Fluid Mechanics & Hydraulic Machinery

AUTHORS

Prof. Suman Chakraborty
Professor,
Department of Mechanical Engineering, IIT Kharagpur

Dr. Sourav Mitra
Assistant Professor,
Department of Mechanical Engineering, IIT Kharagpur

Dr. Aditya Bandopadhyay
Associate Professor,
Department of Mechanical Engineering, IIT Kharagpur

REVIEWER

Prof. Prabhat K S Dikshit
Professor,
Dept. of Civil Engineering (CE), IIT BHU,
Varanasi-221005, Uttar Pradesh

All India Council for Technical Education
Nelson Mandela Marg, Vasant Kunj,
New Delhi, India

Dear Readers,

To prevent the piracy, this book is secured with HIGH SECURITY HOLOGRAM on the front title cover. In case you don't find the hologram on the front cover title, please write us to at contact@khannabooks.com or whatsapp us at +91-99109 09320 and avail special gift voucher for yourself.

Specimen of Hologram on front Cover title:

Moreover, there is a SPECIAL DISCOUNT COUPON for you with EVERY HOLOGRAM.

How to avail this SPECIAL DISCOUNT:
Step 1: Scratch the hologram
Step 2: Under the scratch area, your "coupon code" is available
Step 3: Logon to www.khannabooks.com
Step 4: Use your "coupon code" in the shopping cart and get your copy at a special discount
Step 5: Enjoy your reading!

ISBN: 978-93-5538-951-0
Book Code: DIP458EN

Fluid Mechanics & Hydraulic Machinery
by Prof. Suman Chakraborty, Dr. Sourav Mitra and Dr. Aditya Bandopadhyay

First Edition: 2025

Originally Published by:
All India Council for Technical Education, Ministry of Education, GOI

Exclusively Published & Distributed by:
Khanna Book Publishing Co. (P) Ltd.
4C/4344, Ansari Road, Daryaganj,
New Delhi (INDIA)

Visit us at: **www.khannabooks.com**
Write us at: **contact@khannabooks.com**

To view complete list of books, Please scan the QR Code:

Printed in Bharat.

प्रो. टी. जी. सीताराम
अध्यक्ष
Prof. T. G. Sitharam
Chairman

सत्यमेव जयते

अखिल भारतीय तकनीकी शिक्षा परिषद्
(भारत सरकार का एक सांविधिक निकाय)
(शिक्षा मंत्रालय, भारत सरकार)
नेल्सन मंडेला मार्ग, वसंत कुंज, नई दिल्ली–110070
दूरभाष : 011–26131498
ई–मेल : chairman@aicte-india.org

ALL INDIA COUNCIL FOR TECHNICAL EDUCATION
(A STATUTORY BODY OF THE GOVT. OF INDIA)
(Ministry of Education, Govt. of India)
Nelson Mandela Marg, Vasant Kunj, New Delhi-110070
Phone : 011-26131498
E-mail : chairman@aicte-india.org

FOREWORD

Engineers are the backbone of any modern society. They are the ones responsible for the marvels as well as the improved quality of life across the world. Engineers have driven humanity towards greater heights in a more evolved and unprecedented manner.

The All India Council for Technical Education (AICTE), have spared no efforts towards the strengthening of the technical education in the country. AICTE is always committed towards promoting quality Technical Education to make India a modern developed nation emphasizing on the overall welfare of mankind.

An array of initiatives has been taken by AICTE in last decade which have been accelerated now by the National Education Policy (NEP) 2020. The implementation of NEP under the visionary leadership of Hon'ble Prime Minister of India envisages the provision for education in regional languages to all, thereby ensuring that every graduate becomes competent enough and is in a position to contribute towards the national growth and development through innovation & entrepreneurship.

One of the spheres where AICTE had been relentlessly working since past couple of years is providing high quality original technical contents at Under Graduate & Diploma level prepared and translated by eminent educators in various Indian languages to its aspirants. For students pursuing 2nd year of their Engineering education, AICTE has identified 87 books, which shall be translated into 12 Indian languages - Hindi, Tamil, Gujarati, Odia, Bengali, Kannada, Urdu, Punjabi, Telugu, Marathi, Assamese & Malayalam. In addition to the English medium, books in different Indian Languages are going to support the students to understand the concepts in their respective mother tongue.

On behalf of AICTE, I express sincere gratitude to all distinguished authors, reviewers and translators from the renowned institutions of high repute for their admirable contribution in a record span of time.

AICTE is confident that these outcomes based original contents shall help aspirants to master the subject with comprehension and greater ease.

(Prof. T. G. Sitharam)

ACKNOWLEDGEMENT

The authors would like to thank the All-India Council for Technical Education (AICTE) particularly Prof. T. G. Sitharam, Chairman; Dr. Abhay Jere, Vice-Chairman; Prof. Rajive Kumar, Member-Secretary; and Dr. Sunil Luthra, Director, Training and Learning Bureau for this opportunity to contribute to the Outcome Based Education programme, as per National Education Policy 2020. We also thank Prof. Prabhat Kumar Dikshit, IIT (BHU). SC acknowledges SERB, Department of Science and Technology, Government of India, for J. C. Bose National Fellowship.

The authors would like to extend our gratitude to Mr. Shyamal Biswas for his contribution towards photographic attachment for the preparation of this piece of work. The authors would also like to extend their gratitude to Mr. Chitradittya Barman, Mr. Bikash Mohanty, Dr. Bimalendu Mahapatra, Mr. Mohd. Abdullah and Mr. Prashant Narayan Panday for proofreading the drafts. Lastly, the authors would like to thank their respective family, friends and colleagues for their priceless support.

Suman Chakraborty

Sourav Mitra

Aditya Bandopadhyay

PREFACE

This book introduces students with basic and fundamental understandings of Fluid mechanics and machineries. The objective of the book is to develop the key concepts to solve different fluid mechanics problems and apply the knowledge to real life scenarios, be it in research and development or industrial applications.

We have explained the concepts in an easy and lucid way such that the basic foundational understanding of fluid mechanics will become highly effective for the students. The book contains all the introductory concepts of fluid statics and fluid dynamics. The book elucidates the important governing equations considering a student's expertise in calculus and differential equations. Each chapter has been provided with many solved examples, where we have laid out the mathematical approach towards solving them. Well-designed unsolved problems at the end of each unit will help to improve the problem-solving skills of the students to a great extent. Practical problems given at the end of some units will help the students to gain hands on experience on different aspects of experimental fluid mechanics. The purpose of this book is to provide the students with a comprehensive and well-balanced resource covering all aspects of physical concepts, mathematical derivation, and practical demonstration within the scope of the course. We hope that on sincere completion of the units, the students will be well versed with the principles of fluid mechanics and hydraulic machinery and will also be able to tackle complicated engineering problems.

Suman Chakraborty
Sourav Mitra
Aditya Bandopadhyay

OUTCOME BASED EDUCATION

For the implementation of outcome-based education, the first requirement is to develop an outcome-based curriculum and incorporate an outcome-based assessment in the education system. By going through outcome-based assessments evaluators will be able to evaluate whether the students have achieved the outlined standard, specific and measurable outcomes. With the proper incorporation of outcome-based education there will be a definite commitment to achieve a minimum standard for all learners without giving up at any level. At the end of the program running with the aid of outcome-based education, students will be able to arrive at the following outcomes:

PO-1: Engineering knowledge: Apply the knowledge of mathematics, science, engineering fundamentals, and an engineering specialization in the solution of complex water resources problems.

PO-2: Problem analysis: Identify, formulate, review and analyse complex engineering problems reaching substantiated conclusions using the principles of mathematics, natural sciences, and water resource engineering-based issues.

PO-3: Design/development of solutions: Design solutions for complex water resources engineering problems and design system components or processes that meet the specified needs with appropriate consideration for public and environmental considerations with applications.

PO-4: Conduct investigations of complex problems: Use research-based knowledge and research methods including the design of experiments, analysis, and interpretation of data, and synthesis of the information to provide valid design of Hydraulic structures.

PO-5: Modern tool usage: Create, select, and apply appropriate techniques, resources, and modern engineering and IT tools including prediction and modeling of complex water resources engineering activities with an understanding of the limitations.

PO-6: Environment and sustainability: Understand the impact of the professional engineering solutions in societal and environmental contexts, and demonstrate the knowledge of, and need for sustainable development in water resources engineering.

PO-7: Ethics: Apply ethical principles and commit to professional ethics and responsibilities and norms of engineering practices.

COURSE OUTCOMES (CO)

After completion of the course the students will be able to:

CO-1: Understand basic fluid properties such as density, viscosity, surface tension, and compressibility

CO-2: Apply the principle of hydrostatics and calculate the pressure variation in fluid at static condition

CO-3: Analysing fluid motion using different visualisation technique, like stream lines, streak lines, and path lines.

CO-4: Use the conservation of mass (continuity equation), conservation of momentum (Euler's equation), and energy equation (Bernoulli's equation) to solve various fluid flow problems

CO-5: Learn methods such as Pitot tubes, venturi meters, and orifice plates to measure fluid flow rates

CO-6: Apply the concept of open channel flow and learn about the most economical open channels

CO-7: Applying the fundamental concepts of fluid mechanics to solve practical fluid flow problems.

Course Outcomes	EXPECTED MAPPING WITH PROGRAMME OUTCOMES (1- Weak Correlation; 2- Medium correlation; 3- Strong Correlation)						
	PO-1	PO-2	PO-3	PO-4	PO-5	PO-6	PO-7
CO-1	3	2	2	2	-	-	-
CO-2	3	3	1	2	-	-	-
CO-3	2	2	1	2	-	-	-
CO-4	3	3	1	2	-	-	-
CO-5	2	1	2	1	-	-	-
CO-6	2	2	1	2	-	-	-
CO-7	3	2	2	2	-	-	-

GUIDELINES FOR TEACHERS

To implement Outcome Based Education (OBE) knowledge level and skill set of the students should be enhanced. Teachers should take a major responsibility for the proper implementation of OBE. Some of the responsibilities (not limited to) for the teachers in OBE system may be as follows:

- Within reasonable constraint, they should manoeuvre time to the best advantage of all students.
- They should assess the students only upon certain defined criterion without considering any other potential ineligibility to discriminate them.
- They should try to grow the learning abilities of the students to a certain level before they leave the institute.
- They should try to ensure that all the students are equipped with the quality knowledge as well as competence after they finish their education.
- They should always encourage the students to develop their ultimate performance capabilities.
- They should facilitate and encourage group work and team work to consolidate newer approach.
- They should follow Blooms taxonomy in every part of the assessment.

Bloom's Taxonomy

Level	Teacher should Check	Student should be able to	Possible Mode of Assessment
Creating	Students ability to create	Design or Create	Mini project
Evaluating	Students ability to Justify	Argue or Defend	Assignment
Analysing	Students ability to distinguish	Differentiate or Distinguish	Project/Lab Methodology
Applying	Students ability to use information	Operate or Demonstrate	Technical Presentation/ Demonstration
Understanding	Students ability to explain the ideas	Explain or Classify	Presentation/Seminar
Remembering	Students ability to recall (or remember)	Define or Recall	Quiz

GUIDELINES FOR STUDENTS

Students should take equal responsibility for implementing the Outcome Based Education (OBE). Some of the responsibilities (not limited to) for the students in OBE system are as follows:

- Students should be well aware of each Unit Outcome (UO) before the start of a unit in each and every course.
- Students should be well aware of each Course Outcome (CO) before the start of the course.
- Students should be well aware of each Program Outcome (PO) before the start of the program.
- Students should think critically and reasonably with proper reflection and action.
- Learning of the students should be connected and integrated with practical and real life consequences.
- Students should be well aware of their competency at every level of OBE.

ABBREVIATIONS AND SYMBOLS

List of Abbreviations

General Terms			
Abbreviations	**Full form**	**Abbreviations**	**Full form**
SI	Système International	rms	Root mean square
MEMS	Micro electro mechanical system	AC	Alternating current
		RPM	Revolution per minute
CG	Centre of gravity	BEP	Best efficiency point
CM	Centre of mass	OC	Operating condition
CP	Centre of pressure	NPSH	Net positive suction head

List of Units and Symbols

Symbols	Description	Symbols	Description
τ	Shear stress	K	Compressibility
ρ	Density	F_b	Body force
m	Mass	$\vec{n}$	Unit vector
V	Volume	p_a/p_{atm}	Atmospheric pressure
p	Pressure	p_{vacuum}	Vacuum pressure
n	Number of moles	p_{gage}	Gage pressure
$\bar{R}$	Universal gas constant	I_{xx}	Second moment of area about x – axis
T	Temperature	I_{yy}	Second moment of area about y – axis
M	Molecular weight	W	Weight of the body
γ	Specific weight	$V_{submerged}$	Volume of the submerged body
g	Acceleration due to gravity	V_{total}	Total volume of the body
μ	Dynamic viscosity	ρ_{air}	Density of air
ϑ	Kinematic viscosity	ρ_{water}	Density of water
σ	Surface tension	G	Centre of gravity
F_s	Capillary force	M	Meta centre
F_g	Force due to gravity	B	Centre of buoyancy
p_v	Vapour pressure	t	Time
ΔV	Difference in volume	x, y, z	Cartesian coordinates
Δp	Difference in pressure	u	Velocity field in x – direction

Symbols	Description	Symbols	Description
E	Bulk modulus	v	Velocity field in y- direction
w	Velocity field in z – direction	h_i	Head loss due to sudden exit
$\dot{\epsilon}_x$, $\dot{\epsilon}_y$, and $\dot{\epsilon}_z$	Linear strain rate	h	Minor head loss
$\dot{\epsilon}_{vol}$	Volumetric strain rate	K	Mean loss coefficient
$\dot{\epsilon}_{xy}$	Rate of change of angle between two perpendicular segments	R	Flow resistance
ω	Angular velocity	H	Head loss
Ω	Vorticity	D_H	Hydraulic diameter
Q	Volume flow rate	R_H	Hydraulic radius
$\dot{m}$	Mass flow rate	Fr	Froude number
V_{rel}	Relative velocity	β	Exit angle made by jet
r, θ, z	Polar coordinate	ϕ	Exit angle made by relative velocity and motion of surface
a	Acceleration	V_1	Inlet velocity of the jet
$\forall$	Volume of the control volume	V_{r1}	Relative velocity of the jet with respect to moving surfaces
$\overline{T}^n$	Traction vector	α	Guide blade angle between jet and motion of surface
b_x	Body force per unit volume	V_{w1}	Whirl velocity at the inlet
C_d	Coefficient of discharge	V_{f1}	Flow velocity at the inlet
C_v	Coefficient of viscosity	η	Efficiency of the jet
ρ_m	Density of manometric fluid	N	Rotational speed of the wheel
p_0	Stagnation pressure	D	Diameter of the wheel
C	Correction factor	D_0	Diameter of the runner
C_f	Skin friction coefficient	D_b	Diameter of the boss or hub
τ_w	Shear stress at the wall	η_0	Overall efficiency of the turbine
Re	Reynolds number	P	Power output
f	Darcy friction factor	$\dot{w}$	Weight of the water lifted per second
h_f	Frictional head loss	h_s	Height of the sump
h_e	Head loss due to sudden expansion	h_d	Height of the delivery reservoir
h_c	Head loss due to vena contraction	Q_{th}	Theoretical flow rate

LIST OF FIGURES

LIST OF TABLES

CONTENTS

1 Properties of Fluids

UNIT SPECIFICS

In this unit we will discuss the following aspects:

- A broad perspective of fluid mechanics
- The definition of a fluid
- Fluid properties such as density, viscosity, surface tension
- Concepts of vapour pressure and compressibility

Several examples of real world applications are covered for creating interest in the reader's mind. Several solved examples are given at relevant places in the text. A number of MCQs are given along with numerical questions for practice. Additional topics are given through QR codes as supplementary reading material. A section called 'Know More' is also given for the interested reader to get acquainted with the history of the subject, some real world applications, and case studies.

RATIONALE

This fundamental unit on Properties of Fluids helps student get a basic idea of how to differentiate a fluid to a solid. It is important to understand the fundamental difference arising between the two as many practical applications would require someone to model the items appropriately based on temperature, shear rates etc.

It clearly elucidates the fundamental ideas behind defining pertinent fluid properties such as density, specific volume, etc. We then move on to discuss Newton's law of viscosity and its dependency on temperature. This is important in several industrially relevant applications wherein the motion between two solids is facilitated by liquids.

We then discuss various aspects of surface tension. This is particularly important in the analysis of problems which bear an interface between two fluids. Most commonly seen examples of capillary rise, bubbles, and drops are discussed owing to their practical and natural relevance.

Concepts of vapour pressure and compressibility are required to further understand ideas such as cavitation which are relevant for understanding the operation of turbomachinery and issues arising due to spontaneous vapour formation and collapse, roots of which lie in the concept of vapour pressure.

PRE-REQUISITES

Basic Mathematics (Class XII)

UNIT OUTCOMES

List of outcomes are as follows:

U1-O1 : Define a fluid

U1-O2 : Define density, specific volume, specific gravity

U1-O3 : Describe Newton's law of viscosity

U1-O4 : Describe phenomena observed due to surface tension

U1-O5 : Define concept of vapour pressure and compressibility

Unit-1 Out-come	EXPECTED MAPPING WITH COURSE OUTCOMES (1 - Weak Correlation; 2 - Medium Correlation; 3 - Strong Correlation)						
	CO-1	CO-2	CO-3	CO-4	CO-5	CO-6	CO-7
U1-O1	3	3	1	2	-	-	2
U1-O2	3	2	-	2	-	-	1
U1-O3	2	-	-	2	-	1	3
U1-O4	3	1	-	1	-	-	3
U1-O5	2	2	1	-	1	-	2

1.1 A PERSPECTIVE OF FLUID MECHANICS

Fluids are everywhere around us and also within our body. Many, if not all, processes which govern and affect our lives are in some form or the other related to the flow of fluids. Let us look at some illustrative examples which span across various scales.

From the microscopic length scale, we have the flow of blood and other important fluids in the human body. The heart acts as a biological pump, kidneys act as biological filters, lungs act as a means of oxygen-uptake by the body and so on Fig. 1.1. Understanding the human body as a whole relies on a solid understanding of the fluid mechanical processes associated with it and how each subsystem interacts with each other.

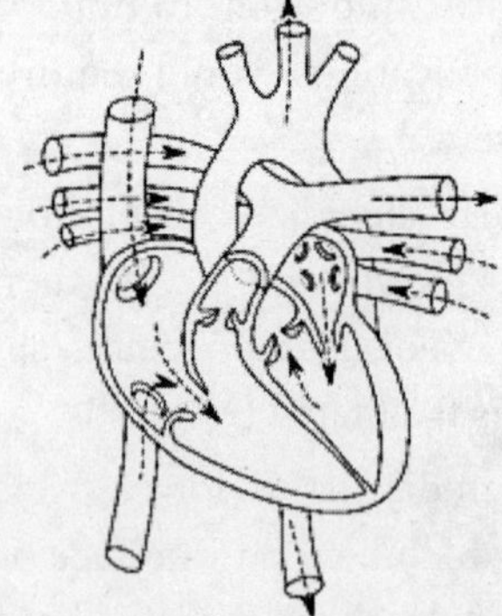

Figure 1.1: A schematic representation of the human heart. It consists of 4 chambers, valves, and a positive displacement pumping achieved through the cardiac muscles. The directions of flow are depicted by the arrows.

The study of fluid mechanics of various birds, fishes, and insects has often given a fresh perspective on engineering and science. There are efforts on the development of drones based on the flight of bats and fireflies. There are also efforts towards aquatic robots which mimic the motion of various fishes. Fluid mechanics is also at play in plants where the interaction of plant tissues causes an uptake of water from roots towards the leaves.

Figure 1.2: The flight of a green bee eater, a kite, and a dragon fly is characterized by completely different aerodynamics. Picture courtesy: Mr. Shyamal Biswas, IIT Kharagpur.

Understanding storage and transmission of fluids is one of the cornerstone of the human civilization as it is central towards many agricultural practices, industrial processes, household water supply, sewage transport and treatment and so on. As a matter of fact, a very basic infrastructure requirement of any region with any population size is an appropriate distribution and storage of water. The design of such pipelines and networks requires a careful understanding of the ideas behind pressure losses and flow rates through pipes; a basic facet of fluid mechanics. Pumps are used practically in all households for raising water to overhead tanks. Fans are used to circulate air to keep our bodies cool. Exhaust fans are used to extract air from spaces and create a ventilated space.

Equally important, in this modern day and age, is understanding the close interaction between climate change, river water levels, and human adaptation to such events. Understanding atmospheric transport also holds the key towards understanding and mitigating air pollution. Similarly, oceanographic transport is important towards studying algal transport in oceans and predicting ocean currents, which have a significant impact on coastal populations. Fig 1.3 shows a snapshot of the process of cloud formation and transport across the Indian subcontinent. Such kinds of predictions are very important from the point of view of the rain dependent Indian agriculture. Weather prediction of hurricanes and cyclones is a very important aspect for population living on the coastal regions of our country. In particular, the IMD forecasts have helped in the evacuation of large masses thanks to the predictions made by them. The formation, motion, and dynamics of such weather patterns is given by a close interplay of fluid mechanics, heat transfer (irradiation, absorption by the atmosphere), and mass transfer (water evaporation and cloud formation).

Figure 1.3: Satellite image showing the cloud coverage. Satellite image taken from https://mausam.imd.gov.in/imd_latest/contents/satellite.php.

Figure 1.4: Power generating units. Top image is that of the Kaveri Engine https://www.drdo.gov.in/labs-and-establishments/gas-turbine-research-establishment-gtre. The lower two images are those of a coal fired power plant and a hydropower plant, https://www.ntpc.co.in/en/about-us/photo-gallery/business-units.

Power generation is critical for the survival of mankind. Power generation in a traditional coal-powered power plant relies on motion of turbines caused due to the flow of high temperature and high-pressure steam over them. On the other hand, power generation in a hydroelectric power plant occurs due to water, which is stored at a large height, flowing over turbines. In both the cases the turbines are used to rotate the shaft of a series of generators, thereby generating electricity. Similarly, wind power plants utilize the kinetic energy of flowing air and convert it into electrical energy through generators. In fact, such means of harnessing the kinetic energy of flowing wind and water have been known for long.

Vehicles utilize the high pressure gases of combustion to generate power, which is eventually transmitted to the wheels. Gears inside the gearbox are lubricated my means of oil, engine is kept at relatively cold by circulation of coolant fluid through various jackets in the engine block, brakes are also actuated through hydraulic fluids. Moreover, the almost universal shape of the car is also designed so that the aerodynamic drag, i.e. the forces causing a retarding force on a moving car, are minimized. Such optimal designs can be seen in the case of high-performance racing cars such as Formula 1. India has made tremendous strides in the development of modern aircraft systems and also space technology. Not only do the propulsion systems involve a close understanding of fluid mechanics of reactive species but also the overall shape and structure of the aircrafts and space crafts at such high velocities depend on their interaction with the atmosphere.

1.1.1 Hydrodynamics

Hydrodynamics refers to the study of the behaviour of fluids in motion, and their interaction with other objects and boundaries. This can be further classified into fluid kinematics and fluid dynamics. Fluid kinematics refers to the study of the motion of the fluid through parameters such as particle displacement, velocity, acceleration, rotationality etc. without due consideration of the forces which act on the fluid. Therefore, in such cases, the role of pressure, viscosity etc. do not appear. On the other hand, fluid dynamics takes into account the various forces acting on the fluid. This can be due to the pressures (or more specifically, gradient in pressure) acting on it or due to interfacial stress at the interface of two fluids. Often in literature, we encounter the term hydrodynamics. While it refers to dynamics of water, it can generally refer to the dynamics of any liquid, whereas aerodynamics refers to the dynamics of air.

1.1.2 Hydraulics

The human endeavour to utilize fluids as a means of power transmission has necessitated a new branch of engineering which is termed as hydraulics. More specifically, hydraulics refers to the study of mechanical properties of liquids. This is to demarcate it from pneumatics, which refers to the study of the mechanical properties of air. There are several applications of hydraulics such as power generation, power control, and transmission of power through high-pressure liquids. One such illustrative image is shown below.

Figure 1.5: A hydraulic crawler dozer makes use of hydraulics to generate tremendous forces which help in earth moving operations. https://www.bemlindia.in/product/bd155/.

1.1.3 Hydrostatics

Hydrostatics is the study of fluids which are not under motion, i.e. under static conditions. Such situations are fairly common in the case of water tanks, dams, reservoirs, and so on. The primary target of the study of hydrostatics is to find out the pressure acting at various surfaces which bear fluid, for example, an overhead tank, or the walls of a dam. Furthermore, floating and transportation of heavy objects across large water bodies by ships is made possible due to the knowledge of hydrostatics. This will form the basis of the study of Unit 2.

Figure 1.6: Chandil Dam, Jharkhand. Picture courtesy: Mr. Shyamal Biswas, IIT Kharagpur.

1.2 WHAT DEFINES A FLUID?

While it is abundantly clear that fluids at rest and motion play a major role in almost all daily activities of mankind, we must first give an outline of what is a fluid. From our high school physics we can recall that all substance exist in three phases: solid, liquid and gases. Liquids and gases together are known as Fluids. What distinguishes a fluid from solid is the behavior under tangential (shear) force. Imagine a rubber block subjected to a tangential force F from top surface, it would deform by a certain angle. This deformation angle will increase with increase in applied force (assuming it doesn't exceed the elastic limit). Upon removal of this force the solid will regain its original shape. Therefore, in case on solids, for a given applied tangential force one can quantify the amount of deformation In contrast, if we were to subject the fluid to a similar shear force, then the fluid would keep on deforming, and it would be found that even after the force is removed, the fluid stays deformed, see Fig. 1.7. In essence, we may say that a fluid is something which flows, and that is quantified by its inability to resist shear forces.

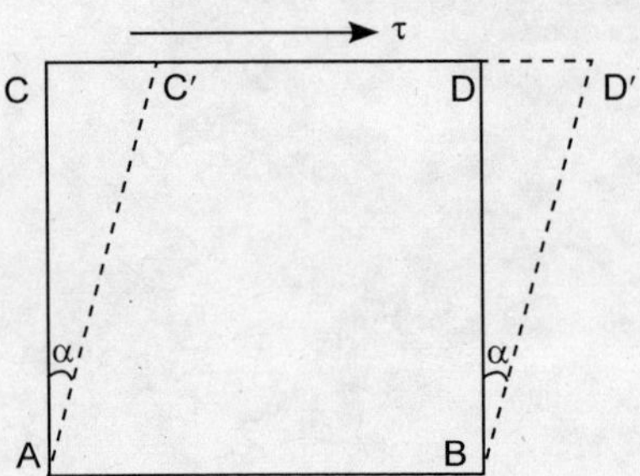

Figure 1.7: An imaginary element ABCD deforms to ABC'D' upon the action of a shear stress, . If the element is that of a perfectly elastic solid, the element regains its original configuration ABCD. However a fluid does not.

Having said this, fluids can further be classified, from an engineering viewpoint, as a liquid or gas. The distinction between a liquid and a gas arises due to the magnitude of intermolecular forces between the molecules. Gas molecules typically have a lower intermolecular interaction than liquids. A direct corollary of this statement is that the gases have a larger mean free path than liquids and correspondingly have a larger average space between each molecule than liquids. This property is useful in the consideration of compressibility of a liquid which we will discuss later. While here the idea of deformation is considered, we have not yet touched upon the total volume of the element. Items such as rubber shrink or expand in volume depending on the forces applied on them; such considerations require our special attention when studied in the context of engineering fluid mechanics. We will touch upon these ideas in later units.

1.3 DIMENSIONS AND UNITS

Before we move on to the description of various fluid properties, it is important to discuss two important concepts which are useful not just for this particular course on fluid mechanics, but for all areas of science and engineering.

A **dimension** is a quantitative way of expressing a physical variable; in other words it is a measure of a physical variable without any numerical value associated with it. Examples of dimensions are mass, length, temperature, current, force, acceleration, density, power, energy and so on. A **unit** is a numerical value which signifies the magnitude of the quantity of the physical dimension. For example, the unit of the dimension of length is meter, the unit of dimension of mass is kg (kilogram) and so on. Over the centuries, there have been several systems of units which have been in use all over the world. However to universalize these system of units in order to enable sharing of knowledge and interoperability of engineering technologies, we have widely adopted the SI (International System of Units) units. However a practicing engineer may still encounter the archaic British system, known as the British gravitational (BG) system.

There are two kinds of dimensions - primary dimensions and secondary dimensions. Primary dimensions refer to those dimensions with the help of which all other dimenions can be derived. For example, length and mass are two examples of primary dimensions. This is so because length cannot be derived from mass, and similarly, mass cannot be derived from length. However, density, a secondary dimension, can be derived with the help of the two primary dimensions of mass and length, i.e. density = mass/length.

We note that while there are seven primary units namely mass, length, time, temperature, electric current, luminous intensity, and amount of matter, for our study of fluid mechanics only four primary dimensions suffice, i.e. mass, length, time and temperature. The dimensions are typically denoted in curly braces such as M, L, T, or θ; these are the dimensions of mass, length, time, and temperature. On similar lines, we can write down various other secondary dimensions by noting their derivations or the formulae associated with it. For example, area is a product of two lengths. For a square of length the area is . For a circle of radius r, the area is $A = \pi r^2$. Therefore, we see that the dimension of area is L^2. Below we tabulate commonly occuring dimensions and their commonly encountered engineering units.

Table 1.1: Primary dimensions and secondary dimensions encountered in fluid mechanics. The relevant SI and British gravitational units are also mentioned.

Primary dimensions				
Dimension	Dimension Representation	SI Unit	BG Unit	Conversion Factor
Mass	{M}	kilogram (kg)	slug	1 slug = 14.5939 kg
Length	{L}	meter (m)	foot (ft)	1 ft = 0.3048 m
Time	{T}	second (s)	second (s)	-
Temperature	{θ}	Kelvin (K)	Rankine (°R)	1 K = 1.8 °R
Secondary dimensions				
Area	$\{L^2\}$	m^2	ft^2	$1\ m^2 = 10.764\ ft^2$
Velocity	$\{LT^{-1}\}$	m/s	ft/s	1 ft/s = 0.3048 m/s
Pressure	$\{ML^{-1}T^{-2}\}$	Pascal (Pa = N/m^2)	lbf/ft^2	$1\ lbf/ft^2 = 47.88$ Pa
Viscosity	$\{ML^{-1}T^{-1}\}$	kg/(m.s)	slug/(ft.s)	1 slug/(ft.s) = 47.88 kg/(m.s)

1.4 DENSITY

Density or mass density, , is defined as the ratio of the total mass of the fluid to the volume occupied by the fluid. The SI unit of density is . At 4 C, the density of water is 999.8395 kg/m. However, in practice we often assume that the density of water is 1000 kg/m. Density may therefore be defined as

$$\rho = \frac{m}{V} \tag{1.1}$$

The density of fluids varies with temperature. For gases such as air, we can estimate the density based on the ideal gas equation,

$$pV = n\bar{R}T \tag{1.2}$$

where represents the pressure, V represents the total volume, n represents the number of moles, represents the universal gas constant, and represents the absolute temperature in Kelvin (K). We can rearrange ideal gas equation by using $n = \frac{m}{M}$, where is the mass of the gas, and is the molecular weight of the gas, to obtain

$$pV = \frac{m}{M}\bar{R}T \Rightarrow p = \frac{m}{V}\frac{\bar{R}}{M}T \Rightarrow p = \rho\frac{\bar{R}}{M}T \tag{1.3}$$

which helps establish that if the pressure remains the same, then an increase in the temperature leads to a decrease in the pressure. This is nothing but another way of putting forward the well known Boyle's law. While the above derivation is true for ideal gases, the trends are also valid for real gases, although the equations become more complicated.

For most fluids, the density reduces as the temperature increases. However, for water, a very common liquid of study, the situation is somewhat different. As the temperature increases beyond 4°C, the density falls. Infact, even as the temperature decreases below 4°C, it is observed that the density falls. This is popularly known as the anomalous behaviour of water.

1.4.1 Specific Weight

The specific weight, is defined as the weight of the fluid per unit volume. It is therefore defined as

$$\gamma = \frac{\rho V g}{V} = \rho g \tag{1.4}$$

where and represent the density and acceleration due to gravity respectively. The SI unit of specific weight is N/m.

1.4.2 Specific Volume

The specific volume of a fluid is defined as the volume of fluid occipied by a unit mass of the fluid. It is therefore defined as

$$v = \frac{V}{m} = \frac{1}{\rho} \tag{1.5}$$

where defines the specific volume. The above definition is more utilized in the context of thermodynamics and when dealing with gases. The SI unit of specific volume is m/kg. The usage of as specific volume must not be confused with the symbol for velocity as it will be obvious from the context.

1.4.3 Specific Gravity

The specific gravity is typically defined as the ratio of a liquid at a given condition to the density of water at a pressure of 101.325 kN/m (which is the standard atmospheric pressure) and at 4C. It is be defined as

$$S = \frac{\rho}{\rho_{water,1atm,4^0 C}} \tag{1.6}$$

Example 1.1

Calculate the density, specific weight, and specific gravity of a fluid which occupies a volume of 1 litre weighs 10N. Assume g = 9.81 m/s^2.

Solution

Let us first convert the volume to m^3.

$$\text{Volume} = 1 \text{ litre} = 10^{-3}\text{m}^3$$

$$\text{Weight} = 10 \text{ N}$$

$$\text{Mass of fluid, } m = \frac{Weight\ of\ fluid}{g} = \frac{10}{9.81} kg = 1.02 kg.$$

$$\text{Density, } \rho = \frac{mass}{Volume} = \frac{1.02}{10^{-3}} = 1020 kg / m^3.$$

$$\text{Specific gravity, S} = \frac{Density\ of\ liquid}{Density\ of\ water} = \frac{1020}{1000} = 1.02$$

$$\text{Specific Weight,} \gamma = \frac{Weight\ of\ fluid}{Volume\ occupied\ by\ fluid} = \frac{10}{10^{-3}} = 10000 N / m^3.$$

1.5 VISCOSITY

Fluid flow is often confined by solid surfaces (e.g. water flowing through a canal or a pipe) and it is important to study the effect of such solid boundaries on the flow behavior. Let us consider fluid flow happening over a stationary plate or over a flat solid non-porous surface. Assume that the fluid approaches this plate at a uniform velocity, U. After striking the surface, the fluid next to the stationary surface comes to complete halt with respect to the stationary plate i.e. $u = 0$. This phenomenon is often called the "no-slip" condition as the fluid sticks to the surface of the solid in contact. This is supported by several experimental observations. Experiments also show that the fluid layer directly in contact with solid surface impedes the flow of fluid layers above it. This results in the formation of a velocity gradient. Fig. 1.8 shows pictorially this no-slip condition of fluid and the developed velocity gradient. The fluid property due to which this phenomenon happens is called the viscosity. Viscosity may be visualized as a frictional force which impedes the relative velocity of fluid layers. As the fluid in direct contact of plate comes to rest, due to viscosity it impedes the relative motion of fluid layer above it, which in-turn impedes the next fluid layer above it and this propagates throughout the fluid flow. Therefore, the uniform fluid flow gets modified to a flow with velocity gradient as depicted in Fig. Fluid layers far away from the solid surface have significantly higher velocity than the fluid near the solid surface and this is a consequence of viscosity.

An additional viewpoint into this is that the cohesion between the fluid molecules cause them to not move easily with respect to each other. Therefore, when two layers of fluids move relative to each other, they experience friction which is due to this cohesiveness. This friction is quantified through the fluid property viscosity, denoted through . A fluid which has a larger cohesion therefore appears as a fluid with a larger viscosity and vice-versa. Let us consider Fig. 1.8, which represents the velocity profile in the direction near a wall. The velocity increases as we move away from the wall. If we consider the velocity at a location to be and the velocity at a location to be , then there is a relative velocity of over a distance . The gradient of velocity, i.e. , is termed as strain rate as well.

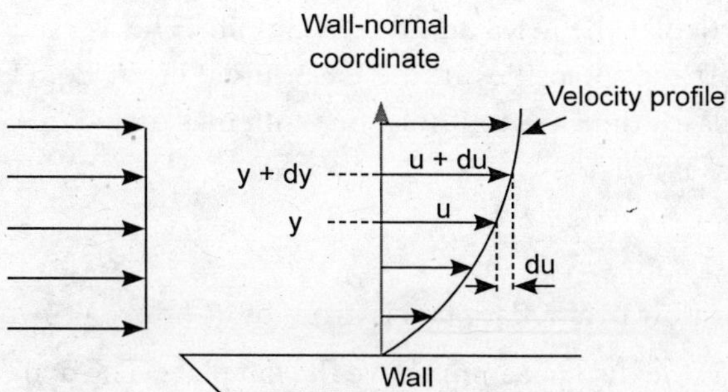

Figure 1.8: Velocity profile for a flow over a solid surface. The incoming flow has a uniform profile. The flow over the solid wall depicts a slowing down of velocity near the wall due to the fluid viscosity. The velocity at the wall becomes zero and is termed as a no-slip boundary.

This leads to the generation of a shear stress at that particular location *y*, τ, between the two layers which has been found experimentally for a certain type of fluid, known as Newtonian fluids to be

$$\tau \propto \frac{du}{dy} \tag{1.7}$$

The proportionality constant can then be written as

$$\tau = \mu \frac{du}{dy} \tag{1.8}$$

where is known as the dynamic viscosity. Essentially, Eq. (1.8) represents the Newton's law of viscosity which states that the shear stress, τ is related to the normal gradient, $\frac{du}{dy}$ (i.e. the gradient perpendicular to the direction of the velocity) through the dynamic viscosity, μ. Similarly, the above discussion can also be used to determine the shear stress at a flat wall for a Newtonian fluid as

$$\tau|_{wall} = \frac{du}{dy}|_{y=0} \tag{1.9}$$

1.5.1 Units of Viscosity

The units of viscosity can be determined using Eq. 1.8 which can be rearranged as $\mu = \frac{\tau}{\frac{du}{dy}}$. This means that the unit of viscosity is

$$Unit\ of\ \mu = \frac{Unit\ of\ stress}{Unit\ of\ velocity\ gradient} = \frac{\frac{N}{m^2}}{\frac{m}{s} \times \frac{1}{m}} = \frac{N \cdot s}{m^2} \tag{1.10}$$

This shows that the SI unit of viscosity is Ns/m^2 which may also be written down as Pa *s*.

While the SI unit is the most used, often in many areas of engineering we will encounter other unit systems for viscosity, in particular the MKS and the CGS units. This is especially true for the selection of oils which typically report the viscosities in the CGS units. From Eq. (1.8), we may also write the MKS unit as kgf-s/m². Similarly, the CGS unit may be written as dyne-s/m². The CGS unit of dynamic viscosity is also referred to as Poise.

Example 1.2

The velocity distribution for flow over a flat plate is given by $u = 4y - y^{\frac{3}{2}}$ where *u* is the velocity at any given location in m/s and *y* is the distance in m above the flat plate. The dynamic viscosity of the fluid is 10^{-3} Pas. Determine the shear stress at y = 4m.

Solution:

Given $$u = 4y - y^{\frac{3}{2}} \Rightarrow \frac{du}{dy} = 4 - \frac{3}{2}y^{\frac{1}{2}}$$

$$\left(\frac{du}{dy}\right)_{y=4} = 4 - \frac{3}{2}4^{\frac{1}{2}} = 1$$

$$\Rightarrow \quad \tau = \mu\left(\frac{du}{dy}\right)_{y=4} = 10^{-3} \times 1 = 10^{-3}\,N/m^2$$

Example 1.3

Fluid is filled between two infinitely long parallel plates. The bottom plate is at rest and the top plate is moving at U = 1 m/s. The distance between these two plates is *h* = 2 cm which is filled with a fluid. It is observed that a force of 10 N has to be applied per unit area of the top plate in order to maintain its speed at 1 m/s. Determine the dynamic viscosity (in Pa·s) of the fluid.

Moving plate U

h

Plate at rest

Solution:

Given: $$u = 1\text{ m/s},\ h = 0.02\text{ m},\ \tau = 10\text{ N/m}^2$$

$$\tau = \mu\frac{du}{dy} = \mu\frac{u-0}{h},$$

$$\Rightarrow \quad 10 = \mu\frac{1}{0.02} \Rightarrow \mu = 0.2\,Pa/s.$$

Example 1.4

Consider a block having a square base area of 0.5 m × 0.5 m having a weight of W = 100 N sliding down an inclined plane at a constant velocity of U = 0.6 m/s. The inclination angle is given as θ = 30°. A thin film of oil having a thickness of *h* = 1.5 mm is used for lubrication between the block and the inclined surface. Calculate the dynamic viscosity of this oil.

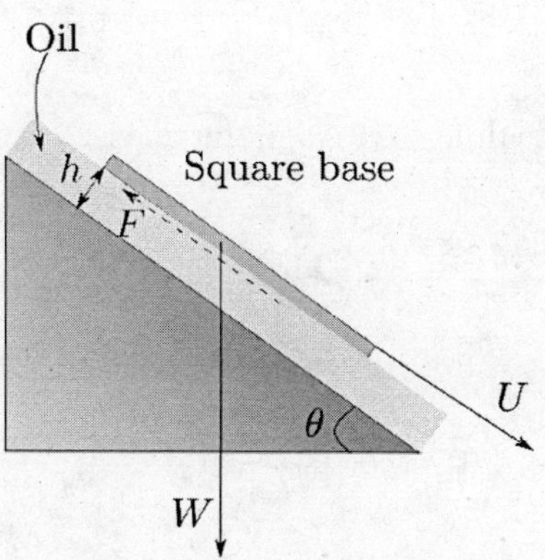

Solution:

Given: $u = 0.6$ m/s, $h = 0.0015$ m, A = 0.25 m², W = 100 N

$\tau = \dfrac{F}{A}$, where F is the tangential force acting on the bottom surface area of the block.

$F = W\sin\theta$

$$\therefore \quad \tau = \frac{100\sin\theta}{0.25} = \frac{100\sin 30^\circ}{0.25} = 200\frac{N}{m^2}.$$

$$\tau = \mu\frac{du}{dy} = \mu\left(\frac{u-0}{h}\right),$$

$$\tau = \mu\left(\frac{0.6}{0.0015}\right) \Rightarrow \mu = 0.5 Pa\cdot s.$$

1.5.2 Kinematic Viscosity

Kinematic viscosity, ν, is defined as the ratio of the dynamic viscosity to the density of the fluid. Mathematically it is defined as

$$\nu = \frac{\mu}{\rho} \tag{1.11}$$

The SI units of kinematic viscosity can be obtained in the following manner

$$Unit\ of\ \nu = \frac{Unit\ of\ \mu}{Unit\ of\ \rho} = \frac{N}{m^2}\cdot s \times \frac{1}{\frac{kg}{m^3}} = \frac{m^2}{s} \tag{1.12}$$

The corresponding MKS unit is also the same as the SI unit which is correspondingly $\dfrac{m^2}{s}$ as kinematic viscosity does not depend on mass. Similarly, the CGS unit of kinematic viscosity is $\dfrac{cm^2}{s}$.

Example 1.5

Determine the dynamic viscosity of a fluid having kinematic viscosity of $6\times10^{-4}\,\frac{m^2}{s}$ and specific gravity of 1.2.

Solution:

Given $\nu = 6 \times 10^{-4}\ m^2/s,\ S = 1.2$

$$S = \rho_{fluid} / \rho_{water} \Rightarrow 1.2 = \frac{\rho}{1000} \Rightarrow \rho = 1200 kg/m^3$$

$$\therefore \quad \nu = \frac{\mu}{\rho} \Rightarrow \mu = \rho \times \nu \Rightarrow \mu = 1200 \times 6 \times 10^{-4} = 0.72 Pa\cdot s.$$

1.5.3 Temperature Dependence on Viscosity

Viscosity depends on the temperature of the fluid. In the case of liquids, it is seen that as the temperature increases, the dynamic viscosity decreases. This can be justified by considering the fact that the intermolecular forces of attraction, that cause a cohesiveness between the fluids, reduces. The cohesiveness appears to us as the tendency of fluid layers to not move relatively to each other, i.e. viscosity. As this cohesive force is overcome, the viscosity appears to have reduced. The situation is different in the case of gases, however. As the temperature of gases increases, the intermolecular collision frequency increases. This leads to a larger loss of energy of the molecules due to collisions. This appears as an increase in the viscosity of the gas as the temperature increases.

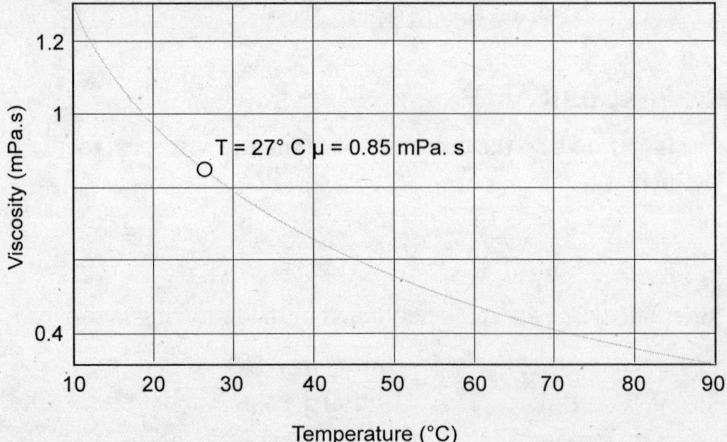

Figure 1.9: Variation of viscosity as a function of temperature for water.

Fig. 1.9 depicts the dynamic viscosity of water as a function of temperature. The tabulated data is obtained from the NIST web-book https://webbook.nist.gov/chemistry/fluid/. The dynamic viscosity of water at room temperature, assumed to be is 27°C, is $\mu = 0.85$ mPas.

1.5.4 Ideal, Real, and Non-Newtonian Behaviour

An ideal fluid is defined as one which does exhibit any viscosity. Moreover, it also exhibits no compressibility, i.e. it does not compress (like a balloon) in response to application of pressure. We may therefore say that an idea fluid is an inviscid and incompressible fluid. Such fluids are never found in practice. However, in many analysis, the influence of the fluid viscosity and compressibility may be neglected. In such situations, the fluid may be *approximated* as an ideal fluid.

Contrary to an ideal fluid, a real fluid is a fluid which has viscosity and which is compressible. All fluids in reality are real fluids.

Real fluids may be further classified into Newtonian and non-Newtonian fluids. Newtonian fluids are fluids which have a linear relationship between the shear stress in the fluid and the velocity gradient in the fluid. Mathematically, this was described in Eq. (1.8). However, the other type of fluids, broadly described by non-Newtonian fluids are more prevalent in several applications. These are fluids, whose behaviour deviates from Newtonian behaviour, either below or above a certain applied shear stress/force or either in response to a high rate velocity gradient (strain rate) or under oscillatory shear rates and so on. Any fluid whose behaviour shows a deviation from the linear relationship in Eq. (1.8) is therefore classified as a non-Newtonian fluid. If, for a simple shear flow, we can denote the stress as a function of the strain rate through a relation $\tau = \mu\left(\frac{du}{dy}\right)^n$, i.e. a power law, then for $n > 1$ we denote the fluid as shear thickneing or dilatant fluid. For $n < 1$, we denote the fluid as a shear thinning or pseudoplastic fluid. The curve, (d), shown in Fig. 1.10 starts with a considerable offset in the shear stress at a fixed rate of strain. In fact, it is only when that τ, i.e. the critical shear stress, exceeds the critical shear stress, i.e. τ_0, that the fluid starts to move. Fluids such as slurries, toothpaste, jams, etc. represent this class of fluid.

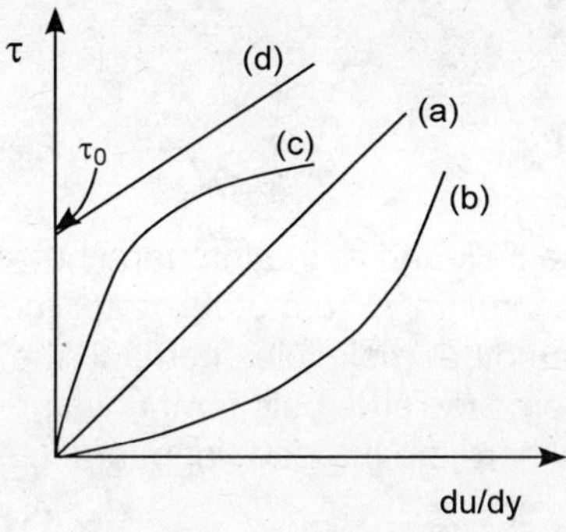

Figure 1.10: Different non-Newtonian behaviours shown for a simple shear flow. The different types of behaviour seen in the curves are: (a) Newtonian fluid, (b) Dilatant or shear thickening fluid, (c) Pseudoplastic or shear thinning fluid, (d) Bingham plastic fluid.

1.6 SURFACE TENSION

Surface tension, , is the inherent property of the interface between two fluids to stay under a state of tension, i.e. as if it is acted upon by a tensile stress. Such interfaces can be formed between a fluid and a gas, such as water and air, or even between two immiscible fluids, such as oil and water. The interface can be thought of as a membrane, such as a rubber membrane, which separates the two fluids. Formally, surface tension can be defined as a force acting at the interface per unit length of the interface. It may also be alternately, and more appropriately, be defined as the surface energy per unit area. The SI unit of surface tension is therefore N/m. The state of tension of the interface can be understood through Fig. 1.11 which shows the interface between air and water as an example. The water molecule in the bulk is attracted by other molecules equally from all sides and hence experience no net force. However, the molecule at the interface is attracted by water molecules from the bulk where as it is weakly attracted by the air molecules on the other side of the interface.

This imbalance of force at the molecular level leads to the collection of molecules at the interface experiencing a net force towards the bulk phase and hence continuously try to contract. Thus, at equilibrium they appear macroscopically to bear resemblance of a membrane trying to contract. Note that the tendency of the interface trying to contract is counteracted by the higher pressure inside the water phase as compared to the air side. Thus, the excess internal pressure balances the surface tension As a matter of fact, this tendency of the interface trying to draw itself inside is responsible for a wide variety of natural occurrences such as drops of water on a lotus leaf etc.

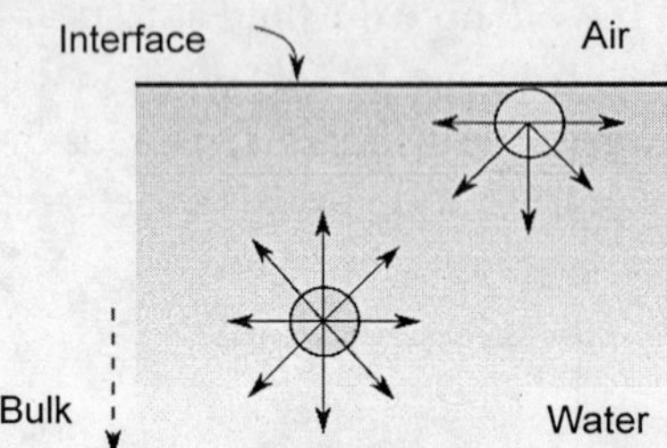

Figure 1.11: Water molecule at the bulk and at the interface between air and water. The molecule at the bulk experiences forces from all directions from other water molecules whereas the water molecule at the interface has a larger pull from other molecules beneath the surface than the pull from air above the interface. This leads to a net force towards bulk for the water molecules at the interface, thereby leading to a situation where the interface appears to be in a state of tension.

1.6.1 Liquid Droplet

With these ideas in mind, we can now determine the excess pressure inside a drop of radius and diameter, , of a liquid in another liquid; the system can be imagined to be a drop of water in air. We assume that the pressure outside is while the pressure inside is . The coefficient of surface tension between the two fluids is . We can imagine a cut in the drop along the centerline. If we consider the equilibrium of the drop, we can write the net force due to the action of surface tension as

$$F_s = \sigma \pi D \tag{1.13}$$

while the net force balancing out the above force due to the pressures acting on the surface of the drop as

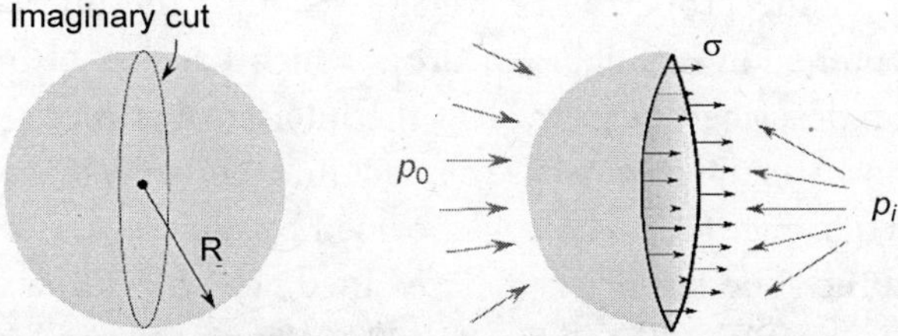

Figure 1.12: Force balance of a droplet

$$F_p = \pi \frac{D^2}{4}(p_i - p_o) \tag{1.14}$$

where we have made use of the fact that the force due to the action of the pressure on a curved surface in the horizontal direction is equal to the pressure times the projected area. Equating the above two, we obtain the equation for the excess pressure as

$$\Delta p = \frac{4\sigma}{D} \tag{1.15}$$

1.6.2 Bubble

Proceeding on the same lines of derivation as that of a liquid drop, we can also determine the excess pressure inside a bubble. Instead of having only one interface, a bubble has two interfaces, one on the outside and one on the inside. Therefore, the force due to the action of surface tension gets altered due to the presence of two circumferences over which the force acts. Thus, we have

$$F_s = 2\sigma\pi D, \quad F_p = \pi \frac{D^2}{4}(p_i - p_o) \Rightarrow \Delta p = \frac{8\sigma}{D} \tag{1.16}$$

Note that the force due to the pressures acting on the outside and inside do not change as they depend only on the projected area which remains constant for the two cases discussed above.

1.6.3 Capillarity

Figure 1.13: (a) Droplets on leaves forming spheres due to the hydrophobic nature of the leaves. (b) Contact angle for a hydrophilic surface. (c) Contact angle for a hydrophobic surface.

Another consequence of the property of surface tension at the interface between two fluids is capillarity. Unlike the phenomenon of bubble or drop, which consists of interactions of two phases, the capillarity phenomenon described here involves the interface between three phases. The interface between three phases is known as the three phase contact line. For convenience, we will discuss a solid-water-air interface as shown in Fig. 1.13. The angle formed by the tangent drawn at the intersection of the air-water interface and the solid surface is denoted by θ, which is known as the contact angle. For hydrophilic walls, i.e. surfaces which alike water, $\theta < 90°$. Whereas for hydrophobic walls, i.e. surfaces which dislike water, $\theta > 90°$. Cellulose, cotton etc. are hydrophilic substrates whereas polymeric surfaces like rubber, and metals are typically hydrophobic. Surfaces can be chemically treated or imparted micro/nanoscopic patterns to alter the contact angle. One such example is self-cleaning glass which is engineered to be super hydrophobic. A natural example where nanoscopic hierarchy of patterns leads to hydrophobicity is a lotus leaf, which exhibits exceptional hydrophobicity.

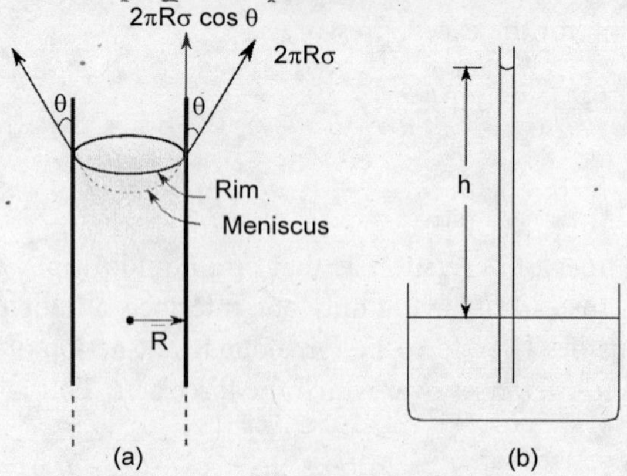

Figure 1.14: Liquid rise and the meniscus in a capillary.

With the general idea of the contact angle, we may now analyze capillarity. Consider a circular capillary of radius, R, with the interface shown in Fig. 1.14. The surface is assumed to be hydrophilic. The meniscus of the surface is also shown in the figure. Let us first determine the *pull* of the interface due to the three phase contact line. The force due to surface tension is along the interface and is equal to

Perimeter of rim × Coefficient of surface tension = $2\pi R\sigma$

The component of force in the direction along the axis of the capillary is therefore,

$$F_s = 2\pi R\sigma\cos\theta$$

Now, if we consider such a situation for the capillary shown in Fig. 1.14(b), then we can see that at equilibrium the upward force due to the surface tension can be balanced by the weight of the column of liquid of height, *h*, pulled by the interface, which can be mathematically expressed as $F_g = \pi R^2 h\rho g$. Note that we have approximated the column of liquid in the capillary as a cylinder by disregarding the dip in the meniscus. Equating the two forces, we can obtain an expression for the height of rise of the capillary as

$$\pi R^2 h = 2\pi R\sigma\cos\theta \Rightarrow h = \frac{2\sigma\cos\theta}{\rho g R} \tag{1.17}$$

A similar derivation can be done for a hydrophobic case wherein the liquid gets pushed further into the capillary instead of rising. This can be seen when a capillary is dipped in a beaker of mercury.

Example 1.6

Calculate the capillary rise in a glass tube of diameter 2 mm when it is immersed vertically into a fluid having surface tension of 0.072 N/m and contact angle of .

Solution:

Given: $\sigma = 0.072\ \text{N/m},\ d = 0.002\ \text{m},\ \theta = 80°$

$$\text{Capillary rise height} = \frac{2\sigma\cos\theta}{\rho g R} = \frac{2\times0.072\times\cos 80^\circ}{1000\times9.81\times0.001} = 0.0025m = 2.5mm.$$

Example 1.7

Determine the excess pressure inside a soap bubble of diameter 5 mm having surface tension of 0.025 N/m.

Solution:

Given: $\sigma = 0.025\ \text{N/m},\ d = 0.005\ \text{m}$

Excess pressure: $$\Delta p = \frac{8\sigma}{R} = \frac{8\times0.025}{0.005} = 40Pa.$$

Example 1.8

What should be the minimum tube diameter so that the maximum capillary rise height is limited to 5 mm when the tube is dipped in water. The surface tension of water is 0.072 N/m and the contact angle between the tube and water is 45°.

Solution:

Given: $\sigma = 0.072\ \text{N/m},\ h = 0.005\ \text{m},\ \theta = 45°,$

Capillary height $$h = \frac{2\sigma\cos\theta}{\rho g R}$$

$$\Rightarrow \quad 0.005 = \frac{2\times0.072\times\cos 45^\circ}{1000\times0.81\times R} \Rightarrow R = \frac{2\times0.072\times\cos 45^\circ}{1000\times9.81\times0.005} = 2.1mm.$$

1.7 VAPOUR PRESSURE

You may recall that all substances can exist in three phases: solid, liquid and gas. The process by which liquid gets converted to gaseous phase is called evaporation and the process by which solid gets converted to gaseous phase is called sublimation. During this phase change process one can have two or even all three phases co-existing. Consider one such case where inside a container mixture of liquid water and water vapor (or steam) is present at T=100 °C. In this state there is continuous evaporation of liquid water and condensation of water vapor at the liquid-vapor interface (see Fig.1.15).

At equilibrium these two rates are equal and the liquid water level inside the container will be stable. The pressure measured by the pressure gauge at any given time instant is known as Vapor Pressure and when "equilibrium" is attained between the Liquid and Vapor phase this Vapor Pressure is equal to the Saturation pressure of water at 100°C. Now let's say we lower the temperature of liquid water to 50°C, then the condensation rate will increase leading to more vapor molecules going into liquid phase (thereby raising the liquid level). This will lead to a reduction in Vapor pressure till a new equilibrium is achieved. This new equilibrium vapor pressure is the saturation pressure of water at 50°C, whereas on increasing the water temperature to 120°C will lead to more evaporation of liquid molecules to vapor phase (thereby lowering the water level). This in turn will increase the vapor pressure and again at equilibrium this vapor pressure will be equal to saturation pressure of water at 120°C. Same phenomenon can be repeated for a closed container having solid ice and water vapor. Only change in this case will be the two processes at equilibrium is known as sublimation (solid to vapor) and deposition (vapor to solid).

Based on the above exercise we can define vapor pressure as the pressure exerted by the vapor column of a substance on its liquid/solid phase (also known as the condensed phase) at any instant in a closed system; whereas the saturation pressure is the vapor pressure when liquid/solid phase is in equilibrium with vapor phase.

The rate of evaporation, which essentially indicates how energetically a molecule can escape the liquid phase and go into the gaseous phase, depends upon the fluid and its temperature. The partial pressure of the vapour phase exerted on the liquid phase at the interface is called as the vapour pressure.

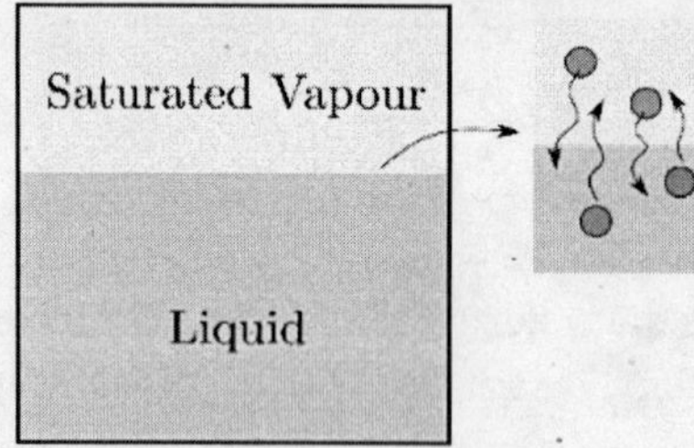

Figure 1.15: Equilibrium of the molecules in the liquid phase with the molecules in the vapour phase.

Let us consider a closed container partially filled with a liquid as shown in the figure. After a sufficiently long time, the vapour molecules will come into equilibrium with the liquid phase, implying that an equal number of molecules will be transiting from the liquid to vapour phase and vice-versa. We then say that the space above the liquid is now saturated with vapour. The vapour pressure is therefore only a function of temperature and is equal to the saturation pressure for boiling at that particular temperature. Accordingly, vapour pressure increases with an increase in the temperature. If we are able to reduce the pressure above the liquid part inside the container, then the temperature at which the liquid boils will also reduce. If the pressure is reduced so that $p_{container} < p_V$, where $p_{container}$ is the pressure above the liquid region, and p_v represents the vapour pressure at the temperature given, then the liquid will spontaneously start boiling at this temperature.

1.7.1 Cavitation

Cavitation is a phenomenon which is closely related to the concept of vapour pressure. We know from intuition that flows occur due to difference in pressure across locations in space. What we mean is that flow occurs from region of high pressure to a region of low pressure. Now if it so happens that the pressure in the liquid during the flow falls below the saturation pressure at that given temperature, then the flowing liquid will spontaneously transition to a vapour phase. This is then seen as bubbles in the flow. These bubbles eventually get carried to regions of pressure which is higher than the vapour pressure. In such regions, the bubble collapses. This sudden transition from a low pressure bubble to high pressure fluid leads to erosion and pitting at the solid surface, if this transition happens near a solid surface confining the liquid flow. This phenomenon is termed as cavitation. This issue requires very keen attention during design and analysis of turbomachinery, submarines etc. where such transition is likely to happen.

1.8 COMPRESSIBILITY

Compressibility of a fluid is the quantification of the change in volume relative to its original volume to the compressive pressure acting on it. Physically, we can relate this to the situation of a balloon filled with air. If we squeeze the balloon from all sides, the volume of the balloon will shrink while the total mass of air inside the balloon remains constant.

Let us assume that a volume of fluid is acted upon by a compressive force so that the change in the pressure, which is the normal force to the surface per unit area of the surface, is $-\Delta p$. The negative sign appears due to the fact that the compressive force acts inwards and by sign convention tensile forces are positive and compressive forces are negative. Let us also assume that in response to the change in pressure, the volume has shrunk by a volume, ΔV, while the original volume is V. Mathematically we can define the bulk modulus of elasticity, E, as

$$E = \lim_{\Delta V \to 0} \frac{-\Delta p}{\dfrac{\Delta V}{V}} = -V\frac{dp}{dV} \tag{1.18}$$

One may interpret the sign appearing in Eq. 1.18 in the following manner. An increase in the pressure, i.e. a positive Δp leads to a reduction in volume, i.e. negative ΔV.

For gases, it is customary to work with compressibility, K, rather than the bulk modulus, E. The compressiblility is defined as

$$K = \frac{1}{\rho}\frac{d\rho}{dp} \tag{1.19}$$

E and K can be related to each other in the following manner. If the mass of fluid, m, then we have

$$m = \rho V \Rightarrow \log m = \log \rho + \log V \Rightarrow dm = \frac{1}{\rho}d\rho + \frac{1}{V}dV \tag{1.20}$$

and if the mass is constant then $dm = 0$ which implies that

$$\frac{1}{\rho}d\rho + \frac{1}{V}dV \Rightarrow -\frac{d\rho}{\rho} = \frac{dV}{V} \tag{1.21}$$

Therefore, from the definition of the bulk modulus, we can write

$$E = -\frac{dp}{\dfrac{dV}{V}} = \rho\frac{dp}{d\rho} = \frac{1}{K} \tag{1.22}$$

Example 1.9

Determine the compressibility of a fluid if its volume changes from 0.0125 m at 100 kPa to 0.0123 m at 150 kPa.

$$P_i = 100 \text{ kPa}, P_f = 50 \text{ kPa}, V_i = 0.0125 \text{ m}^3, V_f = 0.0123 \text{ m}^3$$

$$E = -V\frac{dP}{dV} = -V_i\frac{(P_f - P_i)}{(V_f - V_i)} = -0.0125\times\left(\frac{150\times10^3 - 100\times10^3}{0.0123 - 0.0125}\right)$$

$\Rightarrow$ $$E = 3.125\times10^6 \text{ N/m}^2$$

$\therefore$ $$K = \frac{1}{E} \Rightarrow K = 3.2\times10^{-7}\, m^2/N$$

UNIT SUMMARY

- Ideal gas equation

$$pV = n\bar{R}T$$

- Specific gravity

$$\gamma = \frac{\rho V g}{V} = \rho g$$

- Specific volume

$$v = \frac{V}{m} = \frac{1}{\rho}$$

- Specific gravity

$$S = \frac{\rho}{\rho_{water,1atm,4^0C}}$$

- Newton's law of viscosity

$$\tau = \mu\frac{du}{dy}$$

- Kinematic viscosity

$$\nu = \frac{\mu}{\rho}$$

- Surface tension force for a droplet

$$F_s = \sigma\pi D$$

- Excess pressure inside a droplet

$$\Delta p = \frac{4\sigma}{D}$$

- Surface tension force for a bubble

$$F_s = 2\sigma\pi D$$

- Excess pressure inside a bubble

$$\Delta p = \frac{8\sigma}{D}$$

- Bulk modulus

$$E = \lim_{\Delta V \to 0} \frac{-\Delta p}{\frac{\Delta V}{V}} = -V\frac{dp}{dV}$$

EXERCISES

Multiple Choice Questions

1. Which of the following statements is True?
(a) Water has minimum density at 4 °C.
(b) Water has maximum density at 4 °C.
(c) Density of water doesn't change with temperature.

2. There are two fluids A and B with same dynamic viscosity. Fluid A has higher specific gravity when compared to fluid B. Which of the following statements is TRUE?
(a) Kinematic viscosity for both fluids are same.
(b) Fluid A has higher kinematic viscosity then fluid B.
(c) Fluid A has lower kinematic viscosity then fluid B.

3. For a Newtonian fluid its dynamic viscosity:
(a) Increases with increase in strain rate
(b) Reduces with increase in strain rate
(c) Does not depend on the strain rate
(d) Is Zero

4. Increase in temperature leads to:
(a) Increase in dynamic viscosity for gases
(b) Reduction in dynamic viscosity for gases
(c) Increase in dynamic viscosity for liquids
(d) Reduction in dynamic viscosity for liquids

5. A surface is called "hydrophobic" if the contact angle of water with that surface is:
(a) Zero (b) 90°
(c) > 90° (d) < 90°

6. Two fluids A and B have dynamic viscosities such that $\mu_A > \mu_B$. If they undergo identical strain rate then the relationship between the shear stress τ_A and τ_B is:
(a) $\tau_A > \tau_B$ (b) $\tau_A = \tau_B$
(c) $\tau_A < \tau_B$

7. The capillary rise height for a fluid present inside a tube having contact angle is
(a) Positive (b) Zero
(c) Negative (d) Can't say

8. As a liquid droplet diameter increase the excess pressure inside the droplet

(a) Increases (b) Decreases
(c) Remains constant (d) Can't say

9. The height of capillary rise will reduce in which of the following case(s)?

(a) Increase in surface tension
(b) Increase in radius of a capillary tube
(c) Decrease in acceleration due to gravity
(d) Decrease in surface tension

10. For a liquid in equilibrium with its vapour phase inside a closed container, the saturation pressure of water will increase

(a) As the temperature increases
(b) As the temperature reduces
(c) It remains constant with temperature

ANSWERS															
1.	(b)	2.	(c)	3.	(c)	4.	(a)	5.	(c)	6.	(a)	7.	(b)	8.	(b)
9.	(b)	10.	(a)												

Unsolved Questions

Level - I

1. Two liter of diesel weighs 16.46 N. Calculate its specific weight, density and specific gravity.
2. Determine the dynamic viscosity of a liquid having kinematic viscosity of 10 mm2/s and specific gravity of 1.5.
3. The velocity profile of fluid flow over a stationary plate is given by $u = \frac{3}{2}y - y^2$ where is the velocity in m/s and y is the distance from the plate in m. Determine the shear stress at: (a) $y = 0$ m and (b) $y = 0.1$ m? The dynamic viscosity of this fluid is 1 Pa.s.
4. In the previous problem determine the location at which the shear stress is zero.
5. Determine the mass of air (kg) present inside a room of 27 m³ at a pressure of 101.325 kPa and temperature 30°C. Assume air to be an ideal gas with molar mass of 29 kg/kmol.
6. Consider two parallel horizontal plates separated by 10 mm. The gap is filled with a fluid of viscosity 0.1 Pa.s. The top plate is fixed whereas the bottom plate is moving at a speed of 0.5 m/s. Determine the force per unit area needed to keep the bottom plate in motion.
7. Determine the capillary rise/fall inside a glass tube of diameter d = 2.5 mm when immersed vertically in mercury. The surface tension of mercury σ = 0.52 N/m and contact angle with glass θ = 120°. Mercury has a specific gravity of 13.6.

   The pressure of a liquid is increased from 600 kPa to 800 kPa and its volume decreases by 1%. Determine the bulk modulus of elasticity of the liquid.
8. Consider a water droplet of diameter d = 0.1 mm in air. The atmospheric pressure is 1.01325×10^5 Pa. Determine the pressure inside the droplet if the surface tension of water in contact with air is σ = 0.072 N/m.

9. The compressibility of a fluid is K = 5 $\times 10^{-7}$ m^2/N at 100 kPa pressure. Determine the change in pressure needed to reduce the volume of this fluid by 2
10. Determine the minimum diameter of a glass tube if the capillary fall of mercury has to be limited to 5 mm when it is immersed vertically. Surface tension of mercury N/m and contact angle with glass .

Level - II

1. A square plate of dimensions 0.5 m × 0.5 m is being pushed up an inclined surface with $\theta = 45°$ (see Fig. 1.16) at a constant speed of u = 0.3 m/s. The gap between the plate's bottom surface and the inclined is h = 1 mm and it is filled with an oil of dynamic viscosity 0.6 Pa.s for lubrication. The weight of this plate is 300 N. Determine the force needed to push this square plate up the incline.

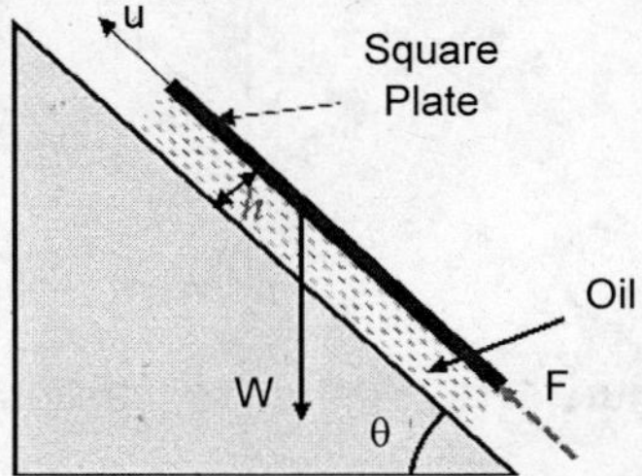

Figure 1.16: Figure for question 1

2. A thin plate of area 0.5 m2 is placed in between two infinitely long parallel stationary plates as shown in Fig. 1.17. The stationary plates are separated by a distance h = 0.1 m. A force of F = 6.25 N is applied on the thin plate in order to maintain its velocity u = 0.5 m/s. The dynamic viscosity of the fluid is 0.5 Pa.s. Determine the distance x of the thin plate from top surface.

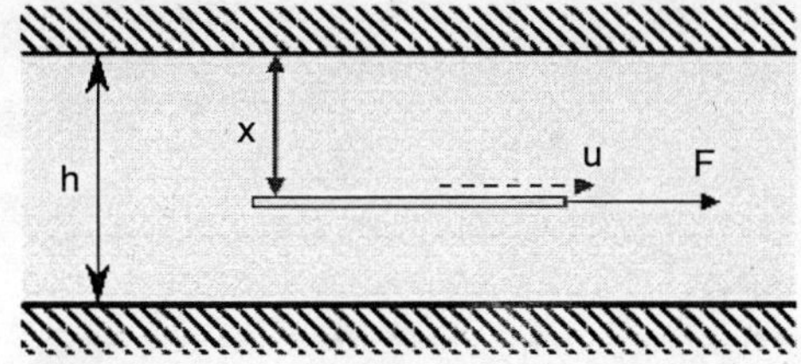

Figure 1.17: Figure for question 2

3. A thin film of oil is used between a shaft and its sleeve for lubrication as shown in Fig. 1.18. The oil has kinematic viscosity $\nu = 5 \times 10^{-4}$ m^2/s and a specific gravity of 1.2. The shaft diameter is d = 200 mm and the film thickness h = 1.5 mm. Determine the shear stress on the shaft due to the oil if it rotates at 1500 r.p.m. (rotations per min).

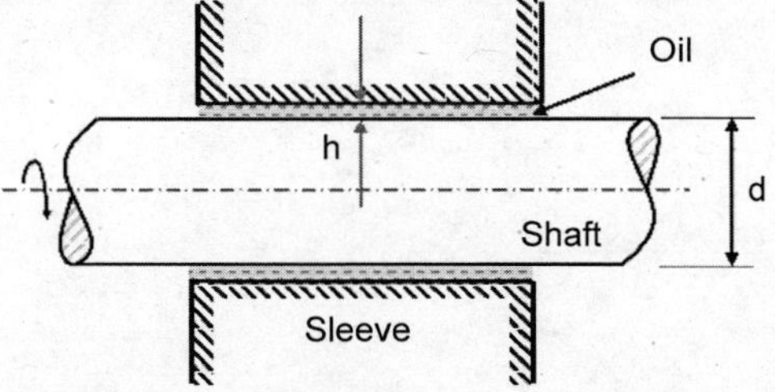

Figure 1.18: Figure for question 3

4. Consider a piston cylinder assembly as shown in Fig. 1.19. The diameter of piston is d_o = 196 mm and its length L = 200 mm. The inner diameter of cylinder is d_i = 200 mm. The annulus space in between the cylinder and piston is filled with an oil of dynamic viscosity μ = 0.2 Pa.s. This piston slides down in vertical position at a constant speed of u = 10 m/s. Determine the mass of this piston.

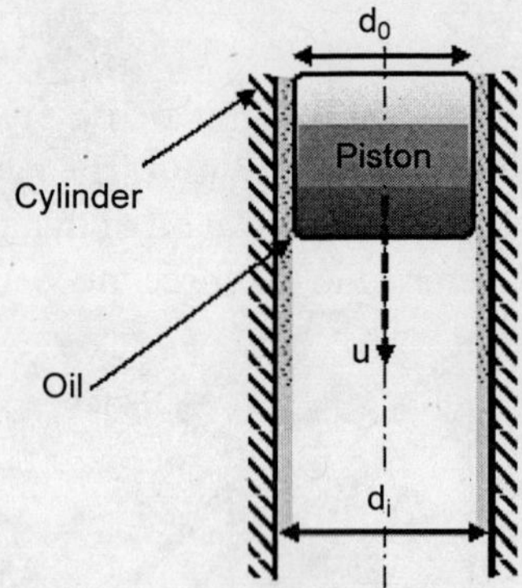

Figure 1.19: Figure for question 4

Answer Key

Level-I

1. Specific weight = 8230 N/m³, Density = 839 kg/m³, Specific gravity = 0.839
2. 1.5×10^{-2} Pa.s
3. (a) 1.5 N/m² (b) 1.3 N/m²
4. 0.75 m
5. 31.48 kg
6. 5 N/m
7. –3.12 mm (– sign indicates capillary fall)
8. 5×10^{-8} m/N
9. 1.02765×10^{5} Pa
10. 40 kPa
11. 2.2 mm

Level - II

1. 257.13 N
2. 0.07236 m or 0.02764 m
3. 6.28 kPa
4. 12.54 kg

To know more about properties of fluids

PRACTICAL

Aim: Experiments on comparing viscosity of some commonly used fluids at home

Apparatus: Two rectangular trays, a piece of white A4 sheet, transparent adhesive tape, a scale, a pencil, cellophane wrap roll, 40-50 ml graduated beaker, a camera phone with video recording app, a phone stand, a table.

Fluids to be tested: Water, Honey, Vegetable Oil, Dish Soap, any other fluid readily available at home.

Theory: The principle for carrying out this comparative study of viscosity for various fluids is that the less viscous fluid will exhibit lower resistance to flow and hence that fluid will have higher velocity when freely flowing over a surface.

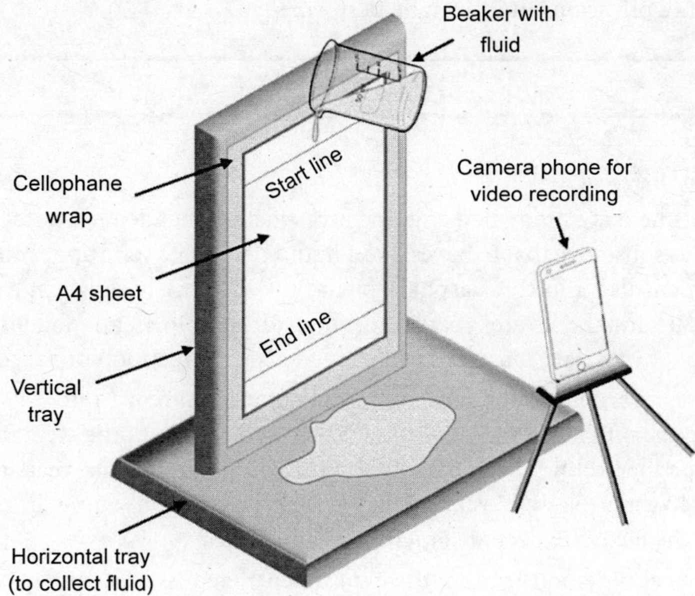

Figure 1.20: Setup for comparison of viscosity across some commonly used house-hold fluids

Procedure

1. Take the A4 sheet and mark two lines, one near the top of the sheet and another one near the bottom. This will denote the "start" and "end" lines for the fluid flow.
2. Paste this A4 sheet on the back surface of one of the rectangular trays using an adhesive tape.
3. Now cover this paper surface completely with the cellophane wrap. Use the adhesive tape (if needed) to stick the cellophane wrap onto the tray surface. Ensure that no portion of the paper is exposed after wrapping.
4. Mount the camera phone on its stand and place this stand on a table.
5. Place the other rectangular tray horizontal on the table. This tray is to collect the various liquids that will fall off the cellophane surface.
6. Hold the cellophane wrapped tray vertically resting on the above mentioned horizontal tray. Ensure that the A4 sheet with pencil markings face the camera of the mobile. This vertical tray with the cellophane wrap will form the final surface on which the fluid flow will occur.

7. Turn ON the video recording of the mobile phone. Ensure that the entire vertical tray with the "start" and "end" line is visible on the screen.
8. Measure around 30-40 ml of each liquid in the beaker and pour it gently near the top start line. While pouring ensure that the liquid flow takes a vertical path.
9. As each liquid flows on the cellophane surface record the time interval it takes to flow from start line and reach the stop line. You may use a video recording app with a time stamp for this purpose.
10. Tabulate these readings for each fluid used in this experiment.

Observations and Conclusion: The fluid with maximum viscosity will face the maximum resistance to fluid flow. Therefore, this fluid will take the maximum time to reach the end line whereas the fluid with lowest viscosity will take the least time. By performing this experiment, a qualitative comparison between the viscosities of various fluids can be performed.

KNOW MORE

Pitch Drop Experiment

The pitch drop experiment is a renowned ongoing investigation that demonstrates the slow movement of pitch, a highly viscous substance that behaves like a liquid despite its solid appearance. Commonly found in the form of bitumen, also known as asphalt, pitch flows extremely slowly at room temperature. Due to its high viscosity, it can take several years for a single drop to form and fall, making this experiment a striking example of how certain materials can challenge our perception of the solid and liquid states.

In 1927, Professor Parnell heated a sample of pitch and poured it into a glass funnel, sealing the stem. After allowing the pitch to cool and settle for three years, he cut the stem in 1930. Since then, the pitch has flowed at an incredibly slow rate, with the first drop taking eight years to fall, followed by five more over the next 40 years. Now, 87 years after the stem was cut, only nine drops have fallen—the last in April 2014—and the next is expected sometime in the 2020s.

Originally set up as a demonstration, the experiment remains in a display cabinet without special environmental controls, so the pitch's flow rate fluctuates with seasonal temperature changes. In 1961, Professor John Mainstone became the experiment's second custodian, overseeing it for 52 years. Like his predecessor, Professor Parnell, he passed away without ever seeing a drop fall. Over the 86 years of the experiment, various technical issues have prevented anyone from witnessing a drop as it falls.

[Source: https://smp.uq.edu.au/pitch-drop-experiment]

REFERENCES AND SUGGESTED READINGS

1. **Introduction to Fluid Mechanics and Fluid Machines** (3rd Edition), S K Som, Gautam Biswas, Suman Chakraborty, McGraw Hill
2. **Fluid Mechanics** (9th Edition), Frank M. White, Henry Xue, McGraw Hill

❑❑

2 Fluids at Rest

UNIT SPECIFICS

In this unit we will discuss about the following aspects:

1. An understanding of fluids at rest
2. Distribution of pressure with depth
3. Notion of surface and body forces
4. Pascal's law of pressure
5. Forces on flat and curved submerged surfaces
6. Buoyancy and Archimedes' principle
7. Concept of floating bodies and their stability

Several real world examples are mentioned for creating interest in the reader. Many solved examples are provided to aid the student in working out problems in hydrostatics. A large number of MCQs are given along with numerical equations for practice. Some additional topics are given through QR codes as supplementary material. A 'Know More' section is given in the end for the reader to get acquainted with the development of fluid statics.

RATIONALE

This fundamental unit on Fluids at Rest helps the student to form a basic concept of describing fluids at rest. Understanding fluids at rest is quite important from the point of view of several key infrastructures at the heart of mankind today such as water storage tanks, reservoirs, dams, ships, etc.

We first discuss about the fundamental concept of the types of forces which can act on a fluid element. Herein in the context of this book we study body forces and surface forces. We then see that the pressure at point in the fluid is isotropic in nature, i.e. it acts equally from all sides at a point a fluid. We then discuss about the fundamental hydrostatics equation wherein we are able to account for the increase in pressure at larger depths of a fluid. We have also elucidated cases where the density is not a constant but varies spatially.

With the help of the ideas formed before, we then discuss about the fundamental ideas behind pressure measurement devices. In particular, we discuss manometry which is a simple and effective method of pressure measurement.

With the help of these basic ideas we study the forces acting on flat and curved surfaces, an idea which is at the heart of complex large-scale infrastructure such as dams. As a practicing engineer, it is very important to understand the various forces and their directions acting on such structures so that the design, including the construction material and shape, can be ascertained. We then move on to the idea of buoyancy and floating bodies. This is quite important for engineers to design items which can float on water, which is useful for several applications relevant to many economic operations such as fishing, off-shore drilling etc. We establish Archimedes' principle using the basic ideas developed earlier.

We ultimately conclude this unit with a detailed discussion on the stability of floating bodies. The stability of objects floating on water is an important parameter for ships and boats. A firm understanding of these ideas sets limits on the maximum weight and the corresponding center of mass that a boat can carry, thereby helping one avoid catastrophic consequences.

PRE-REQUISITES

Elementary calculus

UNIT OUTCOMES

U2-O1: Understand idea of pressure in a fluid and its properties

U2-O2: Various instruments and their principles for pressure measurement

U2-O3: Understand the variation of pressure with depth

U2-O4: Understand the concept of manometry and apply it

U2-O5: Be able to find out forces on submerged surfaces

U2-O6: Find out forces on curved surfaces

U2 - O7: Determine the buoyancy force and general idea of stability

In this Unit, we will consider fluids which are under rest. Essentially it means that under all the conditions mentioned here will consider fluids which are acted upon solely by normal stresses. We have already seen in Unit 1 that fluids deform continuously, i.e. have motion, when they are subjected to shear stress.

Unit-2 Outcome	EXPECTED MAPPING WITH COURSE OUTCOMES (1 - Weak Correlation; 2 - Medium Correlation; 3 - Strong Correlation)						
	CO-1	CO-2	CO-3	CO-4	CO-5	CO-6	CO-7
U2-O1	3	3	-	2	2	-	2
U2-O2	1	3	-	-	3	-	3
U2-O3	2	3	-	2	2	1	2
U2-O4	2	3	-	2	3	-	2
U2-O5	1	3	-	-	-	-	2
U2-O6	-	2	-	-	-	-	1
U2-O7	1	3	-	-	-	-	3

2.1 CONCEPT OF PRESSURE IN A FLUID

We have seen in Unit 1 that fluids cannot sustain shear forces and they deform continuously under the action of a shear. Thus, we can conclude that a fluid which is not acted upon by a shear will not deform and move. A fluid which is at rest must be in a state of zero shear stress. Such a state is known as a *hydrostatic* state. In a hydrostatic state the only force which can act on the fluid is known as pressure, , which is compressive in nature, i.e. it can compress, but will not move the fluid. In other words, we can define pressure mathematically as

$$p = \frac{dF}{dA} \tag{2.1}$$

where it is important to note that the force, dF, is considered to act normally to the area and into the area, i.e. compressive in nature. If we assume that the force, F, is equal and uniform over an area A, then the pressure acting on the area is defined as

$$p = \frac{F}{A} \tag{2.2}$$

Another important property of pressure is that the pressure at a *point* in a fluid acts equally from all sides. We will prove this shortly.

2.1.1 Surface Force and Body Force

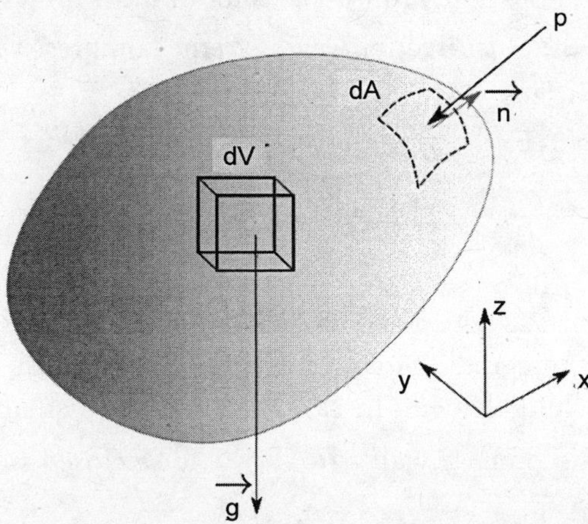

Figure 2.1: Depiction of surface and body force acting on a fluid element.

We first consider an arbitrary fluid element as shown in Fig. 2.1. Such a fluid element can be subjected to two kinds of forces, viz. **surface force** and **body force.** Surface force refers to any force which acts at the surface and the force is proportional to the surface area of the fluid volume. Body force refers to the force which acts on the entire volume of the body and depends on the volume of the body. In the figure, we have shown the surface force as the force due to pressure, and the body force as the force due to the action of gravity on the fluid element.

The body force can be found out as following

$$\text{Force on body} = \text{Mass of body} \times \text{gravity}$$

$$\Rightarrow \quad d\vec{F_b} = \rho\vec{g}dV$$

$$\Rightarrow \quad \vec{F_b} = \int_V \rho\vec{g}dV$$

If we assume that the density and acceleration due to gravity are constant over the entire body, then the integral above may be simplified to obtain the body force as

$$\begin{aligned} \vec{F_b} &= \rho\vec{g}\int_V dV \\ &= \rho\vec{g}V \\ &= m\vec{g} \\ &= -mg\hat{k} \quad \because \vec{g} = -g\hat{k}, \end{aligned} \tag{2.3}$$

where represents the unit vector in the direction. Clearly the body force depends on the volume of the object as seen from the expression above.

On similar lines, the surface force can also be determined. While we know that the force due to the application of a pressure is the pressure times area, we must bear in mind that force is a vector and pressure and area are notionally only scalars. In this regard, the direction of the action of the force due to pressure is normal to the area. In Fig. 2.1, the area vector of the infinitesimal area dA is denoted by $\vec{n}$. Therefore, the force will act in the $-\vec{n}$ direction, owing to the compressive nature of pressure.

$$\begin{aligned} \text{Force on surface} &= \text{Area} \times \text{Pressure} \\ d\vec{F_s} &= -p\, dA\, \vec{n} \\ \vec{F_s} &= -\int_A p\, dA\, \vec{n}. \end{aligned} \tag{2.4}$$

Clearly, the force on the surface depends on the extent of the surface area, and hence it is classified as a surface force. In order to integrate the above expression, we must have the information of how the pressure varies across the fluid element. In case the pressure is equal over the surface of the fluid element, then the integration becomes trivial $\because \int_A dA\, \vec{n} = 0$ for a closed surface. We can imagine this as the integration of the normal vector over a circle, which is a analogous to a surface in 3D. The influence of all direction vectors will be cancelled out by an opposite direction vector. Therefore we have classified the two kinds of forces which we will make use throughout our study.

2.2 UNITS OF PRESSURE

From Eq. (2.2), it is evident that the SI units of pressure are N/m^2. This is also referred to as Pascal, Pa. The MKS unit of pressure is kgf/m^2. In engineering practice we will encounter other units of pressure which are convenient to the problem at hand such as , , and Torr. These units can be related to the SI unit as following:

$$1 \text{ kPa} = 10^3 \text{ Pa} \tag{2.5}$$

$$1 \text{ bar} = 10^5 \text{ Pa} \tag{2.6}$$

$$1 \text{ Torr} = 133.32 \text{ Pa} \tag{2.7}$$

Apart from these pressure is often defined in terms of *head* of a certain liquid. The idea behind such units will be clear below. However, for convenience, the pressure corresponding to a certain head of a fluid can be obtained by $p = \rho gh$, where ρ is the density of the fluid under consideration while reporting the unit, g is the magnitude of acceleration due to gravity, and h is the height reported preceding the unit. Thus, we have

$$\begin{aligned} 1 \text{ mmHg} &= \text{Pressure exerted by 1 mm of Mercury} \\ &= (13.595 \times 10^3) \times 9.80655 \times 10^{-3} \text{ Pa} \\ &= 133.32 \text{ Pa.} \end{aligned} \tag{2.8}$$

We note that 1 mmHg is also equal to 1 Torr. Other such *manometric* units can also be similarly derived. In the above derivation we have used the information that density of mercury is , acceleration due to gravity is 9.80655 m^2/s.

2.3 SCALE AND MEASUREMENT OF PRESSURE

So far we have looked into the various units of pressure. However, a practicing engineer will often encounter pressures relative to some scale. It is important to note that pressure as an absolute quantity holds little importance in many practical examples. It is often reported as relative to the ambient pressure. Therefore we have the following two definitions:

2.3.1 Gage Pressure

Gage pressure is defined as the pressure higher than the atmospheric pressure and can be mathematically represented as

$$p_{gage} = p - p_a \tag{2.9}$$

The atmospheric pressure is p_a = 101.325 kPa. So for example if it is given that the gage pressure is p_{gage} = 200 kPa, we can conclude that the absolute pressure is $p = p_{gage} + p_a$ = 101.325 + 200 kPa = 301.325 kPa.

2.3.2 Vacuum Pressure

On a similar line, for pressures which fall below the atmospheric pressure, we can define the vacuum pressure. It is defined as the pressure below the atmospheric pressure. Mathematically, it can be defined as $p_{vacuum} = p_{a - p}$.

Therefore if it is given that the vacuum pressure is p_{vacuum} = 20 kPa, then the absolute pressure is found out as $p = p_a - p_{vacuum}$ = 101.325–20 = 81.325 kPa.

Note that if pressure is not specified then it may be assumed that the reported pressure is absolute pressure.

2.3.3 Instruments for Pressure Measurement

Pressure is something which cannot be measured directly. This is due to the fact that pressure is fundamentally the measure of the total force exerted by molecules through collision on a surface. So fundamentally, we can measure the force acting on a surface and thereby infer the pressure through the concept behind it. There are multiple ways of pressure measurement which we list below.

1. Weight/gravity based
2. Deformation based
3. Gas-properties based
4. Electronics based

Let us now discuss the fundamental idea behind these.

Weight/gravity based

Manometers are the simplest kind of pressure measuring device. Manometers consist of a bent glass tube. This tube is filled with a fluid known as a manometric fluid. The size of the tube is sufficiently large so that under normal operation the fluid does not come out of the manometer tube. This can be used to measure the total pressure of containers. The fundamental idea is that the pressure of the container will push the manometric fluid inside the manometer tube thereby causing its level to rise on the other side. It is assumed that the manometric fluid inside the manometer is incompressible in nature, i.e. any change in volume across the manometric fluid does not cause the volume occupied by the manometric fluid to change. In practice we may assume that any such change in volume is negligibly small. The rise of manometric fluid is such that the weight of the fluid balances out the force due to the pressure of the container.

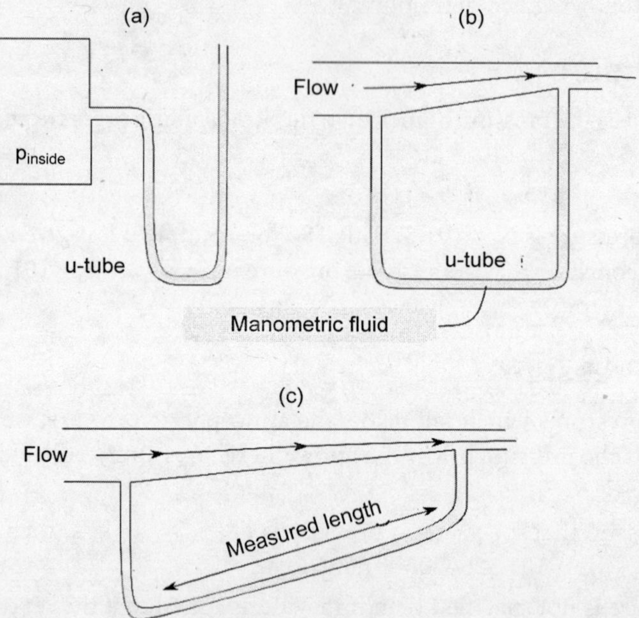

Figure 2.2: Manometer for (a) measurement of pressure in a tank, (b) differential measurement of pressures at two points in a flow through a pipe, (c) differential measurements with a larger sensitivity to measurements of length along the tube rather than the height to which it rises.

It can be also the measures of flowing liquids through holes known as a static port. Typically the hole is kept sufficiently small so that there is no flow directly into the manometer tube.

Accurate measurements of pressure can be done by inclining the manometric tube with respect to the horizontal so that the same change of height causes a longer traversal of the manometric fluid in the tube along the length of the tube. Such measurements are easy and thereby allow a larger sensitivity of the measurement of distance, and hence the pressure.

Differential measurements can also be performed with the help of a manometer tube. A differential measurement refers to the measurement of pressure between two points. This could be the measurement of pressure between two different tanks or the measurement of pressure between two points in a tube. This is easily achieved by simply joining the sources of the two pressures to the two ends of the manometer tube.

Deformation based

A Bourdon tube is one of the most popularly used pressure gages used in modern times. It has a simple construction, it is analog, and can be used across wide pressure changes. It consists of a curved tube with a flattish cross section which deforms outwards when the tube is pressurized. The deformation is due to the internal forces trying to *straighten* out the tube due to mechanical forces. The deformation of the tube causes it to move in such a way that it straightens out. The end of this tube is attached to a mechanical linkage which rotates due to the tube going outward due to the action of the internal pressure. This linkage is kinematically coupled to another gear set which causes a needle deflection.

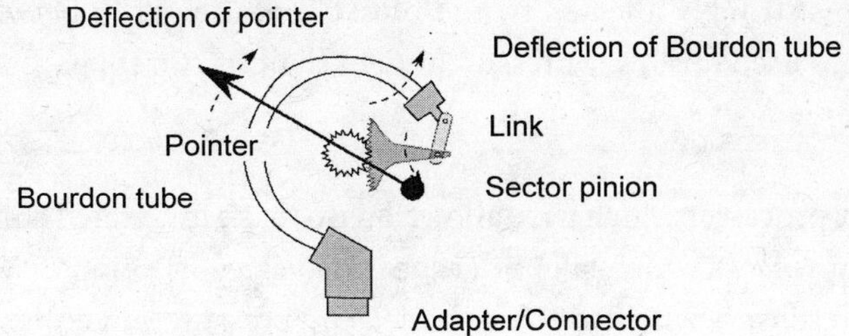

Figure 2.3: Schematic of a Bourdon tube based pressure gage.

The final deflection of the needle gives an analog measurement which can be read off a dial. However, this needle may be replaced by electronic devices which can *transduce* the mechanical movement of the needle to an electric signal.

A diaphragm gage works on similar principles. The construction is also relatively robust. There is a diaphragm which separates two chambers each of which can have a different pressure. In gages which measure relative to the atmospheric pressure, one of the chambers is kept open to the atmosphere with provisions in place to make sure that dust and water do not enter in. When the pressure in the other chamber changes, the diaphragm deforms. This flexure of the diaphragm is the transferred through linkages to the deflection of a needle.

Gas-properties based

Typically such gauges which rely on properties of gas to measure pressure are used for measurement of the vacuum gage pressure, i.e. pressure lower than atmospheric pressures.

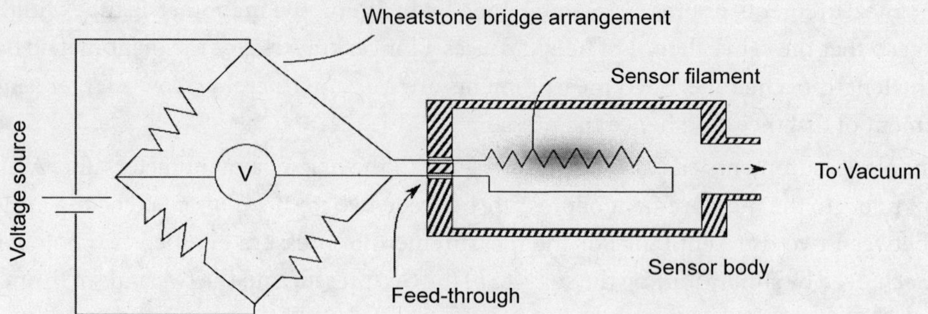

Figure 2.4: Schematic of a Pirani gage which is used to measure vacuum.

A Pirani gage allows for pressure measurements by utilizing the thermal conductivity of gases. A heating circuit is present in which the filament is heated in a closed-loop fashion, i.e. the current through the filament is maintained in such a manner that the temperature of the filament is constant. We all know that the heat transfer to gas increases as pressure increases. In this way with the help of a well-calibrated chart which asserts the information about the heat flux and pressure of the chamber, a very good estimate of the pressure is done.

On the other hand, a Penning gage utilizes high voltage across two electrodes to measure discharge across the filament. The current across the two electrodes is directly related to the discharge between the two electrodes which is strongly affected by the pressure of the gas in between the two electrodes. Typically Penning gages are used for lower pressure ranges than the Pirani gages.

Electronics based

At the dawn of lithography processes which have allowed us to make integrated chips, several technologies were developed to develop sensors which fall under the broad categorization of MEMS (micro-electrical mechanical sensors). Such sensors which are able to measure pressure can either be capacitive in nature or piezoelectric in nature. In capacitive sensors, the microfabricated silicon diaphragm deflects and changes the value of the capacitance. In the piezoelectric type sensors, there are layers of materials which generate charge upon subjecting load to them. Therefore any application of pressure to such layers causes some current fo flow, thereby allowing one to relate the current to the pressure through a calibration curve. These kind of sensors can be used for rapid measurements and storage on computer and are quite versatile due to their small form-factor.

2.4 PASCAL'S LAW OF PRESSURE

Pascal's law states that the pressure at a point in a stationary fluid is equal in all directions. In other words, pressure is *isotropic*. In order to prove this, we consider a wedge shaped fluid element as shown in Fig. 2.5 having dimensions of dx, dy, dz in the x, y, z directions respectively. The wedge makes an angle θ with the xz plane. Since the width dz of the wedge is constant, we can simplify the analysis by considering the 2D figure. We can assume that the pressure acting on the segment OA is p_1 while on segment OB is p_2. Let the pressure acting on the slant face be p_3.

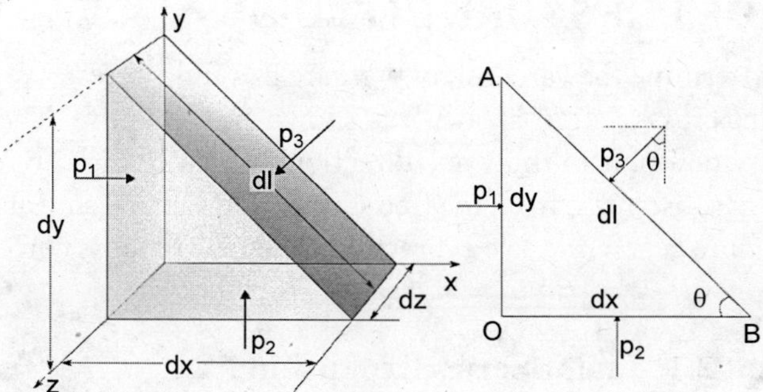

Figure 2.5: A static infinitesimal fluid wedge acted upon by pressure on its faces.

We can now consider the equilibrium of the fluid element by considering a force balance in the x and y directions respectively. Considering first the equilibrium in the x direction we have

$$\Sigma F_x = 0$$

$$\Rightarrow \quad p_1\, dy\, dz - p_3\, dl\, dz \sin\theta = 0 \tag{2.10}$$

$$\Rightarrow \quad p_1\, dy\, dz - p_3\, dy\, dz = 0 \qquad \because dy = dl\sin\theta$$

$$\Rightarrow \quad p_1 = p_3$$

We can now consider the equilibrium of the fluid element in the direction. We assume that gravity acts in the negative direction.

$$\Sigma F_y = 0$$

$$\Rightarrow \quad p_2 dx\, dz - p_3 dl\, dz \cos\theta - \frac{1}{2} dy\, dx\, dz \rho g = 0$$

$$\Rightarrow \quad p_2 dx\, dz - p_3 dx\, dz - \frac{1}{2} dx\, dy\, dz \rho g = 0 \qquad \because dx = dl\cos\theta \tag{2.11}$$

$$\Rightarrow \quad p_2 - p_3 - \frac{1}{2} dy \rho g = 0$$

which in the limit of an infinitesimally small element, i.e. yields

$$p_2 - p_3 = 0 \quad \Rightarrow p_2 = p_3$$

The equilibrium in the z direction is trivial as the pressure on both the front and back faces will cancel each other out in direction.

From Eq. (2.11) and (2.12), we can conclude Pascal's law

$$p_1 = p_2 = p_3 \tag{2.13}$$

for an arbitrary angle θ. This implies that in the limit of an infinitesimally small fluid element, i.e. a point in this case, the pressure acts equally from all directions regardless of any other body force (in this case we have considered gravity) acting on the fluid.

To know more about Pascal's Law

2.5 VARIATION OF PRESSURE IN FLUIDS

We can now proceed to determine the variation of pressure in a fluid at rest. We consider a fluid element of height Δz and constant cross section area ΔA as shown in the figure. The z direction is pointing downwards so that gravity now acts in the +ve z direction as seen in Fig. 2.6. The top face of the fluid element is assumed to be at a distance z from the free surface and the pressure at that face is p. For the top face, the force due to pressure acts in the +ve z direction. The lower face is correspondingly at a location $z + \Delta z$.

Mathematical Concept 2.1: Taylor Series Expansion

A Taylor Series expansion is defined as an infinite series expansion of a function about a given point. To understand this idea, let us consider a one dimensional function $f(x)$ and expand the function in the vicinity of point $x = a$ as seen from the figure. We can write down the value of the function at a point in the neighbourhood of in terms of the function at point $x = a$ as

$$f(x) = f(a) + f(a)(x-a) + \frac{f''(a)}{2!}(x-a)^2 + \frac{f'''(a)}{3!}(x-a)^3 + \ldots \quad (2.14)$$

This is a very convenient way of approximating the function at a point in the neighbourhood of $x = a$ in terms of the value of the function at $x = a$, the slopes at $x = a$, i.e. $f'(a), f''(a)$ etc.

This is shown in the figure below wherein we show the approximations of the functions in the neighbourhood of $x = 1$. The dashed and the dash-dot lines represent the various approximations of the Taylor series as shown in the figure itself. Moreover, the values of the function at $x = 1.25$ as obtained using the exact functional value, and using the Taylor series expansion are also shown.

It is seen that as the number of terms in the expansion are increased, the series value tends towards the exact solution. Moreover, if the point is chosen closer to the point about which the expansion is chosen, the expansion is also better for the same number of terms kept in the series expansion.

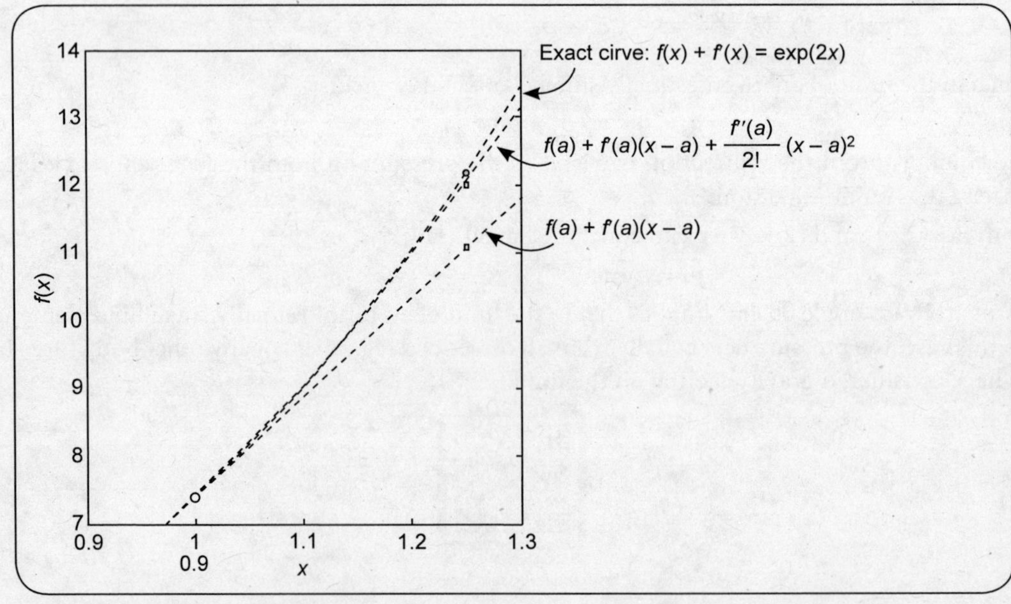

Taylor series expansion about point $x = 1$ for the function $f(x) = \exp(2x)$. The dashed line shows the Taylor series expansion when only the first two terms in the Taylor series in equation (2.14) are considered. The dash-dot line shows the Taylor series expansion when the first three terms in the series in equation (2.14).

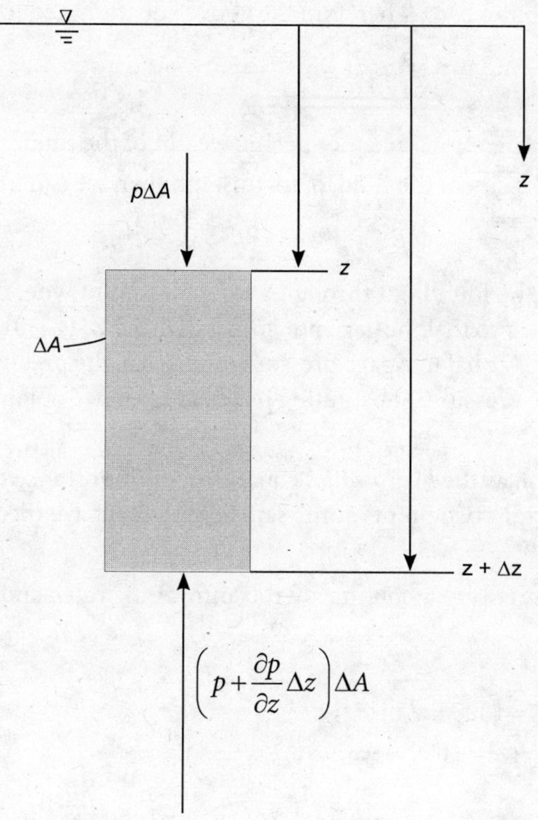

Figure 2.6: Pressure distribution for a fluid element.

Therefore, at this point the pressure can be written in terms of the pressure at through the Taylor series expansion

$$p(z+\Delta z)=p(z)+\frac{\partial p}{\partial z}\Delta z+\frac{1}{2}\frac{\partial^2 p}{\partial z^2}\Delta z^2+\ldots \tag{2.15}$$

In the above Taylor series expansion we only retain the first two terms as the terms appear after these have successively lower orders of magnitude. We can now perform a force balance in the direction.

$$p\Delta A-\left(p+\frac{\partial p}{\partial z}\Delta z+\frac{1}{2}\frac{\partial^2 p}{\partial z^2}\Delta z^2+\ldots\right)\Delta A+\rho g\Delta z\,\Delta A=0$$

$$\frac{\partial p}{\partial z}\Delta z\Delta A+\rho g\Delta A\Delta z+\frac{1}{2}\frac{\partial^2 p}{\partial z^2}\Delta z^2\Delta A+\ldots=0 \tag{2.16}$$

$$\frac{\partial p}{\partial z} + \rho g + \frac{1}{2}\frac{\partial^2 p}{\partial z^2}\Delta z + \ldots = 0$$

In the limit as $\lim_{(\Delta z \to 0)}$, we observe that the last term and all subsequent terms on the left hand side in the above experssion becomes zero. Therefore the above equation, reduces to

$$\frac{\partial p}{\partial z} + \rho g = 0 \tag{2.17}$$

where we recall that $\gamma = \rho g$ represents the specific weight of the fluid.

If we assume that the density of the fluid, ρ, is constant, then we can integrate Eq. (2.17) to obtain

$$\int dp = \int \rho g dz \Rightarrow p = \rho g z + C \tag{2.18}$$

where the constant can be found out through a reference point where p is known. Typically this is chosen as the free surface itself so that the gage pressure is zero, i.e. $p_g\,(z = 0) = 0$ which simply means that the pressure at the interface is equal to a gage pressure of zero, i.e. the pressure at the interface is equal to p_a, which is the atmospheric pressure. Substituting $p = p_a$ at $z = 0$, we obtain

$$p = p_a + \rho g z \tag{2.19}$$

We refer to this equation as the *hydrostatics* equation. An importance consequence of this equation is that in a liquid at equilibrium two points at the same depth will have the same pressure.

Example 2.1

Determine the pressure exerted by a column of 100 mm of a) water and b) mercury having specific gravity of 13.6.

Solution:

Given: $h = 0.1$ m, $\rho_{water} = 1000$ kg/ m³

(a) For water, $p = \rho_{water}\, gh = 1000 \times 9.81 \times 0.1 = 981$ Pa

(b) For mercury,specific gravity $= \dfrac{\rho_{mercury}}{\rho_{water}} \Rightarrow \rho_{mercury} = 13600 kg/m^3$

$$p = \rho_{mercury}\, gh = 13600 \times 9.81 \times 0.1 = 13341.6 \text{ Pa}$$

Example 2.2

Consider a piston cylinder arrangement as shown in the figure. The piston has a diameter of 30 cm. A weight is placed on top of it so that the force acting on the piston is 500 N. Find out the pressure in the hydraulic fluid.

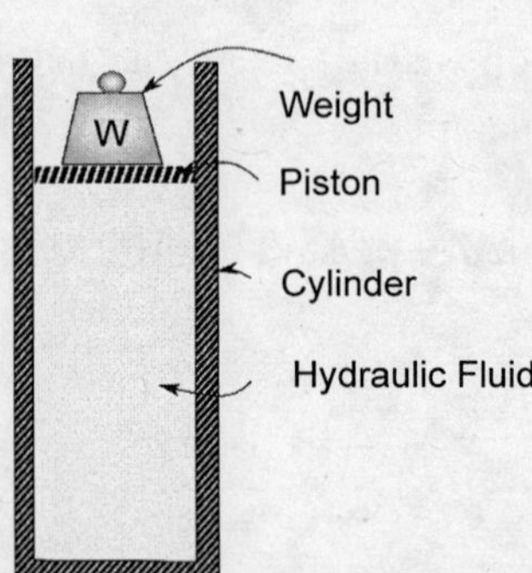

Solution:

The pressure in the liquid is given by

$$P = \frac{Force\ on\ the\ piston}{Area\ of\ the\ piston} = \frac{500}{\pi \times 0.3^2 / 4} = 7073.25N/m^2 = 7073.25Pa$$

Example 2.3

A container has two immiscible fluids A and B placed on top of each other as shown in figure. The column height of fluid A is $h_A = 0.2$ m and for fluid B is $h_B = 0.2$ m. The specific gravities of fluid A and B are 1.5 and 0.8 respectively. Determine the pressure exerted at a) the interface of fluid A and B, b) the bottom of fluid B.

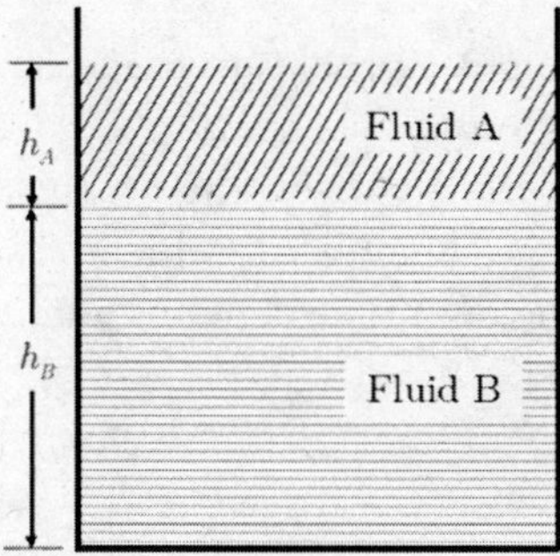

Solution:

Given: $\rho_A = 1500$ kg/ m³, $\rho_B = 800$ kg/m³, $h_A = 0.2$ m, $h_B = 0.5$ m

(a) Pressure at interface of fluids A and B is due to the column of fluid A above it

$$p_{interface} = \rho_A g h_A = 1500 \times 9.81 \times 0.2 = 2943 \text{ Pa}$$

(b) Pressure at the bottom of fluid B is due
to the total column height of fluid A and fluid B above it.

$$p_{bottom} = \rho_A g h_A + \rho_B g h_B = (1500 \times 0.2 + 800 \times 0.5) \times 9.81 = 6867 \text{ Pa}$$

Example 2.4

Consider an open tank which contains water and oil of specific gravity of 0.9. Water has a larger density than oil and hence forms a layer at the bottom of height 1 m. The oil forms a layer of thickness 2 m. Find out the pressure at the bottom of the tank and at the oil-water interface.

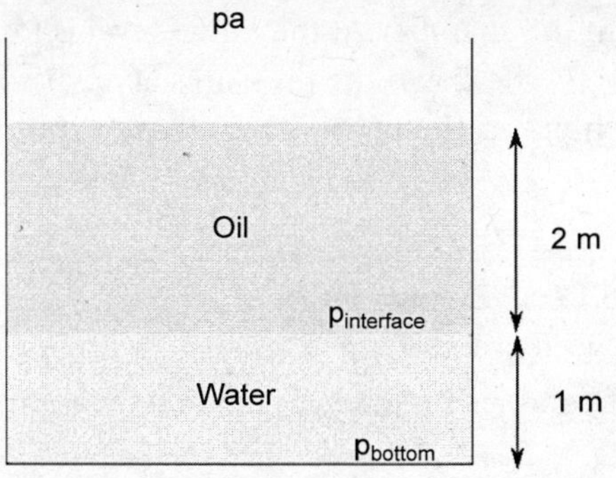

Solution:

Using the hydrostatics equation we can write down the pressure at the oil water interface as

$$p_{\text{interface}} = p_a + \rho_{\text{oil}} \times g \times h_{\text{oil}}$$

$$\Rightarrow \quad p_{\text{interface, gage}} = p_{\text{interface}} - p_a = \rho_{\text{oil}} \times g \times h_{\text{oil}} = 0.9 \times 10^3 \times 9.81 \times 2 = 17658 \text{ Pa}$$

And on similar lines we can write down the pressure at the bottom as

$$p_{\text{bottom}} = p_{\text{interface}} + \rho_{\text{water}} \times g \times h_{\text{water}}$$

$$p_{\text{bottom}} = p_a + 17658 + 10^3 \times 9.81 \times 1$$

$$p_{\text{bottom, gage}} = 17658 + 9810 = 27468 \text{ Pa}$$

2.6 MANOMETRY

The equation derived above, i.e. equation (2.19), helps us to measure pressures at surfaces which support liquid. Manometry is one such application of the equation. Let us first reiterate what the equation tells us. $p = p_a + \rho g z$ tells us that as the depth increases, the pressure increases linearly. Now consider an apparatus shown in the Fig. 2.7. There is a chamber A which has a pressure $p = p_A$ which is connected to the atmosphere with a bent tube. The bent tube contains a liquid of density ρ_2 while the chamber A contains liquid of density ρ_1.

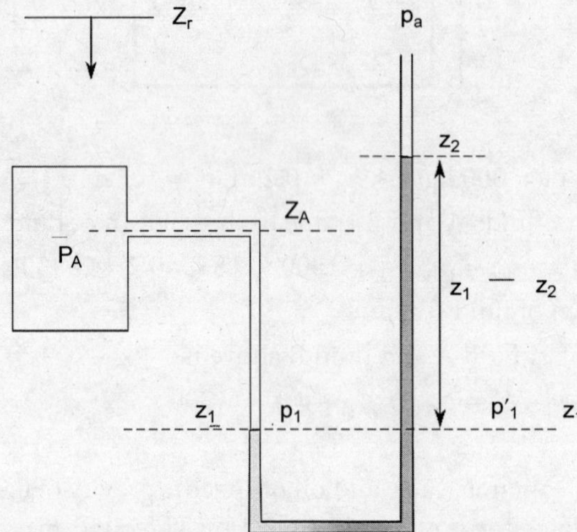

Figure 2.7: Using a u-tube manometer to measure the pressure inside a vessel.

Due to the higher pressure in the chamber A, the liquid level is lower in the left limb as compared to the right limb. We can therefore write down the pressures at any two arbitrarily chosen points 1 and 2. We can assume any reference point in the problem as all the expressions involve differences in two relative heights, i.e.

$$\text{Height of point 1} = z_1 - z_{\text{ref}}$$

$$\text{Height of point 2} = z_2 - z_{\text{ref}}$$

$\Rightarrow$ Difference in heights between points 1 and 2 $= z_1 - z_{\text{ref}} - (z_2 - z_{\text{ref}}) = z_1 - z_2)$

Returning to the problem at hand, using the hydrostatics equation we can write down the pressure at point 1 as

$$p_1 = p_A + \rho_1 g(z_A - z_1) \quad (2.20)$$

Similarly we can write down the pressure at point 1' as

$$p_{1'} = p_a + \rho_2 g(z_2 - z_1) \quad (2.21)$$

where p_a represents the atmospheric pressure.

Now given that there is no flow through the system, both the pressures at point 1 and 1' have to be equal as they both lie at the same level. The hydrostatics equation asserts us two points at the same height will necessarily have the same pressure at equilibrium.

Therefore, we may equate the two pressures above to obtain

$$p_A + \rho_1 g(z_A - z_1) = p_a + \rho_2 g(z_2 - z_1)$$

$$\Rightarrow \quad p_A - p_a = \rho_2 g(z_2 - z_1) - \rho_1(z_A - z_1) \quad (2.22)$$

To know more about manometry

Example 2.5

Consider the flow of a fluid with density ρ_1 through a pipe as shown in the figure. A u-tube manometer containing a liquid with density ρ_2 is connected between two points A and B in order to measure the difference in pressure. The dip in the manometric fluid at location A is h_1 while the dip in the manometric fluid at location B is h_2. The dips are measured with reference to the tube surface where the u-tube is connected. Find out the pressure difference $p_A - p_B$.

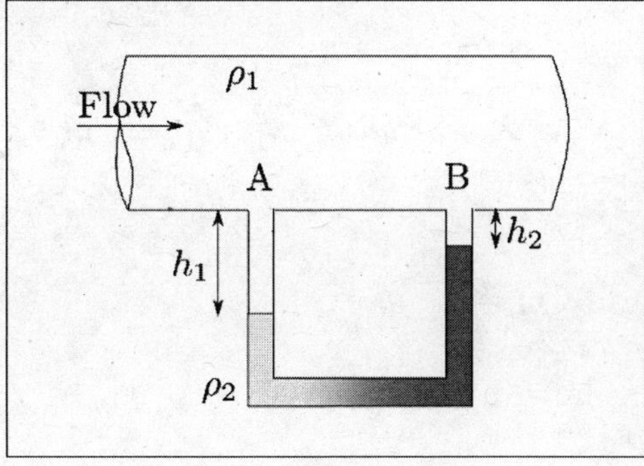

Solution:

We can write down the pressure at $z = h_1$ for both the limbs.

Left limb: $\quad p_\alpha = p_A + \rho_1 g h_1$

Right limb: $\quad p_\alpha = p_B + \rho_1 g h_2 + \rho_2 g(h_1 - h_2)$

$$\Rightarrow \quad p_\alpha + \rho_1 g h_1 = p_B + \rho_1 g h_2 + \rho_2 g(h_1 - h_2)$$

$$\Rightarrow \quad p_\alpha - p_B = \rho_1 g(h_2 - h1) + \rho_2 g(h_1 - h_2)$$

Example 2.6

Consider a fluid media whose density ρ varies with depth *h* and is given by the expression: $\rho = \rho_o e^{(-h)}$ where ρ_o is a constant and equal to the density of fluid at the top surface. Derive an expression for the gauge pressure inside this fluid column of height *h*, taking the pressure on the top surface of fluid as reference.

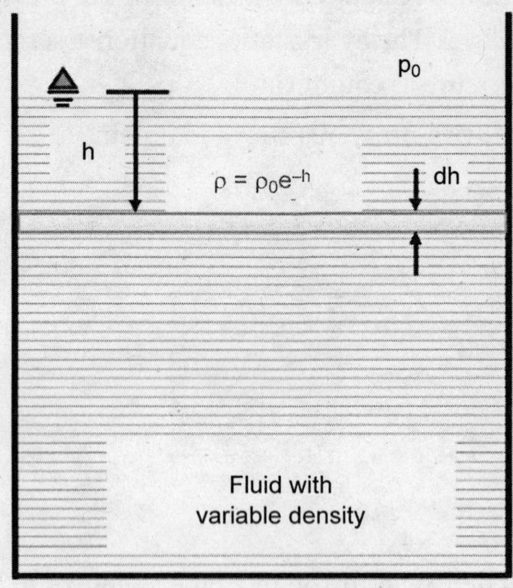

Solution:

Given: The density of fluid varies with depth, $\rho = \rho_o e^{(-h)}$

Consider an infinitesimal fluid column of height *dh* at a depth of *h*

The pressure exerted by this fluid column is given by:

$$d_p = \rho g dh$$

$$\Rightarrow \int_{P_o}^{P} dp = \int_{0}^{h} \rho g dh \Rightarrow \int_{P_o}^{P} dp = \int_{0}^{h} \rho_o g e^{-h} dh$$

$$\Rightarrow p - p_o = \rho_o g \int_{0}^{h} e^{-h} dh$$

$$\text{Gauge pressure} = (p - p_o) = \rho_o g(1 - e^{-h})$$

Example 2.7

Atmosphere exerts a pressure of P_o at the earth's surface. The acceleration due to gravity (*g*) at any height *h* from earth's surface is given by: $g = g_o\left(\frac{R}{R+h}\right)^2$ where g_o represents the acceleration due to gravity on earth's surface and R is the radius of earth. Assume that the density of air ρ doesn't vary with height. Derive an expression for the pressure (*p*) exerted by the atmosphere of total height H assuming the absolute pressure above atmosphere to be zero.

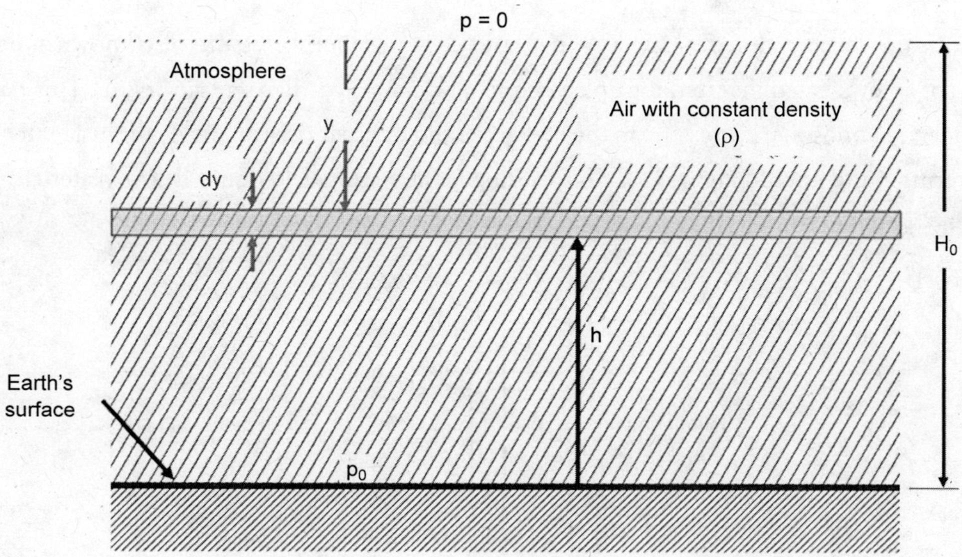

Given: Density (ρ) is constant with height,acceleration due to gravity $g = g_o\left(\frac{R}{R+h}\right)^2$.

Consider a thin element of air of height *dy* at a depth of *y* from the top.

It can be observed that $y + h = H$.

The change in pressure accross this element, $dp = \rho g dy$.

$$dp = \rho g_o\left(\frac{R}{R+h}\right)^2 dy$$

$$\int_0^p dp = \int_0^H \rho g_o\left(\frac{R}{R+h}\right)^2 dy$$

$$\because \qquad y + h = H \Rightarrow dy = -dh$$

$$\therefore \qquad \int_0^p dp = -\int_H^0 \rho g_o\left(\frac{R}{R+h}\right)^2 dh$$

$$\Rightarrow \qquad p = -\rho g_o R^2\left(-\frac{1}{R+h}\right)\Bigg|_H^0$$

$$\Rightarrow \qquad p = \rho g_o R^2\left(\frac{1}{R} - \frac{1}{R+H}\right)$$

Example 2.8

Figure shows the construction of a device to measure the atmospheric pressure, also known as barometer. It consists of an evacuated inverted tube placed into a liquid called "barometric liquid". This barometric liquid is open to atmospheric pressure in the container and hence it rises up inside the evacuated inverted tube. Determine the liquid height h inside this tube if the barometric liquid is: (a) Water (b) Mercury with specific gravity 13.6. The atmospheric pressure is p_{atm} = 101.325 kPa.

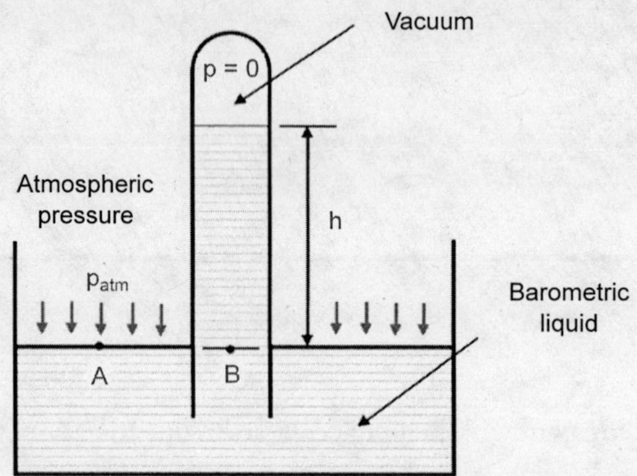

Given: Barometric liquid density

(a) $\rho = 1000$ kg/m³ (b) $\rho = 13600$ kg/m³, p_{atm}=101325 Pa

"Applying Pascal's law between point A and B we get,

$$p_{atm} = 0 + \rho g h \Rightarrow h = \left(\frac{p_{atm}}{\rho g}\right)$$

(a) $$h = \frac{101325}{1000 \times 9.81} = 10.33m$$

(b) $$h = \frac{101325}{13600 \times 9.81} = 0.759m$$

It is therefore observed that the barometric liquid should have high density to keep the liquid rise height within a tolerable range.

Example 2.9

An inclined manometer as shown in the figure is used to measure the pressure inside a channel through which water is flowing. The channel pressure p_i = 125 kPa and atmospheric pressure p_atm = 100 kPa. Assume the water column height h_1 = 10 cm and the manometric fluid is mercury with specific gravity = 13.6. Determine the total length of manometric fluid inside the right limb if (a) $\theta = 60°$ and (b) $\theta = 30°$.

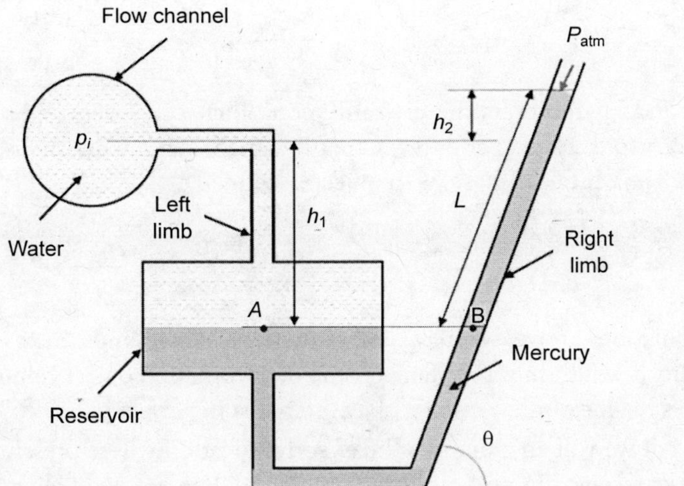

Given: Manometric liquid density

$$\rho_{mercury} = 13600 \text{ kg/m}^3,\ p_{atm} = 10^5 \text{ Pa},\ p_i = 1.25 \times 10^5 \text{ Pa},\ h_1 = 0.1 \text{ m}$$

Applying Pascal ′ s law between point A and B we get,

$$p_{atm} + \rho_{mercury}\, g(h_1 + h_2) = p_i + \rho_{water}\, gh_2$$

$$\Rightarrow \qquad (h_1 + h_2) = \frac{(p_i - p_{atm}) + \rho_{water}\, gh_2}{\rho_{mercury}\, g} \qquad (1)$$

Also we have $\sin\theta = \frac{(h_1 + h_2)}{L} \Rightarrow (h_1 + h_2) = L\sin\theta$ (2)

$\therefore$ From (1) and (2) we get: $L = \frac{(p_i - p_{atm}) + \rho_{water}\, gh_2}{\rho_{mercury}\, g\sin\theta}$

(a) $L = \frac{(1.25 - 1)\times 10^5 + 1000\times 9.81\times 0.1}{13600\times 9.81\times \sin 60^\circ} = 0.224m = 22.4cm$

(b) $L = \frac{(1.25 - 1)\times 10^5 + 1000\times 9.81\times 0.1}{13600\times 9.81\times \sin 30^\circ} = 0.389 = 38.9cm$

Reducing the angle of inclination makes the manometer more sensitive to pressure difference

2.7 FORCES ON SUBMERGED SURFACES

2.7.1 Centroid and Center of Mass

Before discussing the idea of the forces on submerged surfaces, it is instructive to discuss certain points useful throughout. The first one is **centroid**. Centroid refers to the geometric center of a 2D area. It may be imaged as a balance point for an area when the density of an object is constant. Mathematically, we can write this as

$$x_{CG} = \frac{\int x\,dV}{V} \quad y_{CG} = \frac{\int y\,dV}{V} \quad z_{CG} = \frac{\int z\,dV}{V} \tag{2.23}$$

where x_{CG}, y_{CG}, and z_{CG} represent the x, y, and z coordinate of the centroid respectively. The area of the object is V. The centroid and center of mass are the same when the density is uniform for the object under consideration. The **center of mass** of an object is defined as

$$x_{CM} = \frac{\int x\rho dV}{\int \rho dV} \quad y_{CM} = \frac{\int y\rho dV}{\int \rho dV} \quad z_{CM} = \frac{\int z\rho dV}{\int \rho dV}, \textit{where} \int \rho dV = M\,(\textit{mass}) \tag{2.24}$$

In both the calculations above, we may determine the centroid and center of mass for a 2D shape by considering a uniform width into the plane for the determination of the volume. Fundamentally, the center of mass defines the location where the object remains in perfect balance.

Another point that we will encounter is the effective location where pressure acts at a point on an object. This point is known as the center of pressure and will be defined below where we will see how the location can be determined. An astute reader will clearly see the relationship between the centroid, center of mass, and center of pressure.

The foundation of calculating forces on structures which are fully or partially submerged underwater depends on the ability to determine the force acting on a surface which is submerged under water. Typically in these calculations we will consider that one side of the surface is facing the liquid while the other side is simply at atmospheric pressure.

2.7.2 Force Acting on a Horizontal Surface

We first determine the force acting on a horizontal surface with respect to the free surface of water. Let the surface have an area of A. At each point on the surface the pressure is equal to $p = p_A + \rho gz$. Now under the assumption that the other side of the surface has an atmospheric pressure p_a, we can write down the net pressure acting on the surface as $p - p_a = \rho gz$. Therefore the force on a small element of area dA is equal to

$$\text{dF} = (p - p_a)dA = \rho gzdA$$

This force can then be integrated over the entire area as

$$F = \int_A dF = \int_A \rho gzdA = \rho gzA \tag{2.25}$$

where the last step of the integration is trivial owing to the fact that the area in the horizontal direction, density of the fluid, and gravity are not functions of the area coordinates. In fact the above equation may be also interpreted in the following way: $\rho gzA = \rho gV_{\text{over surface}} = m_{\text{over surface}}\,g = W_{\text{over surface}}$ which means that the force acting on the surface due to the gage pressure acting on it is equal to the weight of the liquid over the surface under consideration, which is represented over here by $W_{\text{over surface}}$.

The above force also passes through the centroid of the horizontal surface. The proof of this statement can be inferred from the more general case as shown in the subsection below which considers surfaces oriented at an arbitrary angle with respect to the horizontal free surface.

A list of commonly encountered cross section areas and their moments are listed below for convenience.

Mathematics Concept 2.2: Determining the location of centroid

Below we elucidate the procedure for obtaining centroids of certain commonly encountered shapes. The location of centroids can be found out using

$$x_{CG} = \frac{\int x\,dA}{A},\; y_{CG} = \frac{\int y\,dA}{A} \tag{2.26}$$

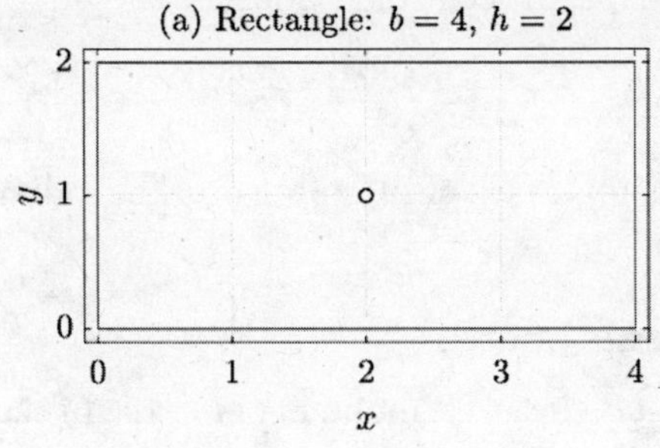

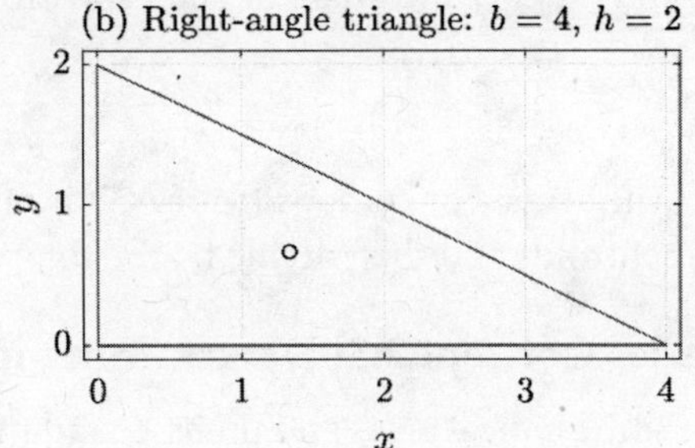

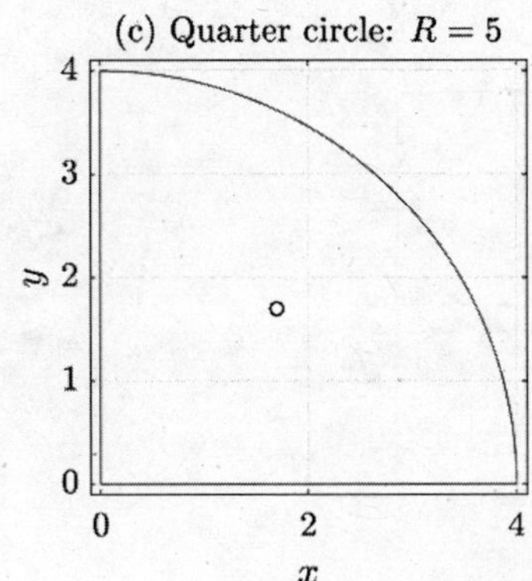

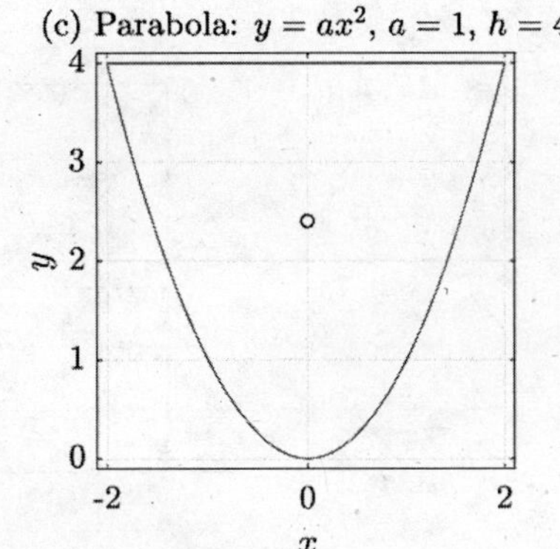

Center of mass for some common shapes and their dimensions

1. **Rectangle:** This is a very commonly encountered shape in hydrostatics, as many gates bear this shape. It is also perhaps the easiest shape for determining the location of the centroid. We can apply equation (2.26) to find out the two locations.
   First, we can write down the location of the centroid as

$$x_{CG} = \frac{\int xdA}{A} = \frac{\iint A^{xdxdy}}{\iint A^{dxdy}} = \frac{\int_0^b xdx \int_0^h dy}{bh} = \frac{b}{2} \tag{2.27}$$

   Similarly we can find out the location of the centroid as

$$y_{CG} = \frac{\int ydA}{A} = \frac{\iint_A ydxdy}{\iint_A dxdy} = \frac{\int_0^b dx \int_0^h ydy}{bh} = \frac{h}{2} \tag{2.28}$$

2. **Right-angled triangle:** The next shape we consider is that of a right angled triangle. We mention below directly the result of the centroid. We leave it as an exercise for the reader to undertake.

$$x_{CG} = \frac{b}{3} \ , \ y_{CG} = \frac{h}{3} \tag{2.29}$$

3. **Quarter circle:** For a quarter circle, the centroid is located at

$$x_{CG} = \frac{4R}{3\pi} \ , \ y_{CG} = \frac{4R}{3\pi} \tag{2.30}$$

4. **Parabola facing upwards:** For such a parabolic shape, the centroid is located as

$$x_{CG} = 0 \ , \ y_{CG} = \frac{3}{5}h \tag{2.31}$$

Composite areas: For composite areas, it is often a simple task of summing over the individual areas times their centroids and then divide by the total composite area.

Mathematics Concept 2.3: Determining moments of area

We list here some typical areas and their second moment of areas, i.e. moment of inertia. In each case the axis about which the moment is found out is indicated.

1. **Rectangle**

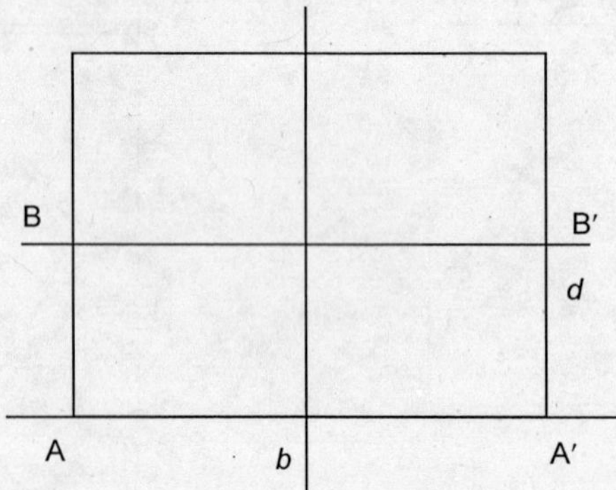

The moment of area of a rectangle of length b and width d is given by

$$About\ AA': \ \frac{bd^3}{3}$$

$$About\ BB': \ \frac{bd^3}{12}$$

2. **Triangle**

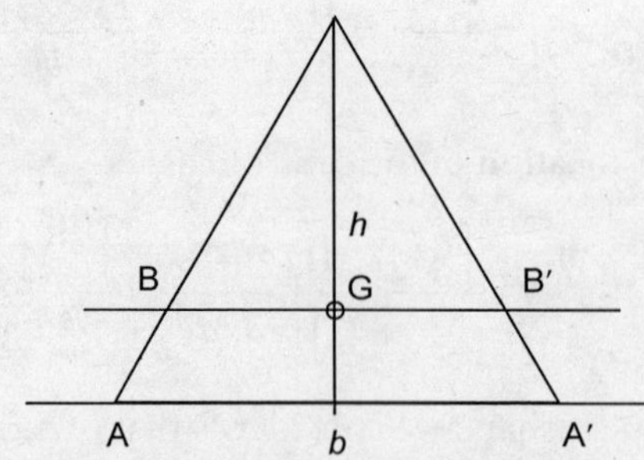

The moment of area of a triangle of base b and height h is given by

$$\text{About } AA': \quad \frac{bd^3}{12}$$

$$\text{About } BB': \quad \frac{bh^3}{36}$$

Note that in the above, axis BB' passes through the centroid, G.

3. **Circle:** One of the most commonly encountered shapes is a circle.

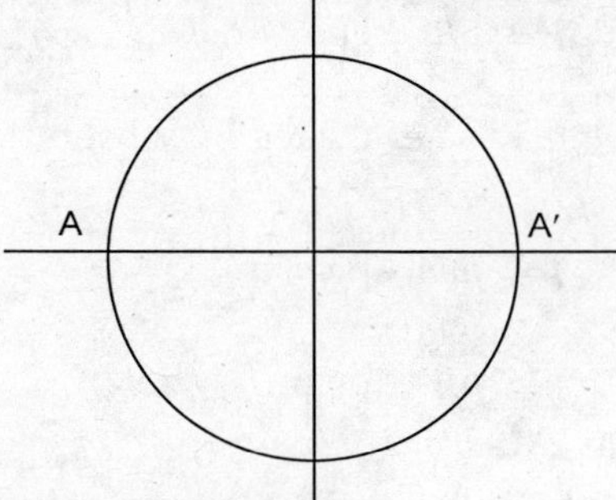

The moment of area of a circle of radius r is given by

About a axis in plane and along diameter: $\frac{1}{4}\pi r^4$

A better idea about such quantities can be obtained from solid mechanics texts such as Srinath, L. S. Advanced mechanics of solids. Tata McGraw-Hill, 2003.

2.7.3 Force on an Arbitrarily Inclined Flat Submerged Surface

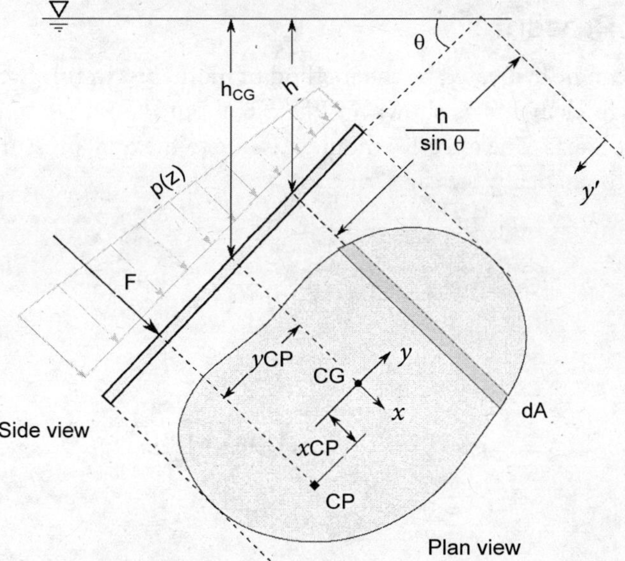

Figure 2.8: Force on a submerged inclined plate.

We first begin by determining the force on a submerged flat plate which makes an angle θ with the horizontal free surface of an arbitrary cross section as shown in Fig. 2.8. Owing to this, various parts of the surface experience a different depth; the part at a larger z experiences a larger pressure than the part at a smaller z as can be deduced through Eq. 2.19. The coordinate system is fixed to the centroid (CG) of the object such that y points along the slant length in the side view. We additionally define an auxilliary coordinate y' which originates at the free surface as shown in Fig. 2.8. Using Eq. (2.19), the total hydrostatic force can be evaluated as

$$dF = p\,dA = (p_a + \rho gh)dA$$

$$\Rightarrow \quad F = \int dF = \int (p_a + \rho gh)\,dA = p_a A + \rho g \int h\,dA \qquad (2.32)$$

The integral appearing above may be evaluated by substituting $h = y'\sin\theta$, thereby allowing us to further simplify

$$F = p_a A + \rho g \int h\,dA$$

$$\Rightarrow \quad = p_a A + \rho g \sin\theta \int y' dA$$

Further utilizing $\int y' dA = y'_{CG} A$.

$$F = \text{pa } A + \rho g \sin\theta \, y'_{CG} A$$

$$\Rightarrow \quad F = pa\, A + \rho g h_{CG}\, A \quad \because h_{CG} = y'_{CG} \sin\theta \qquad (2.33)$$

$$\Rightarrow \quad F = p_{CG} A$$

Force on one side of the plate = Pressure at the centroid × Area of plate"

Notably, the magnitude of the result is independent of the angle at which the plate is immersed in water.

2.7.4 Center of Pressure

The center of pressure can be evaluated by the method of moments. In this, we equate the total moment of the forces on the strip elements as shown in Fig. 2.6 about the x axis and equate it to an effective moment about the same x axis. The effective moment is due to the total pressure or net force determined through Eq. (2.33). Mathematically we can write

$$Fy_{CP} = \int yp\,dA$$

$$= \int y(p_a + \rho g \sin\theta y')\,dA$$

$$= \rho g \sin\theta \int yy'\,dA$$

$$= \rho g \sin\theta \left(y'_{CG} \int y\,dA - \int y^2 dA \right)$$

$$= 0 - \rho g \sin\theta \int y^2 dA$$

$$Fy_{CP} = -\rho g \sin\theta \int y^2 dA.$$

In the above expressions we have made use of the coordinate transform $y' = y'_{CG} - y$ and also made use of the fact that $\int y\, dA = 0$ because the moment of area about the centroid is zero.

Thus, we have an expression for the location of the center of pressure as

$$F_{y_{CP}} = -\rho g \sin\theta \int y^2 dA = -\rho g \sin\theta I_{xx} \tag{2.34}$$

$$y_{CP} = -\rho g \sin\theta \frac{I_{xx}}{F} = -\rho g \sin\theta \frac{I_{xx}}{p_{CG} A} \tag{2.35}$$

In the above expression I_{xx} represents the second moment of area about the x axis. The location of the center of pressure depends on the inclinration of the plate with the horizontal free surface.

All the terms in the expression for y_{CP} are positive and hence the negative sign indicates that the center of pressure is at a larger depth than the location of the centroid. This can be understood by the deeper regions of the plate experiencing a larger pressure thus leading to a shift in the bias of the force towards larger depths. As we move towards a larger h_{CG}, i.e. deeper depth, the magnitude of F increases, thereby reducing the magnitude of y_{CP}, which implies that at larger depths, the center of pressure approaches the centroid.

On similar lines, we can determine the location of the centre of pressure along the x axis.

$$F x_{CP} = \int xp\, dA = \int x\left[p_a + \rho g\left(y'_{CG} - y\right)\sin\theta\right] dA = -\rho g \sin\theta \int xy\, dA$$

$$x_{CP} = -\rho g \sin\theta \frac{I_{xy}}{p_{CG} A} \tag{2.36}$$

In contrast to the expression for y_{CP}, which is necessarily negative, we see that the shift in x_{CP} can be either positive or negative, depending on the sign of I_{xy}, which is the product of inertia of the plate. For objects which have a symmetric about both the and axis, we note that $I_{xy} = 0$.

2.8 FORCES ON CURVED SURFACES

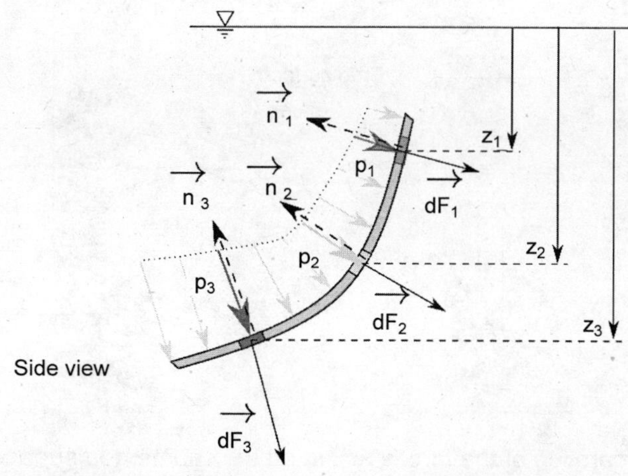

Figure 2.9: Force acting on a curved surface.

The process of determining the force acting on a flat surface can now be extended to curved surfaces. A difficulty in determining force acting on a curved body may be elucidated through Fig. 2.9. Before that, we first describe the problem at hand. We have a curved surface as shown, which may be assumed to have a width, w, into the plane. We have chosen three representative areas, dA_1, dA_2, and dA_3, which are located at a depth of z_1, z_2, and z_3 respectively. Due to the curved nature of the surface, the area normal vector $\vec{n}$ also changes its direction, although it has a unit magnitude. The normal vectors n_1, n_2, and corresponding to areas dA_1, dA_2, and dA_3, are also depicted in Fig. 2.9. The forces acting on the three differential areas can be written as

$$d\vec{F}_1 = p_1 \times dA_1 \times (-\vec{n}_1) \tag{2.37}$$

$$d\vec{F}_2 = p_2 \times dA_2 \times (-\vec{n}_2) \tag{2.38}$$

$$d\vec{F}_3 = p_3 \times dA_3 \times (-\vec{n}_3) \tag{339}$$

While noting that the negative sign in the terms above signify that the force due to pressure is compressive. Also note that we considering the force acting on only one side of the curved interface, as has been already done in the previous derivation for a flat plate. It is clear that the total force acting on the curved surface would be the sum of all such contribution, i.e.

$$\vec{F} = \Sigma_i^N p_i \times dA \times \left(-\vec{n_i}\right)$$

$$\vec{F} = \int_A (-p\vec{n}dA) \tag{2.40}$$

In the above equation, performing the integral on the right hand side would yield the effective force acting on the surface, which may then be broken up into its components along the i, j and k directions. However, the pressure is not a constant; it varies linearly with the depth from the free surface. Moreover, the normal vector also changes its direction as we move along with the surface. This makes evaluating the integral extremely tedious and complicated.

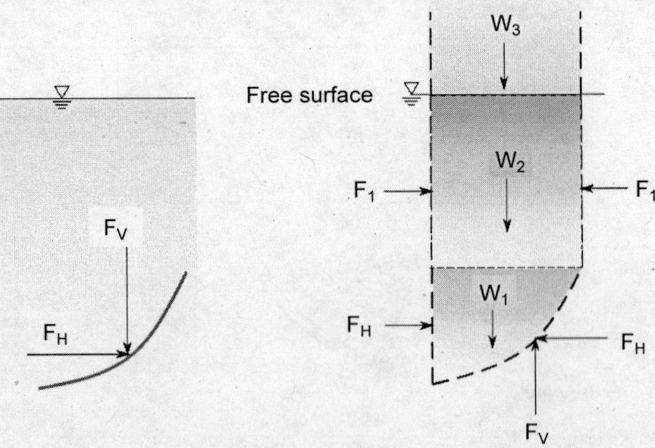

Figure 2.10: (a) Separation of the forces acting on the submerged surface in the horizontal and vertical directions. (b) All the different forces acting on the volume above the curved surface. Note the forces F_H and F_V acting on the opposite direction to that on the surface due to Newton's third law.

Alternately, the determination of the force due to pressure can be readily determined by decomposing it into its horizontal and vertical components. For demonstrating this, we consider Fig. 2.10.

If we consider the free body diagram shown on the right, at equilibrium we can write

$$F_v = \text{Weight of column of water} + \text{Weight of air} \tag{2.41}$$

$$F_v = W_1 + W_2 + W_3 \tag{2.42}$$

From the above, we can conclude that the vertical component of the force due to the pressure acting on the curved plate is equal to the weight of the column of fluid, which includes both the water and air in this case, which is above the curved surface.

Therefore in order to find out the vertical force acting on the curved surface we simply must find out the volume of the fluids on top. This has to be then multiplied with the density of the fluids respectively.

The evaluation of the horizontal force is done in a similar fashion. At equilibrium we can write down the following relationship between the forces:

$$F_H(\text{Curved surface}) = F_H(\text{Vertical area projection of the curved surface}) \tag{2.43}$$

Note that we have not mentioned the forces F_1 in the expression above. This is because the two forces are equal and opposite in nature. The forces are equal because the area on which the pressure acts is exactly the same and is at the same depth. This leads to the F_1 acting on the two horizontal faces equal in magnitude. The direction of the force on the left vertical face is in the positive x direction while the direction of the force on the right vertical face is in the negative x direction, thereby cancelling out.

Example 2.10

Determine the magnitude of the resultant force on the slanted portion of a dam as shown in figure. The height 10 m and angle θ = 45° and the fluid is water. Consider the depth of this dam as unity.

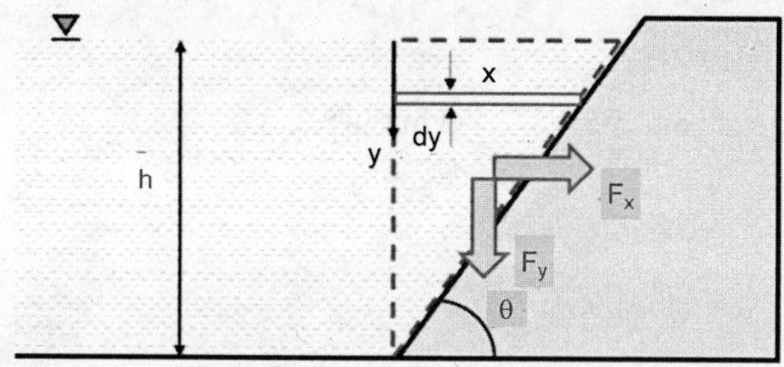

Given: $h = 10$ m, $\theta = 45°$, $\rho = 1000$ kg/m³

$$\text{Horizontal force, } F_x = \int_0^h \rho g y dy = \rho g h\left(\frac{h}{2}\right) = 1000 \times 9.81 \times 10 \times 5 = 490.5 kN$$

Vertical force, F_y = weight of water enclosed in the triangle

Consider a small element of width x and height dy.

Weight of water in this element is given by:

$$F_x = \int_0^h \rho g x dy$$

The relationship between y and x is: $\tan 45° = \frac{y}{x} \Rightarrow y = x$

Substituing $y = x$ in the above equation we obtain:

$$F_y = \int_0^h \rho g y dy = \frac{\rho g h^2}{2} = 490.5kN$$

$$\therefore \quad F_{net} = \sqrt{F_x^2 + F_y^2} = 693.7kN$$

Example 2.11

Determine the magnitude of the resultant force on the curved portion of a dam as shown in figure. The height 9 m and the fluid is water. Consider the depth of this dam as unity.

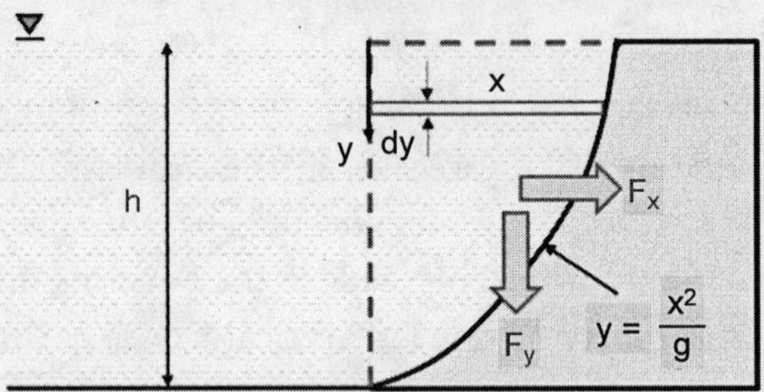

Given: $h = 9$ m, $\rho = 1000$ kg/m³, curved surface is given by $y = \frac{x^2}{9}$

$$\text{Horizontal force, } F_x = \int_0^h \rho g y dy = \rho g h\left(\frac{h}{2}\right) = 1000 \times 9.81 \times 9 \times 4.5 = 397.3kN$$

Vertical force, F_y = weight of water enclosed in the triangle

Consider a small element of width x and height dy.

Weight of water in this element is given by:

$$F_x = \int_0^h \rho g x dy$$

The relationship of the curve is: $y = \frac{x^2}{9}$

Substituting this in the above equation we obtain:

$$F_y = \int_0^h \rho g\left(3\sqrt{y}\right)dy = 3\rho g\left[\frac{h^{3/2}}{3/2}\right] = 2 \times 1000 \times 9.81 \times 9^{3/2} = 529.7kN$$

$$\therefore \quad F_{net} = \sqrt{F_x^2 + F_y^2} = 662.1kN$$

Example 2.12

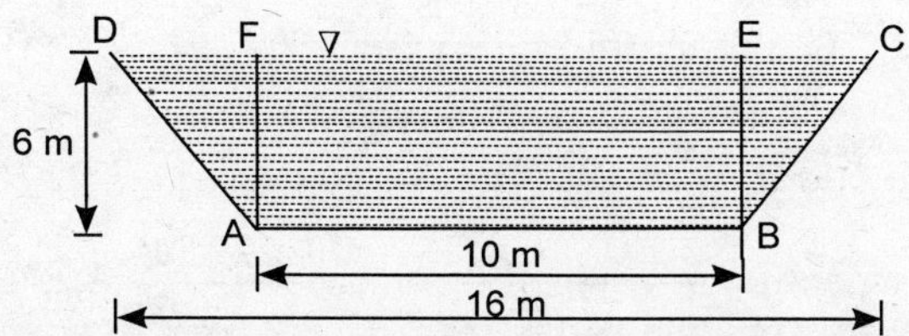

A trapezoidal channel 16 m wide at the top and 10 m wide at the bottom is 6 m deep. Determine the depth of centre of pressure for this channel measured from the top free surface.

Given: Channel is trapezoidal with $a = 10$ m, $b = 16$ m and $h = 6$ m

C.G. of trapezoid is at a depth of:

$$\bar{h} = \frac{C.G.\ of\ \Delta BEC \times ar(\Delta BEC) + C.G.\ of\ \Delta BEC \times ar(\Delta AFD) + C.G.\ of\ ABEF \times ar(ABEF)}{ar(ABCD)}$$

$$\bar{h} = \frac{\left(\frac{h}{3}\right)\times\left[\frac{1}{2}\times\frac{(b-a)}{2}\times h\right] + \left(\frac{h}{3}\right)\times\left[\frac{1}{2}\times\frac{(b-a)}{2}\times h\right] + \frac{h}{2}\times(a\times h)}{\left[\frac{1}{2}\times\frac{(b-a)}{2}\times h\right] + \left[\frac{1}{2}\times\frac{(b-a)}{2}\times h\right] + (a\times h)}$$

$$\bar{h} = \frac{2\times\left(\frac{1}{2}\times 3\times 6\right) + 2\times\left(\frac{1}{2}\times 3\times 6\right) + 3\times(10\times 6)}{\left(\frac{1}{2}\times 3\times 6\right) + \left(\frac{1}{2}\times 3\times 6\right) + (10\times 6)} = \frac{216}{78} = 2.769m$$

(a) M.I of ΔBEC about C.G. of trapezoid

$$= M.I\ of\ \Delta BEC\ about\ its\ C.G. + ar(\Delta BEC)\times\left(h_{C.G} - \bar{h}\right)^2$$

$$\Rightarrow \text{M.I of ΔBEC about C.G.of trapezoid} = \left[\frac{(b-a)}{2}\times\frac{h^3}{36}\right] + \left[\frac{1}{2}\times\frac{(b-a)}{2}\times h\right]\times\left(\frac{h}{3} - \bar{h}\right)^2$$

$$\Rightarrow \text{M.I of ΔBEC about C.G.of trapezoid} = \left[3\times\frac{6^3}{36}\right] + 9\times(2 - 2.769)^2 = 23.32m^4$$

(b) M.I of ΔAFD about C.G. of trapezoid

$$= M.I\ of\ \Delta AFD\ about\ its\ C.G. + ar(\Delta AFD)\times\left(h_{C.G} - \bar{h}\right)^2$$

⇒ M.I of ΔBEC about C.G. of trapezoid = 23.32 m^4

(c) M.I of ABEF about C.G.of trapezoid

$$= M.I\ of\ ABEF\ about\ its\ C.G. + ar(ABEF)\times\left(h_{C\ G} - \bar{h}\right)$$

$$\Rightarrow \text{M.I of ABEF about C.G.of trapezoid} = \left[\frac{a\times h^3}{12}\right] + [a\times h]\times\left(\frac{h}{2} - \bar{h}\right)^2$$

$\Rightarrow$ M.I of ABEF about C.G.of trapezoid $= \left[\dfrac{10\times 6^3}{12}\right] + 60\times(3-2.769)^2 = 183.2m^4$

M.I of ABCD about C.G.of trapezoid, $I_{ABCD} = 23.32 \times 2 + 183.2 = 229.84\ m^4$

Centre of pressure is at a depth $= \dfrac{I_{ABCD}}{ar(ABCD)\times\bar{h}} + \bar{h} = \dfrac{229.84}{78\times 2.769} + 2.769 = 3.83m$

2.9 BUOYANCY

Buoyancy is the phenomenon of hydrostatic lift on a body which is partially or fully submerged in a fluid caused by the net vertical component of the force due to hydrostatic pressure acting on the body.

(a) Completely submerged body

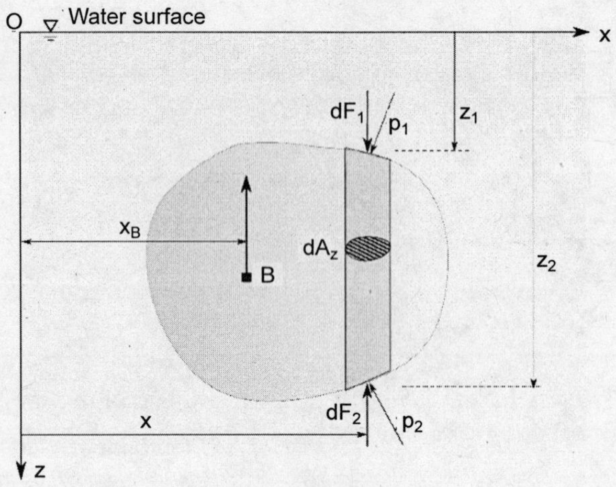

(b) Partially submerged body

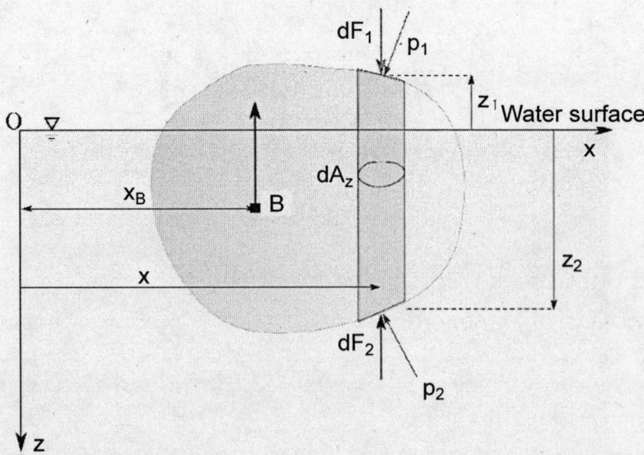

Figure 2.11: An arbitrarily shaped body submerged in a fluid. (a) Where the body is completely submerged in the liquid. (b) Where the body is only partially submerged in the liquid.

2.9.1 Body is Completely Submerged in the Fluid

Let us consider an arbitrarily shaped body submerged in a liquid as shown in Fig. 2.11. It is obvious from the derivation in section 2.8 that the net horizontal force acting on the body will be zero because the surface is a closed surface. The horizontal component of the force due to hydrostatic pressure integrated over the entire closed surface will cancel each other out. We are more interested to determine the force in the vertical direction. We can subdivide the object into elemental blocks as shown in the figure. If we consider one such block, the forces dF_1 and dF_2 can be determined as

$$dF_1 = (p_a + \rho g z_1)dA_z \tag{2.44}$$

$$dF_2 = (p_a + \rho g z_2)dA_z \tag{2.45}$$

where A_z represents the area projected in the xy plane. We can the combine the above two equations to obtain the net force in the z direction as

$$dF_B = dF_2 - dF_1 = \rho g(z_2 - z_1)\, dA_z = \rho g dV \tag{2.46}$$

where dV represents the elementary volume of the segment of the body. The total buoyancy force can then be determined as

$$F_B = \iiint_V \rho g dV = \rho g V \tag{2.47}$$

where it must be kept in mind that the density appearing in the equations above are the density of the medium in which the body is kept. For convenience, the weight of the body which is comprised of a material with density is

$$\mathrm{W} = \rho_b\, gV \tag{2.48}$$

2.9.2 Body is Partially Submerged in the Liquid

In this case where the body is only partially submerged, we can proceed with a similar derivation. We must note, however, that the top surface of the object is exposed to the ambient pressure, i.e. p_a. We can therefore proceed to write down the forces acting on the top face and bottom face as

$$dF_1 = (p_a)\, dA_z, \because \text{ Outer surface is at atmospheric pressure} \tag{2.49}$$

$$dF_2 = (p_a + \rho g z_2)dA_z \tag{2.50}$$

Now, we can add these two together to obtain the net force, i.e. the buoyancy force as

$$dF_B = dF_2 - dF_1 = (\mathrm{pa} + \rho g z_2 - p_a)\, dA_z = \rho g z_2\, dA_Z \tag{2.51}$$

$$\Rightarrow \quad dF_B = \rho g dV_{\mathrm{submerged}} \tag{2.52}$$

It is important to note here that the buoyancy force is obtained solely from the volume that is submerged rather than the total volume of the object.

We can integrate the above expression to obtain the total buoyancy force as

$$F_B = \iiint_{V_{submerged}} \rho g dV_{submerged} = \rho g V_{submerged} \tag{2.53}$$

We can determine the *line of action* along which the effective buoyant force acts by noting that

$$x_B F_B = \int x\, dF_B \Rightarrow x_B = \frac{\int x dF_B}{F_B} = \frac{\iiint_V x\, dV}{\iiint_V dV} = \frac{\iiint_V x\, dV}{V} \tag{2.54}$$

which essentially implies that the point where buoyancy acts is at the centroid of the volume displaced by the body.

This is known as Archimedes principle, which states that the buoyant force is equal to the weight of the liquid displaced by the submerged body. The location through which this buoyant force acts is the centroid of the volume displaced.

2.9.3 Equilibrium of a Floating/submerged Body

While the above discussion is that for a submerged body and the vertical force acting on it due to buoyancy, we can readily extend the idea for a body which is partially submerged, i.e. a body which is partly inside a liquid while the rest of it is above the free surface. Such cases are innumerable, such as ships, boats, and floating icebergs.

For such cases, we can still apply the same idea as above.

$$\text{At equilibrium, FB} = \text{weight of object} \quad (2.55)$$

$$\rho g V_{\text{submerged}} = \rho_m g V_{\text{total}} \quad (2.56)$$

which implies that under equilibrium, the total weight of the floating object, which is given by $\rho_m g V_{\text{total}}$, where ρ_m is the density of the material of the object, is balanced out by the buoyancy force acting on it, i.e. $\rho g V_{\text{submerged}}$, where ρ is the density of the fluid. Moreover, the two forces, i.e. the weight of the object and the buoyancy force have to be collinear as this ensures that there is no net moment acting on the body. If the weight and buoyancy were not to be collinear, then the body would face a net moment, causing it to oscillate, or even overturn. This is discussed in section 2.10.

Example 2.13

A small balloon at atmospheric pressure is filled with 50% (by volume) air and the remaining with water. This balloon when immersed in an oil of unknown specific gravity floats with 75% of the balloon (by volume) inside the oil and remaining outside. Determine the specific gravity of the oil. Neglect the mass of the balloon material. Assume density of air as 1.2 kg/m³.

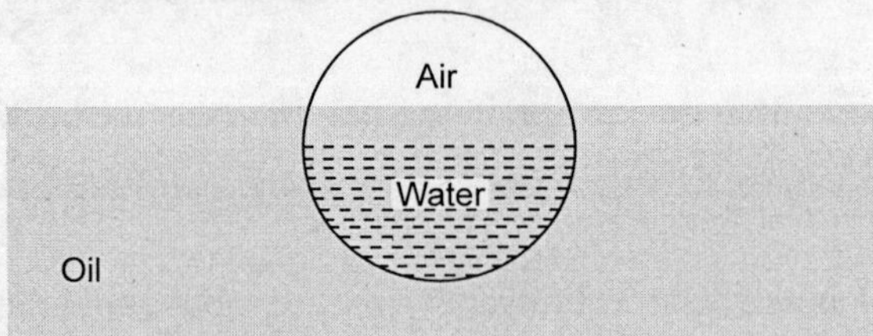

Given: 50% volume of balloon coccupied by air and rest is water,

Volume of oil displaced is 75 % of balloon ´s volume

For the balloon to float, we have:

Mass of fluid inside the balloon = Mass of fluid displaced by balloon

$$0.5V\rho_{\text{air}} + 0.5V\rho_{\text{water}} = 0.75V\rho_{\text{oil}}$$

$$\rho_{\text{oil}} = \frac{0.5\rho_{air} + 0.5\rho_{water}}{0.75} = \frac{0.5\times 1.2 + 0.5\times 1000}{0.75} = 667.5 kg/m^3$$

2.10 STABILITY OF FLOATING BODIES

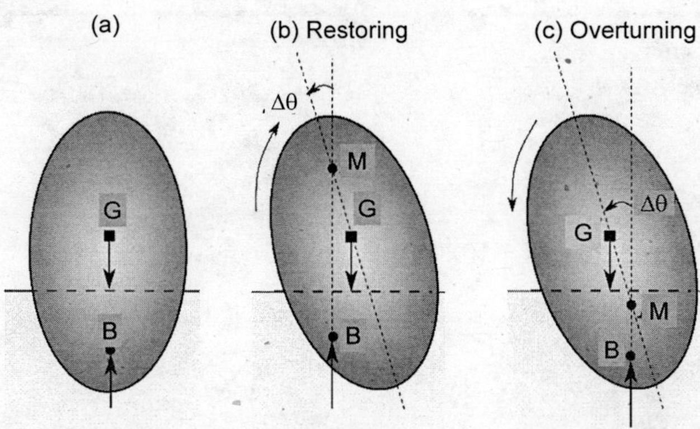

Figure 2.12: Stability of a floating body.

Fig. 2.12 shows a symmetric body floating partially in water. We can now discuss the stability of such floating objects. When the object is not acted upon by any disturbance, we see that the weight and buoyancy act collinear to each other. The weight of the object acts through the center of mass G. Based on the weight of the object, the buoyancy force can be determined and consequently the submerged volume is determined. The submerged volume also gives us the information about the center of buoyancy, B. This condition is the base condition. Now, we displace the object by an angle $\Delta\theta$. Owing to this, the waterline changes and a new submerged volume is achieved. Accordingly a new position of the point of buoyancy is obtained. This could be either subfigure (b) or (c) in Fig. 2.12.

We define the **metacenter**, *M*, as the point of intersection between the vertical line drawn from the center of buoyancy, *B*, and the symmetric axis of the object.

Restoring moment: If the point *M* is found to be above *G*, i.e. the case shown in subfigure (b), then we say that the moment caused by the weight of the object and the buoyancy force create a net moment which is in the opposite direction to the direction of the small disturbance. which causes the body to restore back to its original configuration.

Overturning moment: If the point is found to tbe below , as shown in subfigure (c), then we see that the moment caused by the weight of the object and the buoyancy force creates a net moment which causes the body to further turn in the direction of the disturbance, thereby creating a tendency for the body to overturn from its original equilibrium configuration.

The above situation may be easily be understood by the analogy of a cone kept on a table. If the cone is kept with its base on the table and given a disturbance, then the cone will again come back to its initial configuration. On the other hand, if the cone is somehow, with great difficulty, made to stand on its tip on the table surface and then imparted a small disturbance, then the cone will immediately fall onto its side. The former situation presents the case with a restoring moment, while the latter case presents the case with an overturning moment.

Stability of floating bodies

UNIT SUMMARY

- Pressure acting on a surface

$$p = \frac{F}{A}$$

- Force acting on a body

$$\vec{F_b} = \int_V \rho \vec{g} dV$$

- Force acting on a surface

$$\vec{F_s} = -\int_A p \, dA \, \vec{n}$$

- Gage pressure

$$p_{gage} = p - pa$$

- Hydrostatic equation

$$p = pa + \rho gz$$

- Taylor Series Expansion

$$f(x) = f(a) + f'(a)(x-a) + \frac{f''(a)}{2!}(x-a)^2 + \frac{f'''(a)}{3!}(x-a)^3 + \ldots$$

- General expression for centroids

$$x_{CG} = \frac{\int x\, dV}{V} \quad y_{CG} = \frac{\int y\, dV}{V} \quad z_{CG} = \frac{\int z\, dV}{V}$$

- Force of buoyancy on a submerged body

$$F_B = \iiint_V \rho g dV = \rho g V$$

EXERCISES

Multiple Choice Questions

1. Which of the following statement is TRUE?
   (a) Hydrostatic pressure remains same as we descend deeper into the ocean
   (b) Hydrostatic pressure decreases as we descend deeper into the ocean
   (c) Hydrostatic pressure increases as we descend deeper into the ocean
2. A container has two layers of immiscible fluids A and B with densities $\rho_a > \rho_b$. Consider case I: fluid A is above B and case II: fluid B is above A. Answer which of the following statement(s) are TRUE?
   (a) Hydrostatic pressure at the fluid interface for case I is higher than for case II
   (b) Hydrostatic pressure at the fluid interface for case II is higher than for case I

(c) Hydrostatic pressure at the bottom of the container is same for both cases
(d) Hydrostatic pressure at the bottom of the container is higher for case I

3. The buoyancy force acting on a body of volume V fully submerged into a fluid with density ρ is
(a) Greater than ρV
(b) Less than ρV
(c) Equal to ρV

4. The buoyancy force acting on a body of volume V partially submerged into a fluid with density ρ is:
(a) Greater than ρV
(b) Less than ρV item Equal to ρV

5. Which of the following statement is/are TRUE?
(a) Pascal ′s law is valid only for columns having identical fluids at rest.
(b) Pascal ′s law can be applied across columns containing multiple fluids at rest.
(c) Pascal ′s law cannot be applied for fluids at rest under vacuum.

6. For a body of volume V and mass M to float in a fluid of density ρ, which of the following statements is/are TRUE?
(a) $\frac{M}{V} \geq \rho$
(b) $\frac{M}{V} = \rho$
(c) $\frac{M}{V} \leq \rho$

7. For a thin plate submerged in a fluid, which of the following statement(s) is/are TRUE?
(a) The centre of pressure and gravity are always the same point
(b) The centre of pressure and gravity are the same point when the plate is vertically submerged
(c) The centre of pressure and gravity are the same point when the plate is horizontal

8. A thin plate is immersed vertically inside a fluid. It is unsymmetrical about both the axes passing through its centroid, then which of the following statement(s) is/are TRUE?
(a) The product area moment of inertia = 0
(b) The product area moment of inertia ≠ 0.
(c) The centre of pressure lies along the vertical line joining the free surface and centroid
(d) The centre of pressure lies along the vertical line joining the free surface and centroid

9. For a thin plate submerged in fluid at any arbitrary angle, the centre of pressure is
(a) Can be below or at the centre of gravity
(b) Always at the centre of gravity
(c) Always below the centre of gravity
(d) Can be above or below depending on the shape

ANSWERS															
1.	(c)	2.	(c)	3.	(c)	4.	(b)	5.	(b)	6.	(c)	7.	(d)	8.	(b)
9.	(a)														

Unsolved Questions

Level-I

1. What is the absolute pressure (in Pa) if the gauge pressure is: a) + 20 kPa, b) - 20 kPa? (Atmospheric pressure = 101325 Pa.)
2. Determine the height of mercury (specific gravity 13.6) column inside an inverted tube barometer if the pressure inside the evacuated region is 10 kPa (see figure for example 2.8)
3. A small metal ball of mass 8 gm and volume 1 cm³ is dropped inside a very long fluid column whose density changes with depth as $\rho = 1000(1 + h)$ kg/m³ where h is the depth (in m) from free surface. Determine the depth at which buoyancy force just balances the weight of this ball.
4. A spherical balloon of radius 15 cm is filled with air at density 1.2 kg/m³. Determine the displaced volume (in cm³) when this balloon is floating in a fluid of specific gravity 0.5.
5. A liquid of specific gravity 0.8 is placed inside a measuring cylinder. The height of this liquid column is 300 mm. The cylinder diameter is 60 mm. Determine the net force (in N) exerted by the fluid on the base of this cylinder.
6. For the above problem determine the net force (in N) on the curved surface of the cylinder due to the liquid?
7. Oil of specific gravity 0.8 flows through a pipe. Two pressure tapping (A and B) are made on this pipe. A differential U - tube manometer is connected between these two points (see figure for example 2.5). Mercury is used as the manometric fluid and the difference in mercury level between the two limbs is 20 cm. Determine the pressure difference between A and B (in Pa).
8. A single column manometer is used to measure the pressure inside a pipe with oil of specific gravity 0.9 flowing through it. The height difference between the center of pipe and the oil/mercury interface is 30 cm. The mercury in the right limb is 50 cm above this interface. Determine the gauge pressure (in Pa) in the pipe.

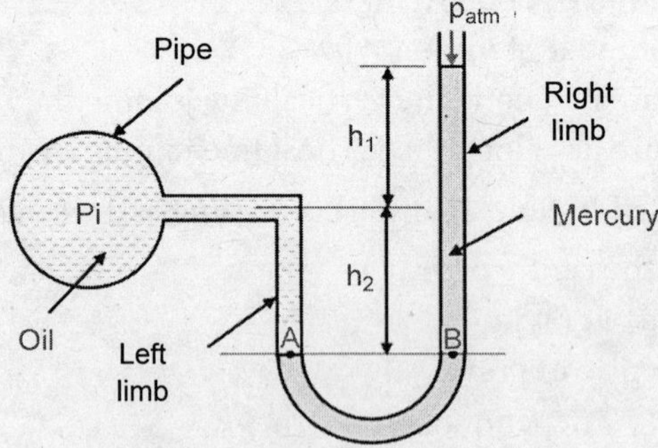

Figure 2.13: Figure for question 8.

9. A hydraulic press is used to lift a mass of 1000 kg. The ram diameter is 42 cm and the plunger diameter is 7 cm. The plunger and ram are at the same height ($h = 0$). Find the force needed (in N) to lift this mass.

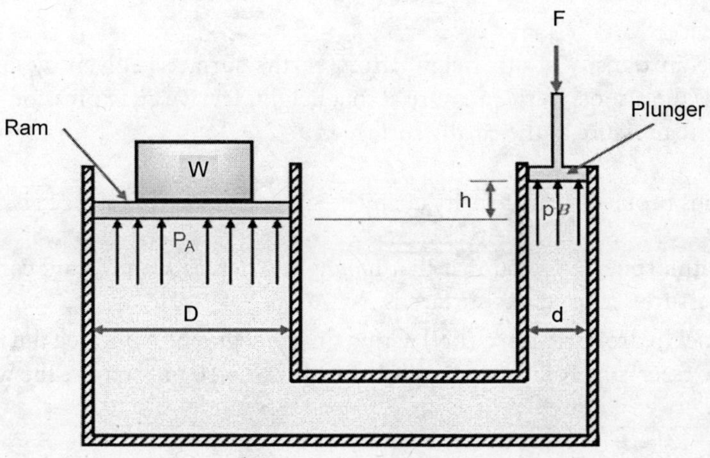

Figure 2.14: Figure for question 9.

10. Redo the above problem if the plunger is at a height = 10 cm above the ram. The density of fluid used: .

11. A manometer is used to measure the pressure of gas inside a channel (see the figure). The gas is air and the manometric fluid is water. Air has density of 1.2 ; and water has density of 1000 . The numerical value of = 10 cm = 5 cm. Determine the absolute pressure of gas if the atmospheric pressure = 101325 Pa.

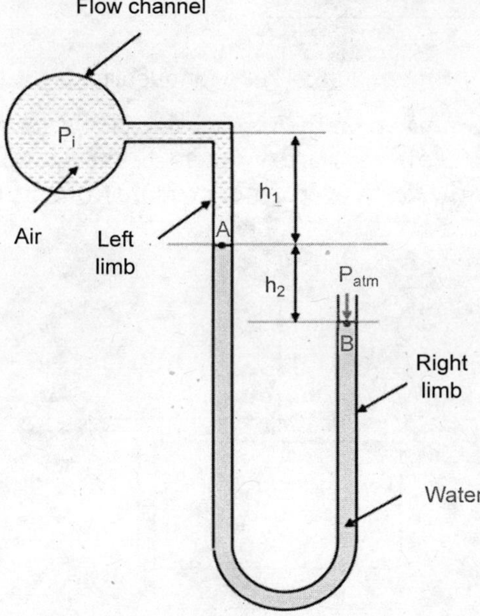

Figure 2.15: Figure for question 11.

Level - II

1. Assume that the air density at any height from earth's surface is given by: $\rho = kp$, where k is a constant and p is the atmospheric pressure at that height. Derive an expression for variation p with if the atmospheric pressure at the earth's surface is p_o.

2. Redo the previous problem if air density at any height h from earth's surface is given by: $\rho = \left(\frac{p}{C}\right)^k$ where p is the atmospheric pressure at that height h, whereas k and C are constants ($k > 1$). The atmospheric pressure at the earth's surface is .

3. Determine the net hydrostatic force (kN) acting on the segment ABC (see the figure) due to water contained in the reservoir. Total water column height, $2h = 10$ m. Assume the width of reservoir as unity.

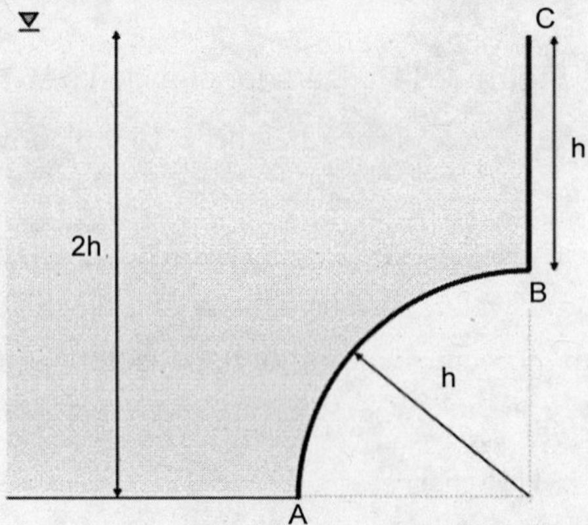

Figure 2.16: Figure for question 3

4. A rectangular sluice gate is situated on the vertical side of a water reservoir. The top edge of this gate is at a depth of h from the free surface of water (see the figure). The height of sluice gate is d and length is b. Determine the depth of center of pressure from the free surface.

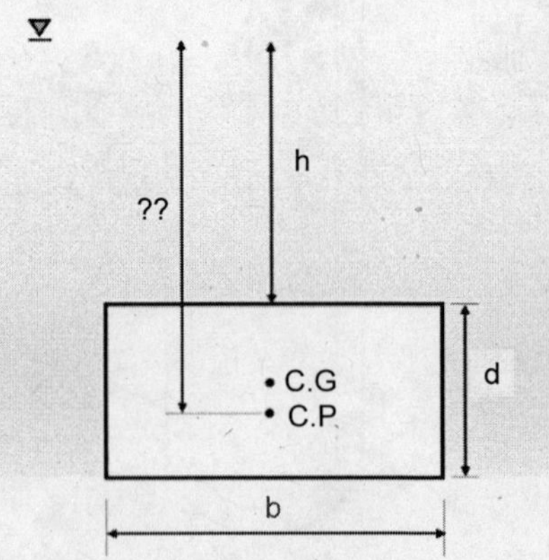

Figure 2.17: Figure for question 4

ANSWER KEY

Level - I

1. (a) 121325 Pa (b) 81325 Pa
2. 0.684 m
3. 7 m
4. 33.9
5. 6.6 N
6. 0
7. 25114 Pa
8. 64059 Pa
9. 272.5 N
10. 266.8 N
11. 10833 Pa

Level - II

1. $p = p_o\, e^{-kgh}$
2. $p = \left[p_o^{\,1-k} + \frac{g}{C^k}(k-1)h \right]^{\frac{1}{1-k}}$
3. 573.9 N
4. $y_{c.p} = \left(h + \frac{d}{2}\right) + \frac{d^2}{12\left(h + \frac{d}{2}\right)}$

To know more about hydrostatics

PRACTICAL

Aim: Estimating the density of commonly available liquids at home using Archimedes principle.

Apparatus: A transparent rectangular plastic container, a graduated beaker (500 ml), a small plastic bowl (it should be light enough to float in water), metal ball bearings, electronic weighing machine, plastic marker.

Fluids to be tested: Water, Vegetable Oil, any other fluid readily available at home.

Theory: This experiment is based on Archimedes principle. An object submerged in fluid with less density will displace more fluid when compared to fluid with higher density.

Procedure:

1. Mark graduations along the height on the plastic container at 5 mm intervals using a scale and marker.
2. Using the graduated beaker pour a known volume of water (ml) into the above mentioned plastic container. Note the water level in this container using the graduations made in step-1.
3. Repeat step-2 for several volumes and note the water level in the plastic box for each volume.
4. Plot the water volume (ml) with water level (mm). This data will provide a calibration of volume vs height for this plastic container (Fig. 1).
5. Now measure the weight of empty plastic bowl (in g) using an electronic weighing machine.
6. Add 20 ball bearings to this plastic bowl and note down the new reading on the electronic weighing machine. This data will provide you the average weight of each ball bearing.
7. Now pour water into the plastic container so that it fills almost half the volume of the plastic container. Note the water level (L_0) using the graduations marked along the height. (Fig. 2)

8. Carefully place the empty plastic bowl in this water. Add 5 ball bearings to this plastic bowl carefully. Ensure that the bowl doesn't sink. Note down the water level (L_1). (Fig. 2)
9. Add 5 more ball bearings and note the new water level (L_2).
10. Continue adding ball bearing for 2 more steps. Ensure that the bowl doesn't sink completely. If needed reduce the number of bearings added in each step to get more readings.
11. Plot the graph of total mass of plastic bowl + metal ball bearings for each case vs the water level measured for each case (Fig. 3).
12. Repeat steps 7-11 for vegetable oil and obtain the plot similar to Fig. 3.

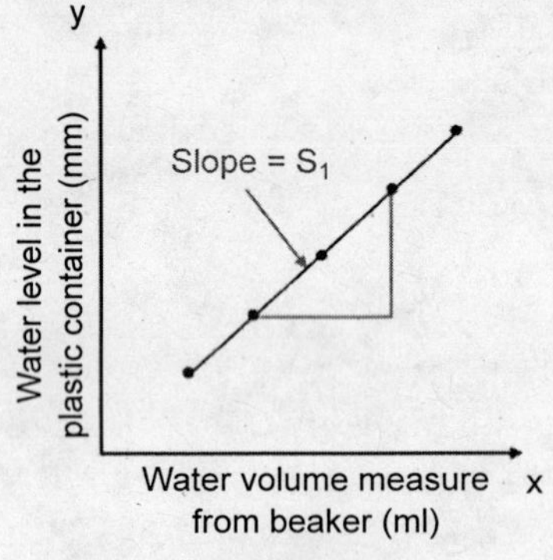

Figure 1

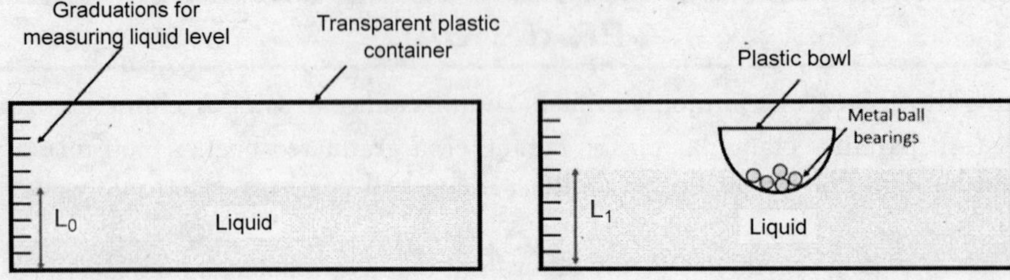

Figure 2

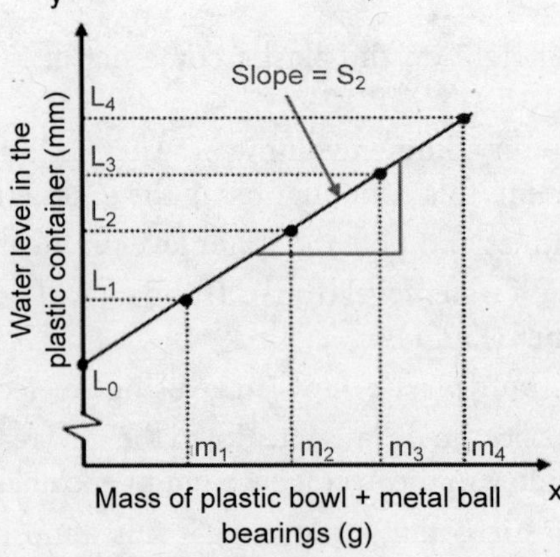

Figure 3

Calculations:

1. **True density calculation:**

   Using the graduated beaker and electronic weighing machine, estimate the true density ρ_l of the liquid by weighing the mass, M, of the unknown volume of liquid, V. The density is given by

$$\rho_l = \frac{M}{V} \tag{2.57}$$

2. Density estimated using Archimedes' principle:

$$Slope, S_1 = \frac{Chamge\ in\ water\ level(mm)}{Change\ in\ volume\ of\ water, ml} = \frac{\Delta L}{\Delta V} \tag{2.58}$$

$$Slope, S_2 = \frac{Change\ in\ water\ level, mm}{Change\ in\ mass\ of\ plastic\ bowl, g} = \frac{\Delta L}{\Delta m} \tag{2.59}$$

   For the plastic bowl and metal ball bearings to float, from Archimedes' principle, we have

$$\Delta mg = \rho_a \Delta V \Rightarrow \rho_a = \frac{\Delta m}{\Delta V} g \quad \therefore \rho_a = \frac{S_1}{S_2} g \times 1000 \tag{2.60}$$

   where ρ_a is the density of liquid estimated using Archimedes' principle.

Observations and Conclusions:

10. Compare the slopes, S_2, obtained from Fig. 3, for water and vegetable oil. Determine which liquid has a higher density.
11. Calculate the density of water and oil using both the methods mentioned above. Determine the density obtained by the two methods and state the reasons for this difference.

KNOW MORE

Archimedes' principle, a key law of buoyancy, explains that any object submerged in a fluid experiences an upward force equal to the weight of the fluid it displaces. For a fully immersed object, the displaced fluid's volume equals the object's volume, while a partially submerged object displaces fluid proportional to its submerged portion. The buoyant force acting on a floating object matches its weight, preventing it from sinking or rising.

For instance, a ship sinks into the ocean until the water it displaces equals its weight, and as cargo is added, it displaces more water. If an object is lighter than the displaced fluid, it rises, like a helium balloon, but if heavier, it sinks with an apparent weight reduction equal to the displaced fluid's weight. This principle even requires adjustments in precise weighings to account for air's buoyant effect.

[Source: https://www.britannica.com/science/Archimedes-principle]

REFERENCES AND SUGGESTED READINGS

1. **Introduction to Fluid Mechanics and Fluid Machines** (3rd Edition), S K Som, Gautam Biswas, Suman Chakraborty, McGraw Hill
2. **Fluid Mechanics** (9th Edition), Frank M. White, Henry Xue, McGraw Hill

❑❑

3 Fluids under Motion 1: Kinematics of Flow

UNIT SPECIFICS

In this unit we will discuss about the following aspects:

1. Particle and field description of fluid flow.
2. Various classifications of flow based on spatial dimensions and time.
3. Material derivative and its role in transforming the conservation equations from the particle description of fluid flow to the field description.
4. Various visualization techniques for flow field.
5. The fundamental kinematic aspects of fluid motion and deformation—rate of translation, rate of rotation, linear strain rate, and shear strain rate.

RATIONALE

Subsequent to discussion on fluid at rest, it is natural to proceed towards analysing fluid under motion. Mankind has been striving for centuries to have thorough understanding of fluid in motion as it governs naturally occurring fluid flow phenomenon. Cyclonic storms, waterfalls and river delta formation are few such examples where fluid motion dictates the natural phenomenon. Furthermore, many man-made fluidic devices, weapons and instruments are also built on principles of fluid in motion. For example, a good understanding of fluid flow is essential in making a plane fly, a ship to stay afloat in an unstable weather or a high speed train to move efficiently. Fig. 3.1 shows the vortex and the eye of cyclone Catarina whereas Fig. 3.2 shows a vortex couple generated by an airplane wing. The scales of both the vortices shown in the two figures are orders of magnitude apart but the mathematical description of such fluid flow patterns are similar.

Figure 3.1: Cyclone Catarina as viewed from the International Space Station; it struck Brazil in late March 2004 (Source: https://pixabay.com).

Broadly, the topics on fluid in motion are divided into: a) Fluid kinematics and b) Fluid dynamics. In fluid kinematics the description of fluid under motion is discussed in detail without delving into the forces that cause these motion. In this chapter on fluid kinematics we primarily focus on the mathematical relationships governing the fluid motion and various ways of classifying and visualising fluid flow. This chapter serves as the precursor to the chapter on fluid dynamics as until we introduce the rigorous mathematics behind description of fluid flow, the various forces leading to these flow patterns cannot be addressed.

Figure 3.2: Airplane wing vortex couple (Source: *https://www.wallpaperflare.com*).

PRE-REQUISITES

Elementary calculus

UNIT OUTCOMES

U3-O1: Describe a flow field using Lagrangian and Eulerian descriptions

U3-O2: Classify flow into steady, unsteady, uniform, and non-uniform flow

U3-O3: Determine the acceleration of flow at a point

U3-O4: Describe a flow using streamline and streakline

U3-O5: Obtain pathline of a particle in a flow

U3-O6: Describe linear, volumetric, and angular deformation of a fluid element due to flow

Kinematics refers to the study of motion without due reference to the forces which are responsible for the motion. In this Unit we will look into various descriptions of flow, classification of flow, ways of visualizing the flow and, finally, the influence of flow on a fluid element.

Unit-3 Outcome	EXPECTED MAPPING WITH COURSE OUTCOMES (1 - Weak Correlation; 2 - Medium Correlation; 3 - Strong Correlation)						
	CO-1	CO-2	CO-3	CO-4	CO-5	CO-6	CO-7
U3-O1	-	-	3	2	-	-	1
U3-O2	-	-	3	3	-	-	2
U3-O3	-	-	1	3	-	-	2

U3-O4	-	-	3	3	-	-	2
U3-O5	-	-	3	1	-	-	1
U3-O6	2	-	2	2	-	-	1

3.1 FLUID MOTION: PARTICLE AND FIELD DESCRIPTION

There are two types of description for describing fluid flow. The first one is the particle based description, i.e. describing the motion of a particle, like we do in classical mechanics. The second one is the description through fixed points in space, i.e. probing a certain fixed location such as a temperature probe fixed at a point, which measures the temperature variation at that location as time passes by. Below, we will describe in more details about the different descriptions

3.1.1 Lagrangian Description

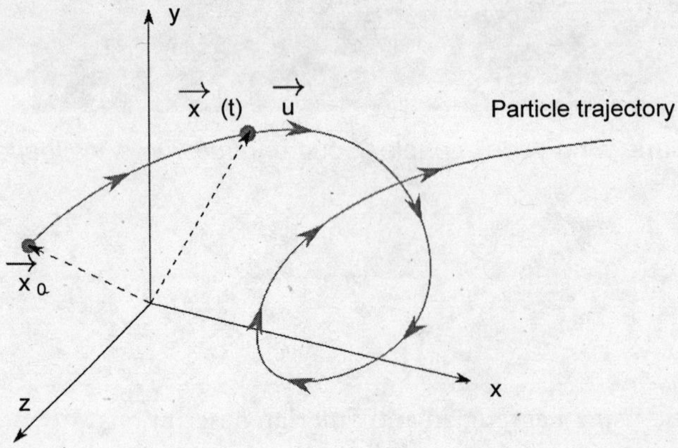

Figure 3.3: Motion of a particle which is at $\vec{x}_0$ at time t_0. At an instant of time, t, the particle is at $\vec{x}$ which experiences a velocity $\vec{u}$, causing it to move ahead in time. The eventual path taken by such a particle is known as the particle trajectory or the particle path.

Lagrangian description is a particle based description wherein we label a certain imaginary particle in the flow field and then look at how the particle location evolves in time. This is shown in Fig. 3.3. In the figure, we track a particle which is initially at a location (x_0, y_0, z_0), which is most conveniently represented by a vector $\vec{x}_0$ at an initial reference time, t_0. Owing to the underlying flow field, the particle is then carried by the flow. We know from classical mechanics that the relationship between the velocity, $\vec{u}$ and acceleration, $\vec{a}$, is given by

$$\frac{d\vec{x}}{dt} = \vec{u}$$

$$\frac{d^2\vec{x}}{dt^2} = \frac{d\vec{u}}{dt} = \vec{a} \tag{3.1}$$

$$at\ t = t_0, \vec{x} = \vec{x}_0\ \vec{u} = \vec{u}_0$$

The last line refers to the initial condition of the particle.

As one can imagine, tracking particles through a flow in order to describe he flow field may often require tracking a very large number of particles which will help in describing the entire flow field over a certain domain of interest. This is not feasible in reality and therefore such a Lagrangian description is not widely adopted here.

3.1.2 Eulerian Description

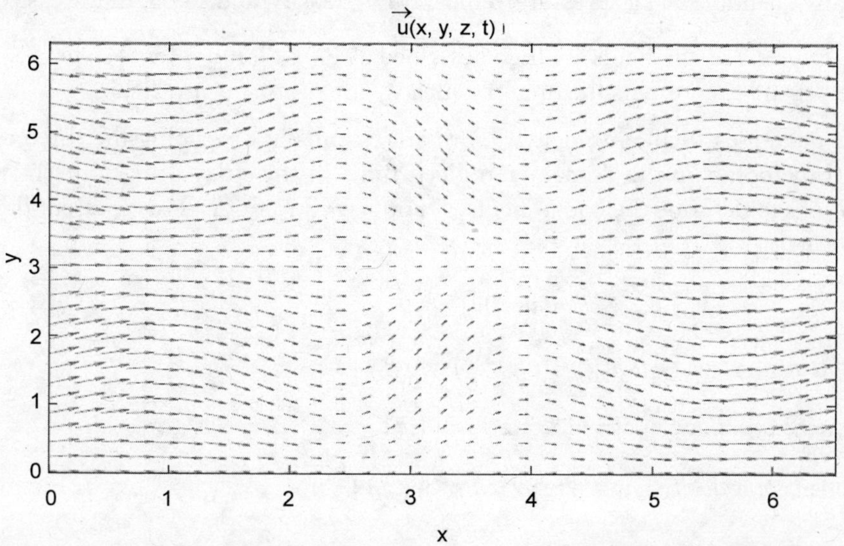

Figure 3.4: Eulerian description of velocity. The velocity vectors are shown throughout the domain as arrows. The arrow points in the direction of the flow while the length of the arrow is an indication of the magnitude of the velocity at that point. The velocity field is shown at a certain snapshot of time. For an unsteady flow, the vector field shown in the figure may change.

In contrast to the Lagrangian description mentioned above, the Eulerian description describes the fluid kinematics at fixed locations in space. Essentially if we focus at a point in the fluid, then the description of the flow at that point is completely defined by the spatial coordinates and time. We can therefore define a property such as velocity in terms of the coordinates and time. Mathematically this may be written as

$$\vec{u} = \vec{u}(\vec{x}, t) \tag{3.2}$$

In certain cases, to avoid confusion between the velocity **vector**, $\vec{u}$ and the x component of the velocity field, u, we represent the velocity vector as

$$\vec{V} = \vec{V}(\vec{x}, t) \tag{3.3}$$

In the theory and questions, we will interchangeably use the two notations and the context often gives a clue as to what we are actually referring to. Notably, the Eulerian description is not just valid for the description. It may be used to describe any other parameter of the flow such as density, temperature at a point in space.

This is the more practical approach to describing fluid flow since describing any properties does not require us to track the particle as it passes through the domain. We can simply focus our attention to a point in space. In the Eulerian approach, we reference to various quantities as fields in space which may or may not evolve in time. This can be seen in Fig. 3.4, which shows the velocity vectors at a given instant of time. The description of the velocity vectors are therefore over the entire domain. To reiterate, this is different from the Lagrangian approach where the velocity of a particle is only known.

The ideas above are not just restricted to the description of velocity but can be readily used for the description of any quantity such as pressure, temperature, density, and so on. In the case of scalars such as temperature or pressure, the fields are then described as $T(\vec{x},t)$ and $p(\vec{x},t)$, where $\vec{x}$ describes the location of interest and t represents the time at which we are *measuring* the field.

For vector quantities such as velocity, $\vec{u}$, we are essentially describing the three components of the vector at all the points in the domain at a given time. In the study of fluid mechanics, we follow certain conventions to describe the components. A popular convention is to represent the velocity with components given by

$$\vec{u} = (u, v, w) \tag{3.4}$$

$$x-\textit{component of velocity} \quad u = u(x, y, z, t) = u(\vec{x}, t) \tag{3.5}$$

$$y-\textit{component of velocity} \quad v = v(x, y, z, t) = v(\vec{x}, t) \tag{3.6}$$

$$z-\textit{component of velocity} \quad w = w(x, y, z, t) = w(\vec{x}, t) \tag{3.7}$$

Different Coordinate Systems

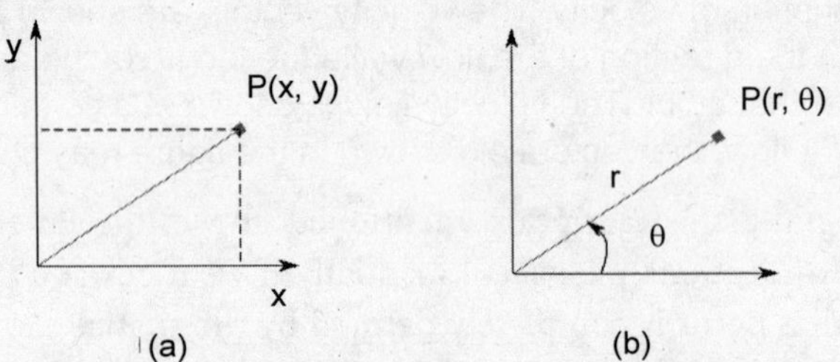

Figure 3.5: (a) A Cartesian coordinate system. (b) A polar coordinate system.

We note that in all the descriptions above, we have explicitly mentioned the location of a point, i.e. the displacement vector, as $\vec{x}$ which we have explicitly mentioned as the triplet (x, y, z). This, however is only true if we are using a Cartesian coordinate system. There are several other practical coordinate systems based on the nature of the problem at hand. For example, a flow which resembles a hurricane, known as a vortex flow, is most readily described in terms of a cylindrical polar coordinate system wherein a general displacement vector $\vec{x}$ is then described as the triplet (r, ϕ, z). Or for flows occuring on a sphere, such as atmospheric scale flows, the systems are most readily described through a spherical polar coordinate system wherein the displacement vector is described as the triplet $\vec{x}$ =(r, θ, ϕ). Such coodinate systems are shown in Fig. 3.6.

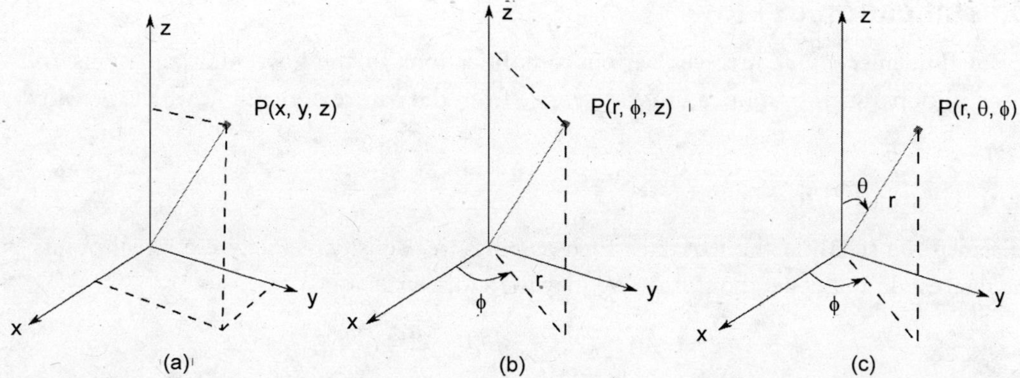

Figure 3.6: (a) A Cartesian coordinate system. (b) A cylindrical polar coordinate system. (c) A spherical polar coordinate system.

Example 3.1

A 2-dimensional flow field is given by $\vec{V} = -y\hat{i} + x\hat{j}$. Determine the magnitude and direction of the velocity vector at (3,4) and (4,3).

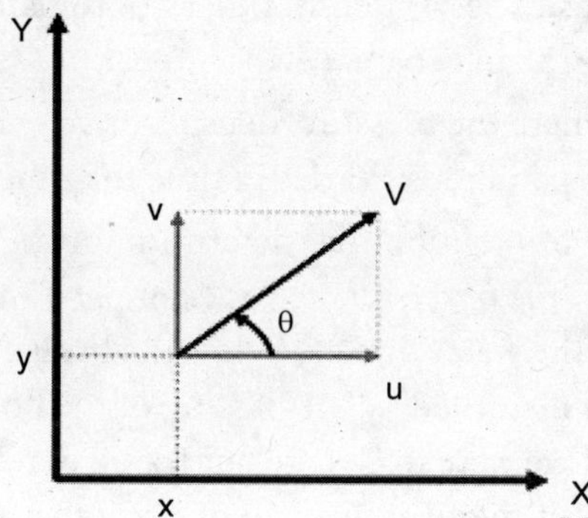

Solution:

Given: Velocity field $V = -y\hat{i} + x\hat{j} \Rightarrow u = -y \quad v = x$

Magnitude of $V = \sqrt{\left(u^2 + v^2\right)} = \sqrt{\left[\left(-y\right)^2 + \left(x\right)^2\right]} = \sqrt{y^2 + x^2}$

Therefore, magnitude of V = 5 units for both coordinates (3,4) and (4,3).

The angle θ made by vector $\vec{V}$ with X-axis is given by: $\tan\theta = \left(\frac{v}{u}\right) = \left(\frac{x}{-y}\right)$

For coordinate (3,4) the angle $\theta = \tan^{-1}\left(\frac{3}{-4}\right) = -36.9^\circ$

For coordinate (4,3) the angle $\theta = \tan^{-1}\left(\frac{4}{-3}\right) = -53.1^\circ$

3.1.3 Classification of Flow

The study of fluid mechanics involves various simplifications in the governing equations and their boundary conditions. Such simplifications often arise from the nature of the flow, which we will discuss below.

Steady and Unsteady Flow

A flow is said to be steady if the variables such as pressure, density, velocity etc. at all points in the domain of interest do not depend on time. We may therefore write

$$\frac{\partial p}{\partial t}=0, \quad \frac{\partial \rho}{\partial t}=0, \text{ and } \quad \frac{\partial \vec{u}}{\partial t}=0 \tag{3.8}$$

for all points in the domain. In such a case, we can simplify the functional form of the variables and write them in a Cartesian coordinate system as

$$u = u(x, y, z),\ v = v(x, y, z),\ w = w(x, y, z) \tag{3.9}$$

$$\text{Pressure: } p = p(x, y, z) \tag{3.10}$$

$$\text{Density: } \rho = \rho(x, y, z) \tag{3.11}$$

The above equation may be easily be written in other coordinate systems by making use of their respective basis triplets such as (r, ϕ, z) and so on.

On the other hand, any flow where the pressure, density, or velocity varies with time is classified as an unsteady flow. Even if one of the parameters varies in time then the flow is said to be unsteady.

It is important in this context to mention that determining whether a flow is steady or unsteady may sometimes determine on the frame of reference. For example if we imagine a boat passing on water as seen from a bridge. If we look at the water directly below the bridge and focus our attention there, we will observe that the river stream is disturbed by the passing of the boat and then again the flow attains that of the river stream. However, if we were to change our frame of reference on the boat, instead of the bridge, and observe the water behind the boat, we will observe that the flow pattern behind the boat, given that the boat is moving with a constant velocity, will remain the same. For the observer on the boat, the flow behind the boat will appear to be steady. For the observer on the bridge, the flow will appear to be unsteady.

Uniform and Non-Uniform Flow

As we have mentioned several times before, a general description of a flow requires the variables in the flow to depend on the spatial coordinates, $(\vec{x})$ and time. We have discussed about steady and unsteady flow in section 3.1.3 and we can now discuss about the various spatial aspects of the flow.

A flow is said to be uniform if its parameters do not vary in space. In Fig. 3.7, we depict some examples of such flow. It may so happen that a flow is uniform in one direction while being non-uniform in the other direction. Therefore, such interpretations must be done on a case to case basis. Unless otherwise specified, a uniform flow implies uniformity in all the directions. A flow which varies in space, which is typical of most engineering flows, is known as a non-uniform flow.

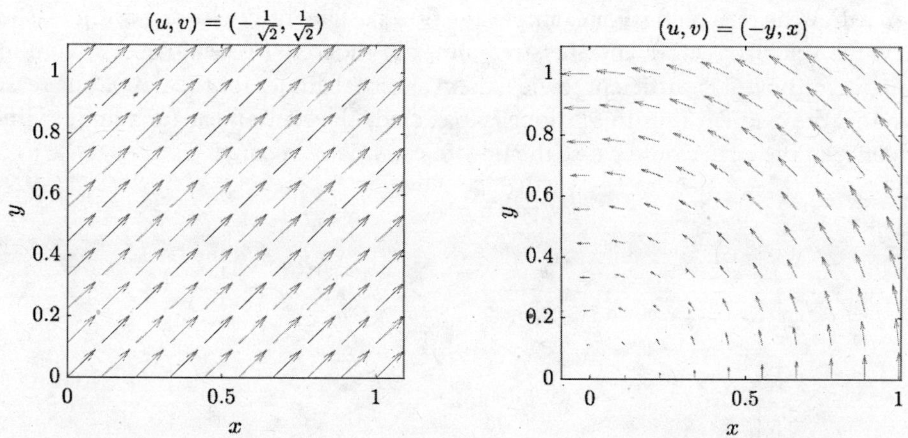

Figure 3.7: (a) A uniform velocity field, (b) A non-uniform velocity field

Example 3.2

Consider two fluid flow fields: a) $\vec{V} = x\hat{i} + 3\hat{j} + 9\hat{k}$ b) $\vec{V} = 2\hat{i} + 3\hat{j} + 9\hat{k}$. Determine which flow field is uniform?

Solution:

Given: (a) $u = x$, $v = 3$, $w = 9$ (b) $u = 2$, $v = 3$, $w = 9$

The flow field described in (b) is uniform as it doesn't depend on the spatial coordinates (x, y, z) whereas the flow field described in (a) depends on the coordinate.

Dimensionality of a flow

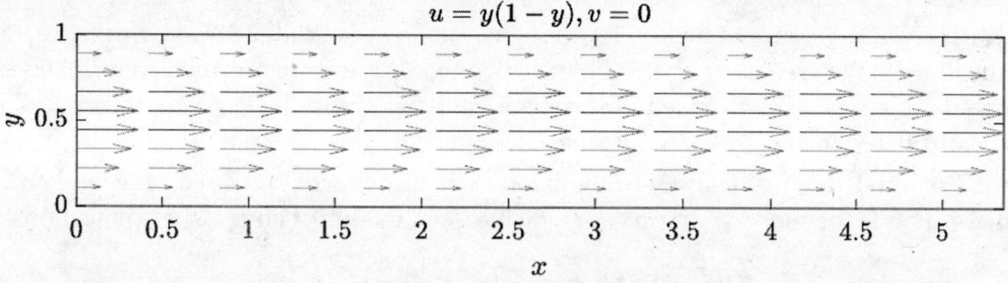

Figure 3.8: A 1D flow in a pipe.

(i) **One-dimensional flow:** A 1D flow is one whose characteristics can be described through only one independent spatial coordinate. Such flows are rarely seen in nature, however some flows can be closely approximated as being 1D in nature. For example, as shown in the Fig. 3.8, we see that the flow in a very long pipe does appear to be such that the flow profile is *uniform* in the direction while being dependent only in the coordinate. Such a flow is a 1D flow. The reason why we have specifically mentioned a long pipe is to exclude the influences of the entry and exit where the flow is no longer only dependent on the coordinate but also depends on the and coordinates as well.

(ii) **Two-dimensional flow:** A 2D flow or a plane flow is a flow whose characteristics are described combpletely by two independent spatial coordinates. Such flows are typically those which can be shown in a drawing on a plane. For example, the flow past an infinitely long cylinder, shown in Fig. 3. depicts the velocity vectors and the streamlines of the flow (we will cover streamlines later in this Unit but for now, it is sufficient to view them as visual guides for the flow). The reason why we mention that the cylinder is infinitely long is to exclude the edge effects for a finite cylinder, where the influence of the edge would cause the flow to essentially become 3D.

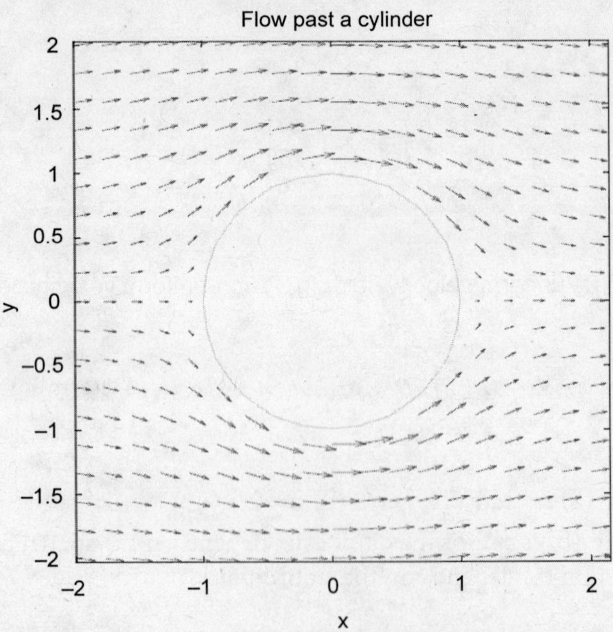

Figure 3.9: Flow past an infinitely long cylinder shows a 2D flow pattern.

(iii) Three-dimensional axisymmetric flow: Although such flows are called 3D axisymmetric flows, in reality, the flow characteristics can be completely defined by means of only two coordinates such as and . Such flows are 2D when viewed in the spherical polar coordinate system, but are essentially 3D when considered in a Cartesian coordinate system.

(iv) Three-dimensional flow: A truly 3D flow is one which depends on three independent spatial coordinates. This is the most common case which is seen in almost all real life applications.

3.1.4 Acceleration in a Flow & Material Derivative

We will now determine the acceleration of a flow in the Eulerian frame of reference. Let us consider a Cartesian coordinate system with the velocity components at a point $\vec{x} = (x, y, z)$ given by

$$u = u(x, y, z, t)$$

$$v = v(x, y, z, t)$$

$$w = w(x, y, z, t)$$

Let us say that there is an imaginary particle at this location and experiences the flow mentioned above as shown in Fig 3.10. After a time Δt, the particle now travels to a new location given by $\vec{x} + \Delta\vec{x} = (x + \Delta x, y + \Delta y, z + \Delta z)$. Let the velocity components in the new position be

$\left(u+\Delta u, v+\Delta v, w+\Delta w\right)$, for which we can write

$$u + \Delta u = u(x + \Delta x, y + \Delta y, z + \Delta z, t + \Delta t) \tag{3.12}$$

$$v + \Delta v = v(x + \Delta x, y + \Delta y, z + \Delta z, t + \Delta t) \tag{3.13}$$

$$w + \Delta w = w(x + \Delta x, y + \Delta y, z + \Delta z, t + \Delta t) \tag{3.14}$$

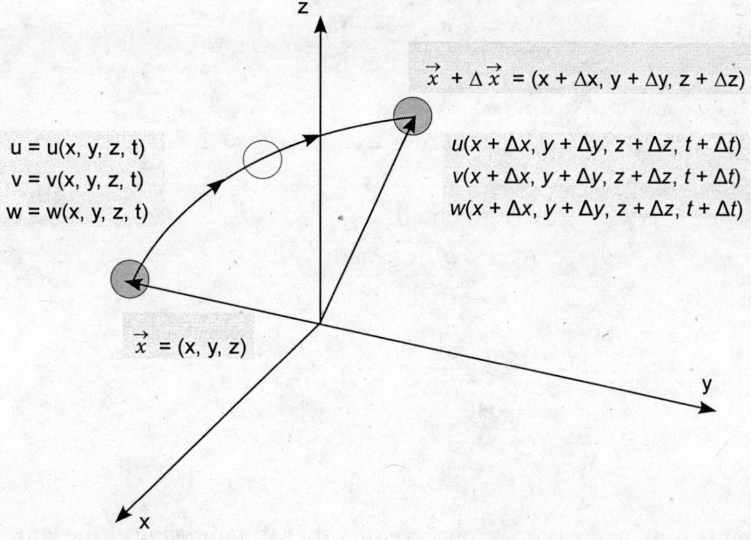

Figure 3.10: The orange particle travels from its location $\vec{x}$ to $\vec{x} + \Delta\vec{x}$ in a time interval of Δt where it experiences a change of velocity from $\vec{u}$ to $\vec{u} + \vec{u}\Delta$.

The components evaluated at $\vec{x}+\Delta\vec{x}$ can be then expanded using a Taylor series expansion about the point as

$$u\left(x+\Delta x, y+\Delta y, z+\Delta z\right) = u\left(x, y, z, t\right) + \frac{\partial u}{\partial x}\Delta x + \frac{\partial u}{\partial y}\Delta y + \frac{\partial u}{\partial z}\Delta z + \frac{\partial u}{\partial t}\Delta t + H.O.T \tag{3.15}$$

$$v\left(x+\Delta x, y+\Delta y, z+\Delta z\right) = v\left(x, y, z, t\right) + \frac{\partial v}{\partial x}\Delta x + \frac{\partial v}{\partial y}\Delta y + \frac{\partial v}{\partial z}\Delta z + \frac{\partial v}{\partial t}\Delta t + H.O.T \tag{3.16}$$

$$w\left(x+\Delta x, y+\Delta y, z+\Delta z\right) = w\left(x, y, z, t\right) + \frac{\partial w}{\partial x}\Delta x + \frac{\partial w}{\partial y}\Delta y + \frac{\partial w}{\partial z}\Delta z + \frac{\partial w}{\partial t}\Delta t + H.O.T \tag{3.17}$$

The H.O.T above represent higher order terms in the Taylor series expansions of each of the terms above which consist of terms such as Δt^2, Δx^2, Δy^2, Δz^2, etc.

We must also bear in mind that the displacements, i.e. $(\Delta x, \Delta y, \Delta z)$ over a time, Δt are caused by presence of the velocity field at point $\vec{x} = (x, y, z)$ at time t. Therefore, we can write

$$\Delta x = u\Delta t \tag{3.18}$$

$$\Delta y = v\Delta t \tag{3.19}$$

$$\Delta z = w\Delta t \tag{3.20}$$

We can now combine equations (3.14) through and (3.20) to obtain

$$\Delta u = u\Delta t\frac{\partial u}{\partial x} + v\Delta t\frac{\partial u}{\partial y} + w\Delta t\frac{\partial u}{\partial z} + \Delta t\frac{\partial u}{\partial t} \tag{3.21}$$

$$\Delta v = u\Delta t\frac{\partial v}{\partial x} + v\Delta t\frac{\partial v}{\partial y} + w\Delta t\frac{\partial v}{\partial z} + \Delta t\frac{\partial v}{\partial t} \tag{3.22}$$

$$\Delta w = u\Delta t\frac{\partial w}{\partial x} + v\Delta t\frac{\partial w}{\partial y} + w\Delta t\frac{\partial w}{\partial z} + \Delta t\frac{\partial w}{\partial t} \tag{3.23}$$

where we have now dropped the higher order terms because their magnitudes will be smaller than the terms that we have retained above.

We can now divide everything by Δt to obtain

$$\frac{\Delta u}{\Delta t} = u\frac{\partial u}{\partial x} + v\frac{\partial u}{\partial y} + w\frac{\partial u}{\partial z} \tag{3.24}$$

$$\frac{\Delta v}{\Delta t} = u\frac{\partial v}{\partial x} + v\frac{\partial v}{\partial y} + w\frac{\partial v}{\partial z} \tag{3.25}$$

$$\frac{\Delta w}{\Delta t} = u\frac{\partial w}{\partial x} + v\frac{\partial w}{\partial y} + w\frac{\partial w}{\partial z} \tag{3.26}$$

If we now take the limit as $\Delta t \to 0$, we observe that the left hand side of the equations above reduce to the Lagrangian description of the accelerations in the x, y, and z directions, i.e.

$$\frac{du}{dt} = \lim_{\Delta t \to 0} \frac{\Delta u}{\Delta t} \tag{3.27}$$

$$\frac{dv}{dt} = \lim_{\Delta t \to 0} \frac{\Delta v}{\Delta t} \tag{3.28}$$

$$\frac{dw}{dt} = \lim_{\Delta t \to 0} \frac{\Delta w}{\Delta t} \tag{3.29}$$

We can now combine equations (3.24), (3.25), and (3.26) along with the definitions in equations (3.27), (3.28), and (3.29), we can relate the acceleration obtained through the Lagrangian viewpoint to the description based on the Eulerian viewpoint as

$$\begin{aligned} &Acceleration\ in x direction: \frac{du}{dt} \equiv \frac{Du}{Dt} = \frac{\partial u}{\partial t} + u\frac{\partial u}{\partial x} + v\frac{\partial u}{\partial y} + w\frac{\partial u}{\partial z} \\ &Acceleration\ in y direction: \frac{dv}{dt} \equiv \frac{Dv}{Dt} = \frac{\partial v}{\partial t} + u\frac{\partial v}{\partial x} + v\frac{\partial v}{\partial y} + w\frac{\partial v}{\partial z} \\ &Acceleration\ in z direction: \frac{dw}{dt} \equiv \frac{Dw}{Dt} = \frac{\partial w}{\partial t} + u\frac{\partial w}{\partial x} + v\frac{\partial w}{\partial y} + w\frac{\partial w}{\partial z} \end{aligned} \tag{3.30}$$

Notice that we have used D as the ordinary derivate operator instead of d because we would like to emphasize on the fact that the derivative connects the Lagrangian description to the Eulerian description.

The operator $\frac{D}{Dt}$ is known as the total derivative or the material derivative. This means that at a given point in the domain, the total derivative of the x-component of velocity is given by $\frac{Du}{Dt}$ which is described component-wise in equation (3.30) In general for a given field variable, ϕ, we can define its total derivative as

$$\frac{D}{Dt}\phi = \frac{\partial \phi}{\partial t} + \vec{V}\cdot\nabla\phi \tag{3.31}$$

Example 3.3

Determine the acceleration for the flow field: $V = xy\hat{i} - 4z\hat{j} + y^2\hat{k}$ at the coordinate .

Solution:

Given:

$$u = xy,\ v = -2z,\ w = y^2$$

$$a_x = \frac{\partial u}{\partial t} + u\frac{\partial u}{\partial x} + v\frac{\partial u}{\partial y} + w\frac{\partial u}{\partial z} = -1 + xy.y + (-3z).x + y^2.0 = xy^2 - 3xz \tag{3.32}$$

$$a_y = \frac{\partial v}{\partial t} + u\frac{\partial v}{\partial x} + v\frac{\partial v}{\partial y} + w\frac{\partial v}{\partial z} = -1 + xy.0 + (-3z).0 + y^2.(-3) = -3y^2 \tag{3.33}$$

$$a_z = \frac{\partial w}{\partial t} + u\frac{\partial w}{\partial x} + v\frac{\partial w}{\partial y} + w\frac{\partial w}{\partial z} = -1 + xy.0 + (-3z).(2y) + y^2.0 = -6yz \tag{3.34}$$

Acceleration vector is given by:

$$\vec{a} = (xy^1 - 3xz)\hat{i} - 3y^2\hat{j} - 6yz\hat{k}$$

Therefore, acceleration at coordinate (0,1,2) is given by:

$$\vec{a} = -5\hat{i} - 3\hat{j} - 12\hat{k}$$

Example 3.4

Determine the acceleration for the flow field: $\vec{V} = xy\hat{i} - (3z - 10e^{-t})\hat{j} + y^2\hat{k}$ at the coordinate at time $t = 0$. Is the acceleration constant with respect to time? Given: $u = xy$, $v = (-3z + 10\ e^{-t})$, $w = y^2$

$$a_x = \frac{\partial u}{\partial t} + u\frac{\partial u}{\partial x} + v\frac{\partial u}{\partial y} + w\frac{\partial u}{\partial z} = 0 + xy.y + (-3z + 10e^{-t}).x + y^2.0 = xy^2 - 3xz \tag{3.35}$$

$$a_y = \frac{\partial v}{\partial t} + u\frac{\partial v}{\partial x} + v\frac{\partial v}{\partial y} + w\frac{\partial v}{\partial z} = 10e^{-t} + xy.0 + (-3z + 10e^{-t}).0 + y^2.(-3) = 10e^{-t} - 3y^2 \tag{3.36}$$

$$a_z = \frac{\partial w}{\partial t} + u\frac{\partial w}{\partial x} + v\frac{\partial w}{\partial y} + w\frac{\partial w}{\partial z} = 0 + xy.0 + (-3z + 10e^{-t}).(2y) + y^2.0 = -6yz \tag{3.37}$$

Acceleration vector is given by: $\vec{a} = \left(xy^2 - 3xz\right)\hat{i} + \left(10e^{-t} - 3y^2\right)\hat{j} - 6yz\hat{k}$ and it is evident that acceleration is dependent on time. Thus, acceleration at the coordinate (1, 1, 2) at t = 0 is given by: $\vec{a} = -5\hat{i} + 7\hat{j} - 12\hat{k}$

Example 3.5

Determine the magnitude of acceleration for the 2-dimensional flow field $V = -y\hat{i} + x\hat{j}$ along a circle of radius 5 units and centre (0, 0).

Solution:

Given: Velocity field is $\vec{V} = -y\hat{i} + x\hat{j} \Rightarrow u = -y \,\&\, v = x$

$$a_x = \frac{\partial u}{\partial t} + u\frac{\partial u}{\partial x} + v\frac{\partial u}{\partial y} = 0 + (-y).0 + x.(-1) = (-x) \qquad (3.38)$$

$$a_y = \frac{\partial v}{\partial t} + u\frac{\partial v}{\partial x} + v\frac{\partial v}{\partial y} = 0 + (-y).1 + x.0 = (-y) \qquad (3.39)$$

Magnitude of acceleration = $\sqrt{(-x)^2 + (-y)^2} = \sqrt{x^2 + y^2}$. The equation of circle of radius 5 units is given by: $x^2 + y^2 = 5^2$. Therefore, the magnitude of acceleration of fluid element along this circle is $\sqrt{x^2 + y^2} = 5$ units.

3.2 FLOW VISUALIZATION

In the above sections, we have discussed about the Lagrangian and Eulerian frame of references. In what follows, we discuss various ways in which a flow may be visualized which are very useful for analytically plotting flow fields, and also for extracting descriptions of flow fields from experimental images.

3.2.1 Streamline

Streamlines are geometric ways of describing the velocity in a domain. In an Eulerian description of velocity, we have the information of velocity throughout the domain. If we freeze time, then a curve which is tangential to velocity vectors is known as a streamline. We first define what *freezing time* means. Freezing time refers to setting time to a particular value, and not treating time as an independent variable. In the definition of streamline, time is not relevant. We now define what it means to say *a curve which is tangential to velocity vectors.*

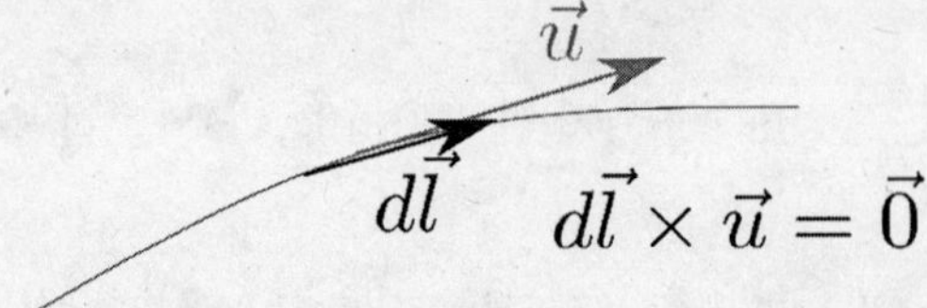

Figure 3.11: Schematic of a streamline on which the velocity vector at a point and the displacement along the streamline are parallel to each other.

Let us consider Fig. 3.11 where we have shown a velocity vector, $\vec{u}$. We have also shown a curve whose differential element along that particular curve is shown as $d\vec{l}$. If we were to describe the curve in Cartesian coordinates, then we can write down the components of velocity and curve length as

$$\vec{u} = (u, v, w) \equiv u\hat{i} + v\hat{j} + w\hat{k} \tag{3.40}$$

$$d\vec{l} = (dx, dy, dz) \equiv dx\hat{i} + dy\hat{j} + dz\hat{k} \tag{3.41}$$

where $\hat{i}$, $\hat{j}$, and $\hat{k}$ represent the three unit vectors in the x, y, and z directions respectively.

A necessary condition for two arbitrary vectors, $\vec{a}$ and $\vec{b}$ to be parallel can be given by the vector identity

$$\vec{a} \times \vec{b} = \vec{0}.$$

Utilizing the vector identity for the case of $d\vec{l}$ and $\vec{u}$, we can write the cross product as

$$\begin{matrix} \hat{i} & \hat{j} & \hat{k} \\ dx & dy & dz \\ u & v & w \end{matrix} = \hat{i}(wdy - vdz) + \hat{j}(udz - wdx) + \hat{k}(vdx - udy) = \vec{0} \tag{3.42}$$

Setting each of the components of $\hat{i}$, $\hat{j}$, $\hat{k}$ to 0 in the expressions above, we obtain

$$wdy - vdz = 0 \Rightarrow \frac{dy}{v} = \frac{dz}{w} \tag{3.43}$$

$$udz - wdx = 0 \Rightarrow \frac{dz}{w} = \frac{dx}{u} \tag{3.44}$$

$$vdx - udy = 0 \Rightarrow \frac{dx}{u} = \frac{dy}{v} \tag{3.45}$$

$$\Rightarrow \quad \frac{dx}{u} = \frac{dy}{v} = \frac{dz}{w} \tag{3.46}$$

The above equation describes the differential equation for a streamline in a Cartesian frame of reference.

Example 3.6

Find the equation for a streamline passing through (5,0) for the 2-dimensional velocity vector field $\vec{V} = -y\hat{i} + x\hat{j}$.

Solution:

Given: Velocity field is $\vec{V} = -y\hat{i} + x\hat{j}$. $\Rightarrow u = (-y)$ & $v = x$. For the streamline we have: $\frac{dx}{u} = \frac{dy}{v}$

$$\Rightarrow \qquad \frac{dx}{-y} = \frac{dy}{x} \Rightarrow xdx + ydy = 0 \tag{3.47}$$

$$\Rightarrow \qquad \int xdx + \int ydy = C \tag{3.48}$$

$$\Rightarrow \qquad x^2 + y^2 = 2C \tag{3.49}$$

For the streamline passing through (5,0) we get $5^2 + 0^2 = 2C$ Therefore, the equation of streamline passing through (5,0) is $x^2 + y^2 = 25$.

Example 3.7

Find the equation for a streamline passing through (1,3) for the 2-dimensional velocity vector field $\vec{V} = y^2\hat{i} + 3x\hat{j}$.

Solution:

Given: Velocity field is $\vec{V} = y^2\hat{i} + 3x\hat{j} \Rightarrow u = y^2 \;\&\; v = 3x$

For the streamline we have: $\frac{dx}{u} = \frac{dy}{v}$

$$\Rightarrow \qquad \frac{dx}{y^2} = \frac{dy}{3x} \Rightarrow 3xdx - y^2dy = 0 \tag{3.50}$$

$$\Rightarrow \qquad 3\int xdx - \int y^2dy = C \tag{3.51}$$

$$\Rightarrow \qquad 3x^2 - \frac{y^3}{3} = 2C \tag{3.52}$$

For the streamline passing through (1,3) we get $3 \times 1^2 - \frac{3^3}{3} = 2C$ Therefore, the equation of streamline passing through (1, 3) is $3x^2 - \frac{y^3}{3} = (-6)$.

3.2.2 Pathline

While the above method of flow visualization was based on an Eulerian method, we can also track only a single particle as it goes through the flow. A pathline is defined as the trajectory of a fluid particle as it moves through the flow. In the definition of a pathline, the identity of a particle is kept fixed. This is to be expected since in a Lagrangian description, we always define everything with respect to an object of fixed identity. This is depicted in Fig. 3.12.

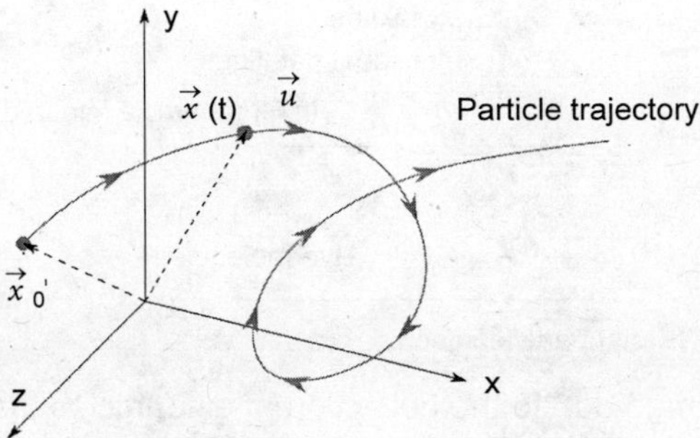

Fig. 3.1.2: The pathline of a single particle as it moves through the flow. The initial condition is also shown in the figure. It is clear that the pathline and the Lagrangian description refer to the same thing.

Mathematically, we can determine the trajectory of a particle by noting that the velocity of the particle at a the location of the point in the domain, say (x_p, y_p, z_p) and time t, will be equal to the Eulerian velocity at that same point. The subscript is to denote the fact that the coordinates are infact those of the particle in equation. In that case we can write

$$\frac{dx_p}{dt} = u\left(x_p, y_p, z_p, t\right) \tag{3.53}$$

$$\frac{dy_p}{dt} = v\left(x_p, y_p, z_p, t\right) \tag{3.54}$$

$$\frac{dz_p}{dt} = w\left(x_p, y_p, z_p, t\right) \tag{3.55}$$

where we note that the above ordinary differential equations represent an initial value problem. An initial value problem is one which requires a valid initial condition, which in this case represents the location of particle at a reference time, typically chosen as $t = 0$. We can therefore write down the initial conditions as

$$\text{At } t = 0, \quad x_p = x_{p0}, \quad y_p = y_{p0}, \quad z_p = z_{p0} \tag{3.56}$$

3.2.3 Streakline

The concept of streakline is particularly useful for experimental observations. Typically fluid flow experiments are done with the help of some kind of a tracer material, i.e. something which helps in aiding in visualizing how the flow is taking place. Examples of tracers can be dye such as ink, or smoke in air. One is familiar with the flow that takes place above a lit incense (agarbatti) stick. We are able to qualitatively understand the nature of the flow due to such a source of flow. Similarly, we are able to judge the direction of flow in the atmosphere if we look at the smoke coming out from a chimney. In a similar way, if we take an injection syringe and continuously inject a colored dye in flowing water, we observe that the dye is *flown* away with the water. These are all examples of tracers used to visualize flow.

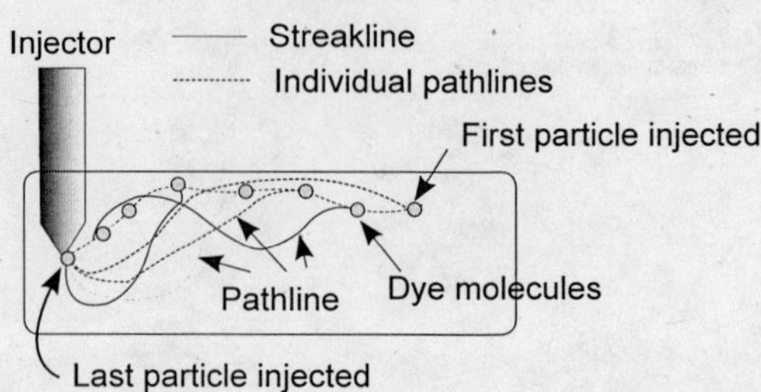

Figure 3.13: Streakline formed due to the collection of all particles that have been injected into the domain at some earlier time at a given point through an injector. All the individual pathlines are depicted in translucency. The streakline is the line that is comprised of all the dye molecules and is represented by the red line in the figure.

Let us now mathematically describe the curve that we see the tracer makes in the flow. A streakline at any given instant of time is the locus of all particles that have passed through a fixed point in the flow field at all previous times. It is essentially the curve which is obtained by connecting the locations of all fluid particles that have passed through a given point in space at all previous times to the time under consideration. The above definition can be best understood by means of the visual example given in Fig. 3.13.

Example 3.8

The velocity components in a flow field is given as follows: $u = \dfrac{x}{t + t_0}$, $v = v_0$, $w = 0$, where t_0 and v_0 are constants. A colored dye is injected at the point A(x_0, y_0, z_0) in the flow field. Find the equation of a colored line visible in the flow field at $t = 2t_0$, as a consequence of the dye injection, given that the dye injection starts at $t = t_0$.

Solution

The velocity components in a flow field is given as follows: $u = \dfrac{x}{t + t_0}$, $v = v_0$, $w = 0$ where t_0 and v_0 are constants. A colored dye is injected at the point A(x_0, y_0, z_0) in the flow field. Find the equation of a colored line visible in the flow field at , as a consequence of the dye injection, given that the dye injection starts at $t = t_0$.

Solution: Here, essentially we need to obtain the equation of streakline at $t = 2t_0$. In order to obtain the same, we consider that the fluid particles are injected through the point at instants of time symbolically denoted by t_i, where $t_i \leq 2t_0$. Importantly, t_i is a variable and not a fixed instant of time. Now based on the given velocity field, we have

$$u = \frac{dx}{dt} = \frac{x}{t + t_0}$$

$$\Rightarrow \quad \int_{x_0}^{x} \frac{dx}{x} = \int_{t_i}^{2t_0} \frac{dt}{t + t_0}$$

$$\Rightarrow \quad \ln\frac{x}{x_0} = \ln\frac{2t_0 + t_0}{t_i + t_0}$$

$$\Rightarrow \quad \frac{x}{x_0} = \frac{3t_0}{t_i + t_0} \tag{3.57}$$

Similarly,

$$v = \frac{dy}{dt} = v_0$$

$$\Rightarrow \quad \int_{y_0}^{y} dy = \int_{t_i}^{2t_0} v_0 dt$$

$$\Rightarrow \quad y - y_0 = v_0\,(2t_0 - t_i)$$

$$\Rightarrow \quad t_i = 2t_0 + \frac{y - y_0}{v_0} \tag{3.58}$$

Eliminating from Eqs (3.57) and (3.58), one can write

$$\frac{x}{x_0} = \frac{3t_0}{2t_0 + \frac{y - y_0}{v_0} + t_0} = \frac{3t_0}{3t_0 + \frac{y - y_0}{v_0}} = \frac{1}{1 + \frac{y - y_0}{3t_0 v_0}}$$

$$\frac{x}{x_0}\left(1 + \frac{y - y_0}{3t_0 v_0}\right) = 1 \tag{3.59}$$

Equation (3.59) is the equation of a colored line visible in the flow field at $t = 2t_0$.

Example 3.9

The velocity components in a flow field is given as follows: $u = \frac{x}{t + t_0}$, $v = v_0$, $w = 0$, where t_0 and v_0 are constants. A colored dye is injected at the point A(x_0, y_0, z_0) in the flow field. Find the locus of a fluid particle that passes through the point A at $t = t_0$.

Solution

Here, essentially we need to obtain the equation of pathline, corresponding to a fluid particle that at t = t_0 passed through (x_0, y_0)

Now,

$$u = \frac{dx}{dt} = \frac{x}{t + t_0}$$

$$\Rightarrow \quad \int_{x_0}^{x} \frac{dx}{x} = \int_{t_0}^{t} \frac{dt}{t + t_0}$$

$$\Rightarrow \quad \ln\frac{x}{x_0} = \ln\frac{t + t_0}{t_0 + t_0} = \ln\frac{t + t_0}{2t_0}$$

$$\Rightarrow \quad \frac{x}{x_0} = \frac{t + t_0}{2t_0} \tag{3.60}$$

Similarly,

$$v = \frac{dy}{dt} = v_0$$

$$\Rightarrow \quad \int_{y_0}^{y} dy = \int_{t_0}^{t} v_0 dt$$

$$\Rightarrow \quad y - y_0 = v_0 (t - t_0)$$

$$\Rightarrow \quad t - t_0 = \frac{y - y_0}{v_0} \tag{3.61}$$

Eliminating from Eqs (3.60) and (3.61), one can write

$$\frac{x}{x_0} = \frac{t_0 + \frac{y - y_0}{v_0} + t_0}{2t_0} = 1 + \frac{y - y_0}{2t_0 v_0}$$

$$\Rightarrow \quad \frac{x}{x_0} - \frac{y - y_0}{2t_0 v_0} \tag{3.62}$$

Equation (3.62) is the equation of the locus of a fluid particle that passes through the point A at $t = t_0$.

3.3 DEFORMATION OF FLUID ELEMENTS

In general in an arbitrary flow, the motion of an imaginary fluid element may be imagined as being composed of three distinct properties namely

1. Translation
2. Rotation
3. Rate of deformation

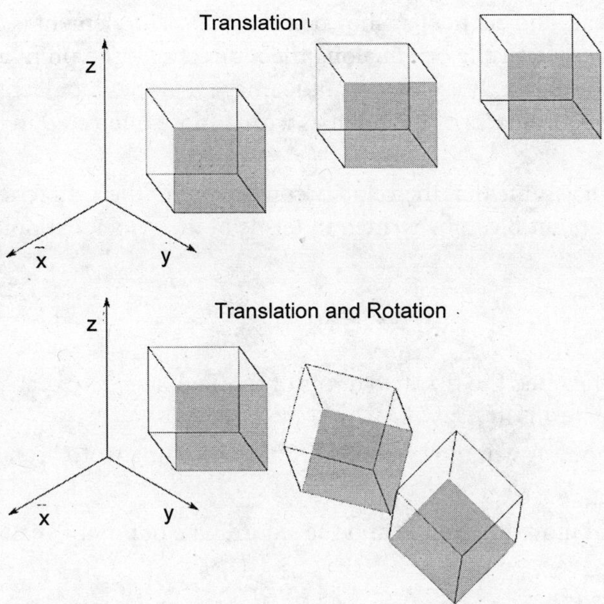

Figure 3.14: (a) Translation of a cubic fluid element, (b) Translation and rotation of a cubic fluid element.

Translation and rotation are common to the motion of a rigid body as well wherein the motion does not cause any strain or rate of strain in the body. Such a motion is shown in Fig. 3.14. More generally, we can say that the fluid element has undergone only translation and/or rotation if the included angles between the sides of the fluid element do not change. Essentially, two sides which are initially parallel will be displaced in such a manner that the sides will remain parallel to each other.

In what follows we will now look at the different conditions under which fluid deformation may occur. We shall first look into the rate of linear deformation followed by volumetric deformation, followed by rate of angular deformation.

3.3.1 Rate of Linear Deformation

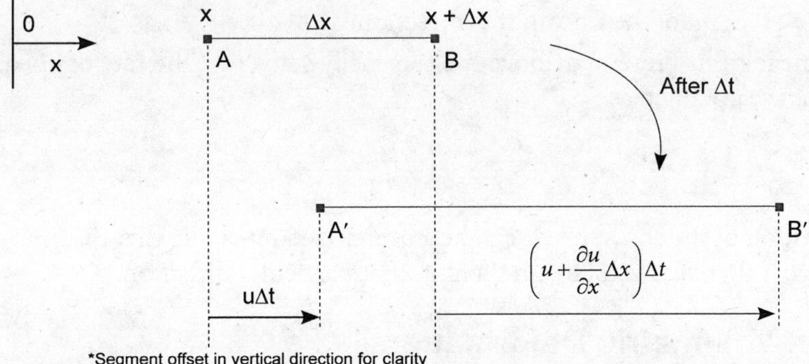

Figure 3.15: Linear deformation of a segment AB. After a time the segment now becomes A'B'. For sake of clarity the segment A'B' is shown at a vertical offset. The displacements of points A and B are shown in the figure. The new length of the segment can then be found out.

Let us consider a linear fluid element as shown in Fig. 3.15. The element is denoted by AB where the left end, A, is at a distance x from the origin along the x direction. The point at the right end, B, is at a distance of $x + \Delta x$ from the origin. It is clear from the image that the initial length of this element is Δx. We do not consider the y coordinates of the point as we are more interested in linear deformation along the direction.

In such a case, we can assume that the velocity component in the x direction is $u(x)$ at point A. The corresponding velocity at point B can be written in terms of the velocity at point A in terms of a Taylor series expansion, i.e.

$$u(x+\Delta x)=u(x)+\frac{\partial u}{\partial x}\Delta x+\ldots=u(x)+\frac{\partial u}{\partial x}\Delta x \tag{3.63}$$

where we have only retained the first two terms of the Taylor series expansion as the other terms in the series are higher order terms in Δx.

At the end of time point A will move to point A'. The coordinate of A' can then be written as.

$$A' = x + u(x)\Delta t \tag{3.64}$$

Similarly, point B will move to point B' and the coordinate of point B' can be written as

$$B' = x+\Delta x+\left(u(x)+\frac{\partial u}{\partial x}\Delta x\right)\Delta t \tag{3.65}$$

Using the two coordinates of points A' and B' we can then write down the new length of the fluid element as

$$\text{Old length} = B - A = \Delta x \tag{3.66}$$

$$\text{New length} = B' - A' = \Delta x+\frac{\partial u}{\partial x}\Delta x\Delta t \tag{3.67}$$

We are now in a position to determine the rate of change of length of the fluid element with respect to the original length in the direction. This is mathematically written as

$$\dot{\epsilon}_x = \frac{\textit{Rate of change of length}}{\textit{Original length}} = \frac{1}{\Delta t}\frac{B'A'-BA}{BA} = \frac{1}{\Delta t}\frac{\Delta x+\frac{\partial u}{\partial x}\Delta x\Delta t-\Delta x}{\Delta x} = \frac{\partial u}{\partial x} \tag{3.68}$$

The above expression shows that the length of a line element in a flow will change along the direction if there is a velocity gradient of the flow in the direction.

Using the same logic as presented above, we can easily show that the rate of linear deformations in the and directions are obtained as

$$\dot{\epsilon}_x = \frac{\partial u}{\partial x} \quad \dot{\epsilon}_y = \frac{\partial v}{\partial y} \quad \dot{\epsilon}_z = \frac{\partial w}{\partial z} \tag{3.69}$$

The interpretation of the above result can be summarized by saying that the linear dimension of an element will change if there is a velocity gradient in that particular direction.

3.3.2 Rate of Volumetric Deformation

Using the same arguments made for a linear element deforming in a flow, the ideas can be extended to the deformation in a 3d flow field. In the following derivation we refer to the figure of the linear deformation in the previous section and also Fig. 3.16.

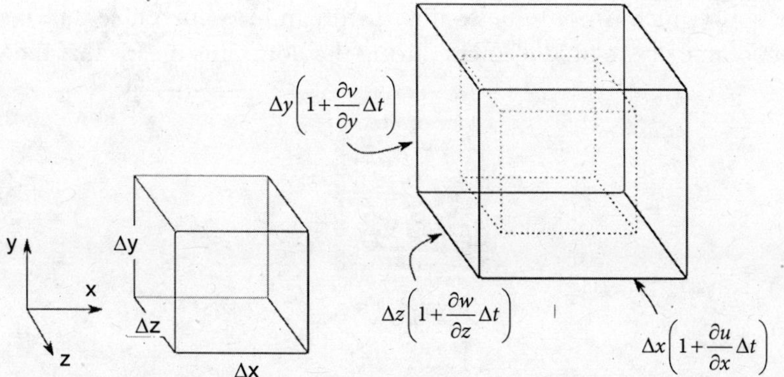

Figure 3.16: Deformation of a cubic volume in the presence of a flow. The original lengths and the final lengths are depicted in the schematic.

We have already seen that due to velocity gradients, the lengths which were initially

Initial lengths: $\Delta x, \quad \Delta y, \quad \Delta z,$

get changed to the new values due to the gradient in the velocity field:

Final lengths :

$$x \text{ direction}: \left(\Delta x + \frac{\partial u}{\partial x}\Delta x \Delta t\right)$$

$$y \text{ direction}: \left(\Delta y + \frac{\partial v}{\partial y}\Delta y \Delta t\right)$$

$$z \text{ direction}: \left(\Delta z + \frac{\partial w}{\partial z}\Delta z \Delta t\right)$$

The change in the volume of the fluid element can then be figured out using

$$New - Initial\ Vol = \left(\Delta x + \frac{\partial u}{\partial x}\Delta x \Delta t\right)\left(\Delta y + \frac{\partial v}{\partial y}\Delta y \Delta t\right)\left(\Delta z + \frac{\partial w}{\partial z}\Delta z \Delta t\right) - \Delta x \Delta y \Delta z \tag{3.70}$$

$$= \Delta x \Delta y \Delta z \left[\left(\frac{\partial u}{\partial x} + \frac{\partial v}{\partial y} + \frac{\partial w}{\partial z}\right)\Delta t\right] \tag{3.71}$$

which implies that the rate of change of volume per initial volume can be written down as

$$\frac{1}{\Delta t}\frac{New - old\ vol}{old\ vol} = \frac{\Delta x \Delta y \Delta z \left[\left(\frac{\partial u}{\partial x} + \frac{\partial v}{\partial y} + \frac{\partial w}{\partial z}\right)\Delta t\right]}{\Delta x \Delta y \Delta z \Delta t} \tag{3.72}$$

$$\frac{1}{V}\frac{DV}{Dt} = \dot{\epsilon}_{vol} = \frac{\partial u}{\partial x} + \frac{\partial v}{\partial y} + \frac{\partial w}{\partial z} \tag{3.73}$$

The above equation states that the rate of change of volume relative to the initial volume is given by the divergence of the velocity field, i.e. $\nabla \cdot \vec{u}$.

Incompressible flow (which is not to be confused with an incompressible fluid) is one that which suffers no change of volume of the fluid element during the flow. This means that the velocity fields in such flows satisfies $\nabla \cdot \vec{u} = 0$.

To know more about deformation of fluids

Example 3.10

Check if the velocity field $\vec{V} = x^2 y\hat{i} + y^2 z\hat{j} - (2xyz + yz^2)\hat{k}$ satisfies the steady incompressible flow criterion.

Solution:

Given: Velocity field is $\vec{V} = x^2 y\hat{i} + y^2 z\hat{j} - (2xyz + yz^2)\hat{k}$. ⇒ u = $x_2\, y$, $v = y^2\, z$ & $w = -(2xyz + yz^2)$

For steady incompressible fluid flow: $\frac{\partial u}{\partial x} + \frac{\partial v}{\partial y} + \frac{\partial w}{\partial z} = 0$.

$$\frac{\partial u}{\partial x} = 2xy \tag{3.74}$$

$$\frac{\partial v}{\partial y} = 2yz \tag{3.75}$$

$$\frac{\partial w}{\partial z} = -(2xy + 2yz) \tag{3.76}$$

$$\Rightarrow \quad \frac{\partial u}{\partial x} + \frac{\partial v}{\partial y} + \frac{\partial w}{\partial z} = 2xy + 2yz - (2xy + 2yz) = 0$$

Hence, the above mentioned flow field represents steady incompressible fluid flow.

3.3.3 Rate of Angular Deformation

Physically, we have already seen that the gradient of velocity which is aligned in the same direction as that of a line segment leads to the elongation or contraction of that element, depending on whether the gradient is positive or negative. However, such kinds of flows do not lead to any angular deformation, i.e. a pair of segments with an included angle in between them, flow in such a way that the angle between them does not change. This can stem from either translation or rotation.

On the other hand, there are often cases where the flow is such that an element with an initial angle between the two line segments changes with the flow. Either the angle reduces, resulting in a coming together of the two segments, or the angle increases, resulting in an increased separation between the two line segments.

Let us now understand the basis for such angular deformation of fluid elements. We consider, for convenience a two dimensional velocity field given by

$$u = u(x, y) \quad v = v(x, y) \tag{3.77}$$

Although the velocity field chosen above is steady, the derivation shown below is equally valid for unsteady fields as well as the instantaneous deformation does not depend on time derivatives of the velocity components.

We consider a rectangular fluid element ABCD as shown in Fig. 3.17. The lengths of the sides along the x and y axis are Δx and Δy respectively. The initial angle between the sides AB and AD is $\frac{\pi}{2}$.

Now, let us consider what happens physically due to the variation of the component of the velocity as a function of x. At point A, say we have a velocity v_A. Now at point B, we can write down the velocity as v_B. Now clearly, the points A and B are separated by a distance of Δx. If we say that the velocity at point B is larger than the velocity at point A, i.e. $\frac{\partial v}{\partial x} > 0$, then after a time Δt, point B would have moved a larger distance in the y direction than point A. On the other hand, if we were to have a situation where $\frac{\partial v}{\partial x} < 0$, then point A would move a larger distance than point B. All other things remaining the same, we can conclude that if the segment AD does not deform at all, then the segment AB would either turn counter clockwise if $\frac{\partial v}{\partial x} > 0$ and vice-versa.

Therefore, we have a very important assessment that the orthogonal gradient of the velocity field leads to *deformation* of the fluid element. Let us now formally derive the result using Fig 3.17 as the reference.

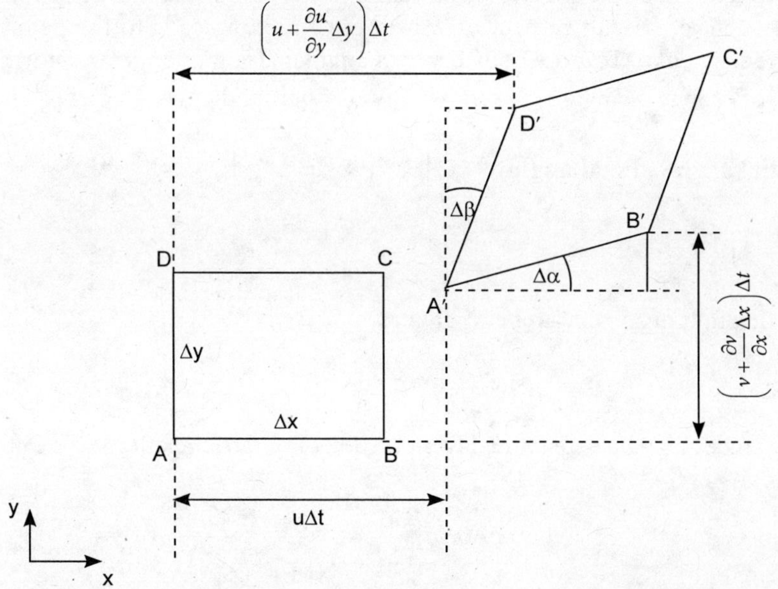

Figure 3.17: Angular deformation of a fluid element in a flow.

We can write down the displacement of points A and B in the x direction during a time Δt as

Displacement for point A in the x direction: $u\Delta t$ (3.78)

Displacement for point B in the x direction: $\left(u+\frac{\partial u}{\partial x}\Delta x\right)\Delta t$ (3.79)

Similarly, we can write down the displacement of points A and B in the y direction during a time Δt as

Displacement for point A in y direction: $v\Delta t$ (3.80)

Displacement for point B in the y direction: $\left(v+\frac{\partial v}{\partial x}\Delta x\right)\Delta t$ (3.81)

With the above equations, we can write down the relative displacement of point B with respect to point A as:

Relative displacement of B with respect to A in x direction: $\frac{\partial u}{\partial x}\Delta x\Delta t$ (3.82)

Relative displacement of B with respect to A in y direction: $\frac{\partial v}{\partial x}\Delta x\Delta t$ (3.83)

By the same arguments, we can now write down the displacement of point D with respect to point A as:

Relative displacement of D with respect to A in x direction: $\frac{\partial u}{\partial y}\Delta y\Delta t$ (3.84)

Relative displacement of D with respect to A in y direction: $\frac{\partial v}{\partial y}\Delta y\Delta t$ (3.85)

We again reiterate that the relative displacement of B with respect to A in the x direction and the relative displacement of D with respect to A in the y direction leads to the deformation of the angle between the line segments AB and AD. The rate of angular strain or angular deformation is defined as:

$\dot{\epsilon}_{xy}$ = Rate of change of angle between two initially perpendicular segments (3.86)

$$= \frac{1}{\Delta t}\text{[Initial angle – Final angle]} = \frac{1}{\Delta t}\left[\frac{\pi}{2}-\left(\frac{\pi}{2}-\Delta\alpha-\Delta\beta\right)\right] \quad (3.87)$$

$$= \frac{1}{\Delta t}[\Delta\alpha+\Delta\beta] \quad (3.88)$$

which, in the limit of $\Delta t \to 0$ can be written as

$$\dot{\epsilon}_{xy} = \frac{d\alpha}{dt}+\frac{d\beta}{dt} \quad (3.89)$$

The above angles may be determined from the trigonometric relationships found from Fig. 3..

$$\tan\Delta\alpha = \frac{B'M}{A'M} = \frac{\frac{\partial v}{\partial x}\delta x\delta t}{\delta x\left(1+\frac{\partial u}{\partial x}\delta t\right)} \quad (3.90)$$

$$\text{For small}\Delta\alpha, \tan\Delta\alpha \approx \Delta\alpha = \frac{\frac{\partial v}{\partial x}\Delta x\Delta t}{\Delta x\left(1+\frac{\partial u}{\partial x}\Delta t\right)} \tag{3.91}$$

$$\Rightarrow \qquad \frac{\Delta\alpha}{\Delta t} = \frac{\frac{\partial v}{\partial x}\Delta x}{\Delta x\left(1+\frac{\partial u}{\partial x}\Delta t\right)} \tag{3.92}$$

which, in the limit of $\Delta t \to 0$, can be simplified to

$$\frac{d\alpha}{dt} = \frac{\partial v}{\partial x} \tag{3.93}$$

On similar lines, we can derive the expression for $\frac{d\beta}{dt}$. The reader is strongly encouraged to derive the expression on their own. From the same figure, we can write

$$\frac{d\beta}{dt} = \lim_{\Delta t\to 0}\frac{\Delta\beta}{\Delta t} = \lim_{\Delta t\to 0}\frac{\frac{\partial u}{\partial y}\Delta y}{\Delta x\left(1+\frac{\partial v}{\partial y}\Delta t\right)} = \frac{\partial u}{\partial y} \tag{3.94}$$

We can now combine equations (3.93) and (3.) to obtain the rate of angular deformation as

$$\dot{\in}_{xy} = \frac{\partial v}{\partial x} + \frac{\partial u}{\partial y} \tag{3.95}$$

Therefore we have derived how fluid elements deform in arbitrary flows. The same derivation may be easily extended to three dimensional flow fields through a pairwise consideration of the deformations seen through the yz and xz planes (a note of caution - this statement is true only for Cartesian coordinate systems and not generally valid for other non-orthogonal coordinate systems). However, it may be easily derived using the basic principles highlighted above.

3.3.4 Rotation of Fluid Elements

While we have derived the rate of angular deformation of two initially orthogonal fluid elements, we are now in a position to determine the rate of rotation of the fluid elements.

Given the impending motion of the two line segments AB and AD, it is seen that the sense of direction of the two line segments are in the opposite directions. Line segment AB has a tendency to rotate counter-clockwise when the velocity gradient, $\frac{\partial v}{\partial x}$, is larger than zero. Similarly, line segment AD has a tendancy to rotate in the clockwise direction when the velocity gradient, $\frac{\partial u}{\partial y}$ is larger than zero. In that case, if we consider the xy plane as that shown in the figure, any counter-clockwise rotation is considered to be a positive angular velocity and hence the average angular velocity of the segment is given by

$$\omega_{xy} = \frac{1}{2}\left(\frac{d\alpha}{dt} - \frac{d\beta}{dt}\right) = \frac{1}{2}\left(\frac{\partial v}{\partial x} - \frac{\partial u}{\partial y}\right) \tag{3.96}$$

On similar lines, the average rotation rates as seen in the other two planes can also be found out. We now attempt to connect the result (3.96) with the concept of vorticity.

3.3.5 Vorticity

We define vorticity as the curl of the velocity field. Mathematically it is written as

$$\vec{\Omega} = \nabla \times \vec{u} = \begin{vmatrix} \hat{i} & \hat{j} & \hat{k} \\ \frac{\partial}{\partial x} & \frac{\partial}{\partial y} & \frac{\partial}{\partial z} \\ u & v & w \end{vmatrix} \tag{3.97}$$

$$= \hat{i}\left(\frac{\partial w}{\partial y} - \frac{\partial v}{\partial z}\right) + \hat{j}\left(\frac{\partial u}{\partial z} - \frac{\partial w}{\partial x}\right) + \hat{k}\left(\frac{\partial v}{\partial x} - \frac{\partial u}{\partial y}\right) \tag{3.98}$$

$$\text{where, } \Omega_x = \left(\frac{\partial w}{\partial y} - \frac{\partial v}{\partial z}\right) \tag{3.99}$$

$$\Omega_y = \left(\frac{\partial u}{\partial z} - \frac{\partial w}{\partial x}\right) \tag{3.100}$$

$$\Omega_z = \left(\frac{\partial v}{\partial x} - \frac{\partial u}{\partial y}\right) \tag{3.101}$$

It is clear from equations (3.101) and (3.96) that

$$\omega_{xy} = \frac{1}{2}\Omega_z \tag{3.102}$$

Example 3.11

Determine the angular deformation and vorticity for velocity field: $\vec{V} = y^2\hat{i} + 3x\hat{k}$.

Solution:

Given: Velocity field is $\vec{V} = y^2\hat{i} + 3x\hat{k}$. $\Rightarrow u = y^2$ & $v = 3x$

Angular deformation $\epsilon_{xy} = \frac{1}{2}\left(\frac{\partial v}{\partial x} + \frac{\partial u}{\partial y}\right) = \frac{1}{2}(3 + 2y)$

Vorticity $\Omega_z = \left(\frac{\partial v}{\partial x} - \frac{\partial u}{\partial y}\right) = (3 - 2y)$

Example 3.12

Determine if the following flow field is irrotational or not: $u = \frac{y^3}{3} + 2x - x^2 y$ and $v = xy^2 - 2y - \frac{x^3}{3}$.

Solution:

Vorticity

$$\Omega_z = \left(\frac{\partial v}{\partial x} - \frac{\partial u}{\partial y} \right)$$

$$\frac{\partial v}{\partial x} = y^2 - x^2 \tag{3.103}$$

$$\frac{\partial u}{\partial y} = y^2 - x^2 \tag{3.104}$$

$$\therefore \quad \Omega_z = 0 \tag{3.105}$$

Thus, the flow field is irrotational.

UNIT SUMMARY

- Acceleration

  Acceleration in x direction: $\frac{du}{dt} \equiv \frac{Du}{Dt} = \frac{\partial u}{\partial t} + u\frac{\partial u}{\partial x} + v\frac{\partial u}{\partial y} + w\frac{\partial u}{\partial z}$

  Acceleration in y direction: $\frac{dv}{dt} \equiv \frac{Dv}{Dt} = \frac{\partial v}{\partial t} + u\frac{\partial v}{\partial x} + v\frac{\partial v}{\partial y} + w\frac{\partial v}{\partial z}$

  Acceleration in z direction: $\frac{dw}{dt} \equiv \frac{Dw}{Dt} = \frac{\partial w}{\partial t} + u\frac{\partial w}{\partial x} + v\frac{\partial w}{\partial y} + w\frac{\partial w}{\partial z}$

- Total derivative

$$\frac{D}{Dt}\phi = \frac{\partial \phi}{\partial t} + \vec{V} \cdot \nabla \phi$$

- Equation of streamline

$$\frac{dx}{u} = \frac{dy}{v} = \frac{dz}{w}$$

- Linear deformation rate in x, y, and z direction

$$\dot{\epsilon}_x = \frac{\partial u}{\partial x} \quad \dot{\epsilon}_y = \frac{\partial v}{\partial y} \quad \dot{\epsilon}_z = \frac{\partial w}{\partial z}$$

- Volumetric strain rate

$$\frac{1}{V}\frac{DV}{Dt} = \dot{\epsilon}_{vol} = \frac{\partial u}{\partial x} + \frac{\partial v}{\partial y} + \frac{\partial w}{\partial z}$$

- Rate of angular deformation

$$\dot{\epsilon}_{xy} = \frac{\partial v}{\partial x} + \frac{\partial u}{\partial y}$$

- Vorticity of a flow field

$$\vec{\Omega} = \nabla \times \vec{u} = \begin{vmatrix} \hat{i} & \hat{j} & \hat{k} \\ \frac{\partial}{\partial x} & \frac{\partial}{\partial y} & \frac{\partial}{\partial z} \\ u & v & w \end{vmatrix}$$

EXERCISES

Multiple Choice Questions

1. A 2-dimensional flow field is given by $\vec{V} = x\hat{i} + y\hat{j}$. The flow field is:

(a) Uniform
(b) Non-uniform
(c) Steady
(d) Cannot be determined

2. For the 2-dimensional flow field $\vec{V} = 3t\hat{i} + y\hat{j}$, where t is time, the acceleration vector is given by

(a) Time dependent
(b) Time independent
(c) Spatial coordinate independent
(d) Spatial coordinate dependent

3. For the 2-dimensional flow field $\vec{V} = xt\hat{i} + y\hat{j}$, where t represents time, the acceleration vector is

(a) Time dependent
(b) Time independent
(c) Spatial coordinate independent
(d) Spatial coordinate dependent

4. A 2-dimensional flow field is given by $\vec{V} = -y\hat{i} + x\hat{j}$. The magnitude of the velocity $\vec{V}$ along a circulat path of any radius R is

(a) Dependent on the radius R of the circle
(b) Constant irrespective of the radius of the circle
(c) Cannot be determined from the above information

5. Pathlines, streaklines, and streamlines are the same for which of the following case?

(a) Unsteady uniform flow
(b) Unsteady irrotational flow
(c) Steady incompressible flow
(d) Steady non-uniform flow

6. The equation of a streamline passing through the point (0, 0) for a flow field $\vec{V} = 5\hat{i} + x\hat{j}$ is given by

(a) $x^2 + 10y = 0$
(b) $x^2 - 10y = 0$
(c) $y^2 - 10x = 0$
(d) $y^2 + 10x = 0$

7. For the flow field $\vec{V} = x^2 y\hat{i} - y^2 x\hat{j}$, the vorticity vector is in the

(a) x direction
(b) y direction
(c) z direction
(d) Any arbitrary direction in the 3-dimensional space

8. The flow field $\vec{V} = x\hat{i} - y\hat{j}$ represents a:

(a) Steady compressible flow
(b) Unsteady compressible flow
(c) Steady incompressible flow
(d) Steady irrotational flow

9. The flow field $\vec{V} = x^2\hat{i} - y^2\hat{j}$ represents an

(a) Steady compressible flow
(b) Steady incompressible flow
(c) Steady rotational flow
(d) Steady irrotational flow

ANSWERS															
1.	(b), (c)	2.	(b), (d)	3.	(a)	4.	(a)	5.	(c), (d)	6.	(b)	7.	(c)	8.	(c), (d)
9.	(d)														

Unsolved Questions

Level - I

1. For the velocity field $\vec{V} = Ax^2 y\hat{i} + Bxy^2\hat{j}$, where A = 2 m^{-2}s^{-1} and B = 1 m^{-2}s^{-1}, and the coordinates are measured in meters, obtain an equation for the flow streamlines.

2. The velocity field $\vec{V} = Ax\hat{i} - Ay\hat{j}$, where A = 2s^{-1}, can be interpreted to represent flow in a corner. Find an equation for the flow streamlines.

3. A velocity field is given by $\vec{V} = ax\hat{i} - bty\hat{j}$, where a = 1s^{-1} and b = 1s^{-2}. Find the equation of the streamlines at anytime t.

4. A velocity field is given by $\vec{V} = ax^3\hat{i} - bxy^3\hat{j}$, where a = 1m^{-2}s^{-1} and b = 1 m^{-3}s^{-1}. Find the equation of the streamlines.

5. Consider the flow field given by $\vec{V} = xy^2\hat{i} - y^3/3\hat{j} + xy\hat{k}$. Determine the acceleration of a fluid particle at point $(x, y, z) = (1, 2, 3)$.

6. Consider the velocity field $\vec{V} = A\left(x^4 - 6x^2y^2 + y^4\right)\hat{i} + A\left(4xy^3 - 4x^3y\right)\hat{j}$ in the xy plane, where A = 0.25 m^{-3}s^{-1}, and the coordinates are measured in meters. Calculate the acceleration of a fluid particle at point $(x, y) = (2, 1)$.

7. Consider the flow field given by $\vec{V} = ax^2 y\hat{i} - by\hat{j} + cz^2\hat{k}$. where, a = 2 $m^{-2}s^{-1}$, b = $2s^{-1}$ and c = 1 $m^{-1}s^{-1}$ Determine if it is a possible incompressible flow, and the acceleration of a fluid particle at point $(x, y, z) = (2, 1, 3)$.
8. Given the incompressible flow $\vec{V} = 3y\hat{i} + 2x\hat{j}$, find the stream function $\psi(x, y)$.
9. A two-dimensional unsteady velocity field is given by $u = x(1 + 2t)$, $v = y$. Find the time-varying streamlines which pass through some reference point (x_0, y_0).
10. Consider the two-dimensional velocity distribution $u = -By$, $v = +Bx$, where B is a constant. If this flow possesses a stream function, find its form. Compute the local angular velocity of the flow, if any, and describe what the flow might represent.

Level - II

1. The x component of velocity in a steady, incompressible flow field in the xy plane is $u = A/x$, where A = 2 m^2s^{-1}, and x is measured in meters. Find the simplest y component of velocity for this flow field.
2. A two-dimensional incompressible flow is defined by $u = -\dfrac{Ky}{x^2 + y^2}$ and $v = \dfrac{Kx}{x^2 + y^2}$, where K = constant. Discuss whether this flow is rotational or irrotational and how?
3. Consider the flow described by the velocity field $\vec{V} = A(1 + Bt)\hat{i} + Cty\hat{j}$, with A = 1 m/s, B = $1s^{-1}$, and C = $1s^{-2}$. Coordinates are measured in meters. Plot the pathline traced out by the particle that passes through the point (1, 1) at time $t = 0$.
4. Consider the flow field given in Eulerian description by the expression $\vec{V} = A\hat{i} - Bt\hat{j}$, where A = 2 m/s, B = 2 m/s^2, and the coordinates are measured in meters. Derive the Lagrangian position functions for the fluid particle that was located at the point $(x, y) = (1, 1)$ at the instant $t = 0$. Obtain an algebraic expression for the pathline followed by this particle.
5. Given the velocity field $u = \dfrac{x}{1+t}$, $v = y$. (a) Find the stream lines at an instant of time t. (b) Pathline from $t = 0$ to time t. (c) Streaklines at an instant of time t for particle which have passed through a point x_0, y_0 over a time period from $t' = 0$ to $t' = t$.

To know more about kinematics of fluids

Answer Key

Level – I

1. $y = C/\sqrt{x}$

2. $\ln(xy) = C$

3. $y = cx^{\frac{-bt}{a}}$

4. $y = \dfrac{1}{\sqrt{2\left(\dfrac{b}{ax} + c\right)}}$

5. $\vec{a} = \dfrac{16}{3}\hat{i} + \dfrac{32}{3}\hat{j} + \dfrac{16}{3}\hat{k}$

6. $a = 69.9\ \text{m/s}^2$

7. $\vec{a} = 48\hat{i} + 4\hat{j} + 54\hat{k}$

8. $\psi = \dfrac{3}{2}y^2 - x^2$

9. $y = y_0\,(x/x_0)^{1/(1+2t)}$

10. $\psi = -\dfrac{B}{2}\left(x^2 + y^2 + constant\right)$ and angular velocity = B

Level – II

1. $v = \dfrac{Ay}{x^2}$

2. Irrotational

3. $x = A\left(t + B\dfrac{t^2}{2}\right) + 1$

4. $y = 1 - \dfrac{(x-1)^2}{4}$

5. Streamline: $y = Cx^{1+t}$, Pathline: $y = y_0 e^{\left(\frac{x}{x_0} - 1\right)}$, Streakline: $y = y_0 e^{\left[t - \frac{x}{x_0}(1+t) + 1\right]}$

PRACTICAL

Aim: Visualizing the flow of air using a smoke-wire apparatus.

Apparatus: A small exhaust fan (for example, a CPU cooler fan), nichrome wires, a DC power supply (a 12V source would work), building materials for the stand, used toothbrush, black chart paper, light source, camera (a smart phone camera will also work)

Theory: Air flow can be visualized with the help of smoke. A very easy way of generating smoke is to flash it with the help of a high temperature surface. In the example described here, we make use of a heated wire to generate the smoke. The points where the smoke gets generated acts as the points where the streakline originates. The streaklines can be easily obtained by taking a single snapshot of the smoke particles in the flow at a chosen later time.

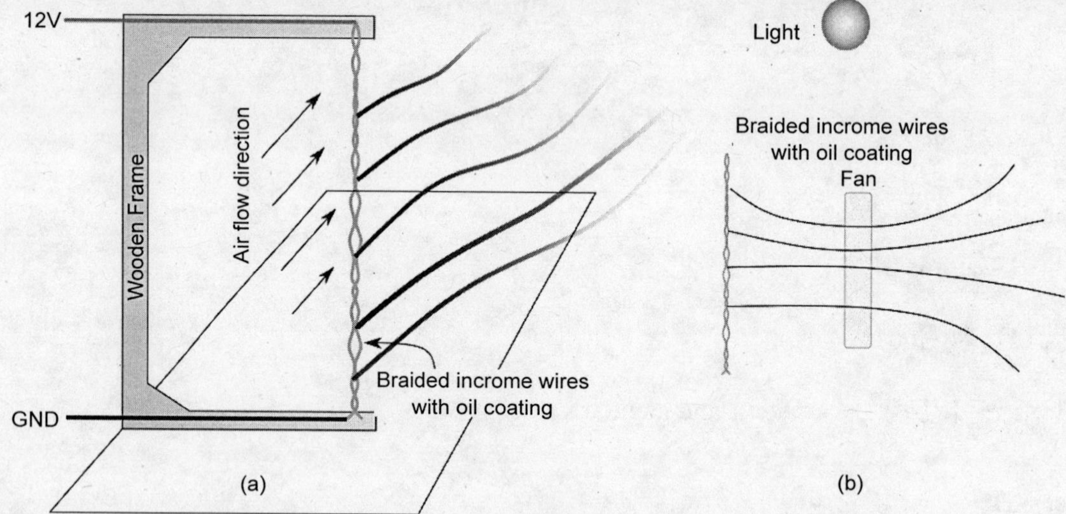

Figure 3.18: Schematic of the smoke-wire apparatus in a flow field. At a certain instant of time, the smoke emanating from the heated wire will be visible against a dark background. (a) An isometric view of the experiment showing the frame which holds the wire and the streaklines due to the flow. (b) A side view of the demonstration. The air is drawn in by the fan from the wire towards it. An interested student may try various locations and observe the streaklines.

Procedure:

1. Two Nichrome wires of length 10 cm each are twisted like a braid.
2. The wires are fixed at the two ends of a stand which is made with the help of a non-conducting base such as wood.
3. The ends of the wires are then connected to the DC power supply. The current passing through the Nichrome wires causes heating of the wires.
4. With the help of a old toothbrush, kerosene oil can be coated onto the wires.
5. This apparatus can be placed in the suction region of the CPU cooler fan which is drawing air from one direction.

6. A black background can be used in conjunction with a strong light placed orthogonal to the apparatus as shown in the figure.
7. The apparatus can be placed at various points in the domain to understand the nature of the fluid flow.

Observations: The person performing the experiment will observe that when the apparatus is placed in the suction region of the fan, then the streaklines do not fluctuate a lot and the flow upstream of the fan appears to be rather stationary and well-behaved. On the contrary when the apparatus is placed in the region downstream of the fan, the disturbances are such that it is not possible to clearly elucidate the flow. There appears to be a rapid dispersion of the smoke particles in such kinds of flow. The pattern of the smoke captured by a single photograph represents the streaklines formed due to the smoke originating from various points on the heated wire.

KNOW MORE

In fluid dynamics, Leonard Euler and Joseph-Louis Lagrange independently formulated distinct mathematical methods for analyzing fluid movement, each offering valuable insights into the phenomena. The Eulerian approach examines changes in velocity (speed and direction) at fixed points, while the Lagrangian approach tracks the positions of individual fluid particles over time, with both methods complementing each other effectively.

These methods employ different mathematical techniques, which makes integration challenging. Euler's method is particularly suited for analyzing data collected at stationary locations, whereas Lagrange's method excels in measurements from drifters and floats. This distinction underscores the unique strengths of each approach in the study of fluid dynamics.

[Source: https://incois.gov.in/Tutor/science+society/lectures/illustrations/lecture21/lagrangian.html#:~:text=Leonard%20Euler%20and%20Joseph%2DLouis,are%20complementary%20in%20their%20use]

REFERENCES AND SUGGESTED READINGS

1. **Introduction to Fluid Mechanics and Fluid Machines** (3rd Edition), S K Som, Gautam Biswas, Suman Chakraborty, McGraw Hill
2. **Fluid Mechanics** (9th Edition), Frank M. White, Henry Xue, McGraw Hill

❑❑

4 Fluids under Motion 2: Conservation Laws

UNIT SPECIFICS

In this unit we will discuss about the following topics:

1. Conservation of mass and various approaches towards it.
2. Conservation of momentum and the inviscid equation which describes it.
3. An integral approach to the conservation of momentum
4. Bernoulli's equation as obtained through the conceptual framework of the conservation of momentum

RATIONALE

While the previous unit delved deep into the understanding of the various descriptions of fluid flow, there was no connection to the actual forces which cause the flow. A very natural and important aspect of fluid mechanics is how we can describe the connection between the flow itself and the nature of forces acting upon it. While we have briefly touched up on the idea of a pressure gradient in Unit 2, we are now able to formally introduce it through the idea of conservation laws.

The most elementary conservation law that must hold true, in systems beyond fluid mechanical systems, is the conservation of mass. In what follows we have first described general frameworks to address this conservation law and shown that regardless of the approach, we naturally arrive at the same result.

The other elementary conservation law is that of momentum. The basis of the conservation of momentum lies in the Newton's second law which states that the rate of change of momentum of a body is equal to the sum of all forces acting on it; an idea which has been introduced at a high school level. Extending this to fluid systems is not straightforward as fluids are not discrete objects such as billiard balls. In this Unit, we will address these issues and obtain some very important results such as the Euler's equation and the Bernoulli's equation which are extremely useful in many day-to-day applications and observations. In this sense, this chapter lays down useful ideas which the practicing engineer should be very conversant in.

PRE-REQUISITES

Elementary calculus

UNIT OUTCOMES

U4-O1: Conservation of mass

U4-O2: Conservation of momentum

U4-O3: Euler's equation

U4-O4: Bernoulli's equation

Unit-4 Outcome	EXPECTED MAPPING WITH PROGRAMME OUTCOMES (1 - Weak Correlation; 2 - Medium Correlation; 3 - Strong Correlation)						
	CO-1	CO-2	CO-3	CO-4	CO-5	CO-6	CO-7
U4-O1	2	-	1	3	-	2	3
U4-O2	2	-	1	3	-	2	3
U4-O3	2	-	-	3	-	-	2
U4-O4	2	-	2	3	3	2	2

4.1 CONSERVATION OF MASS

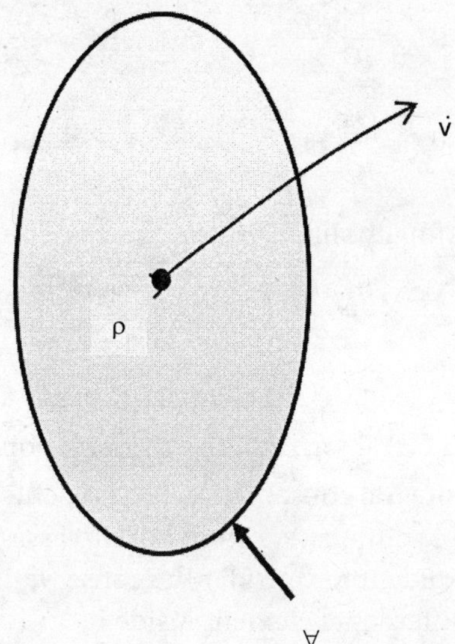

Figure 4.1: Control mass moving with a velocity moving with a velocity $\vec{V}$.

Let us first begin by looking at the most fundamental premise of any transport process, i.e. conservation of mass. We all know that mass can neither be created nor be destroyed. The case of fluid mechanics is no different. In this case, let us begin by consider a control mass of a fluid as shown in Fig. 4.1. We have shown a control volume having a volume $\forall$ and density ρ. The total mass of the system can be written as

$$m = \rho\forall \Rightarrow \ln m = \ln \rho + \ln\forall \tag{4.1}$$

Now, if we track this control mass, the volume may deform in time but the deformation will be such that the total mass inside the chosen control volume will be constant. This implies

$$\frac{D}{Dt}\ln m=\frac{D}{Dt}\ln\rho+\frac{D}{Dt}\ln\forall \tag{4.2}$$

$$0=\frac{D}{Dt}\ln\rho+\frac{D}{Dt}\ln\forall \tag{4.3}$$

We can now expand the above expression to write

$$\frac{1}{\rho}\frac{D}{Dt}\rho+\frac{1}{\forall}\frac{D}{Dt}\forall=0 \tag{4.4}$$

Making use of the definition of the total derivative given in the previous Unit, equation Eq. (3.31) and the definition of the volumetric strain rate derived also in the previous chapter, Eq. (3.73), we can write down the above equation as

$$0=\frac{1}{\rho}\left(\frac{\partial\rho}{\partial t}+u\frac{\partial\rho}{\partial x}+v\frac{\partial\rho}{\partial y}+w\frac{\partial\rho}{\partial z}\right)+\nabla\cdot\vec{V} \tag{4.5}$$

The above equation can now be simplified as

$$0 \quad =\frac{1}{\rho}\left(\frac{\partial\rho}{\partial t}+u\frac{\partial\rho}{\partial x}+v\frac{\partial\rho}{\partial y}+w\frac{\partial\rho}{\partial z}\right)+\frac{\partial u}{\partial x}+\frac{\partial v}{\partial y}+\frac{\partial w}{\partial z}$$

$$0 \quad =\frac{\partial\rho}{\partial t}+u\frac{\partial\rho}{\partial x}+v\frac{\partial\rho}{\partial y}+w\frac{\partial\rho}{\partial z}+\rho\left(\frac{\partial u}{\partial x}+\frac{\partial v}{\partial y}+\frac{\partial w}{\partial z}\right)$$

The above equation can thus be finally simplified to

$$\frac{\partial\rho}{\partial t}+\frac{\partial(\rho u)}{\partial x}+\frac{\partial(\rho v)}{\partial y}+\frac{\partial(\rho w)}{\partial z}=0 \tag{4.6}$$

The above equation represents the differential form of the conservation of equation. To this book, it is more suitable for us to work with integral forms of the conservation of mass.

Let us first look at why such an integral conservation law is useful to us. If we consider a simple tank which has multiple inputs and multiple outputs, we would like to assess how the total mass in the tank is varying in time, for such a case it is prudent to consider the conservation of mass contained in the entire tank, rather than analyzing a small differential element inside it.

Similarly, if we consider syringe pushing a fluid due to the user pushing the injector, we are interested to know the rate at which fluid is coming out of the needle in terms of gram per hour. We do not bother much about the state of a differential element inside the syringe.

The differential approach established through Eq. (4.6) helps us in assessing very fundamental properties of flow fields and density fields because as the equation reveals, the density and velocity field must be such that the Eq. (4.6) must be identically satisfied. We cannot have an *inconsistent* velocity field in conjunction with a density field which does not satisfy the conservation of mass. Notwithstanding this importance of a differential field, for the purpose of this Unit, it is more important to us to analyze larger systems.

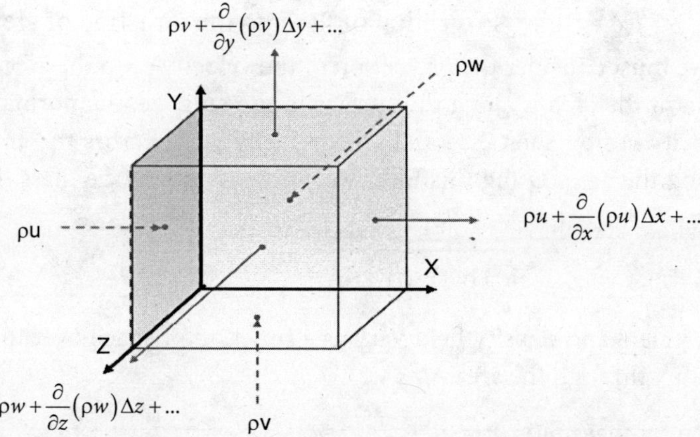

Figure 4.2: A control volume depicting mass flow in and out of a control volume.

We can now work with a control volume rather than working with a control mass and obtain a simpler physical understanding of Eq. (4.6). Let us consider a control volume as shown in Fig. 4.2. The control volume is fixed while the flow *crosses* the control volume. It is clear from the schematic that flow enters and exits the control volume. We can now proceed to perform a balance of mass on the control volume. We can say that the rate of increase of mass inside the control volume is equal to the net mass brought into the control volume by the flow. When written in the form of an equation we can write

Rate of change of mass inside a CV = Rate of influx into the CV (4.7)

We can write down the rate of change of mass inside the control volume as

$$\text{Rate of change of mass inside a CV}=\frac{\partial}{\partial t}\left(\int \rho d\forall\right) \tag{4.8}$$

$$\text{If density is constant inside the control volume }=\frac{\partial}{\partial t}(\rho\forall)=\frac{\partial}{\partial t}(\rho\Delta x\Delta y\Delta z) \tag{4.9}$$

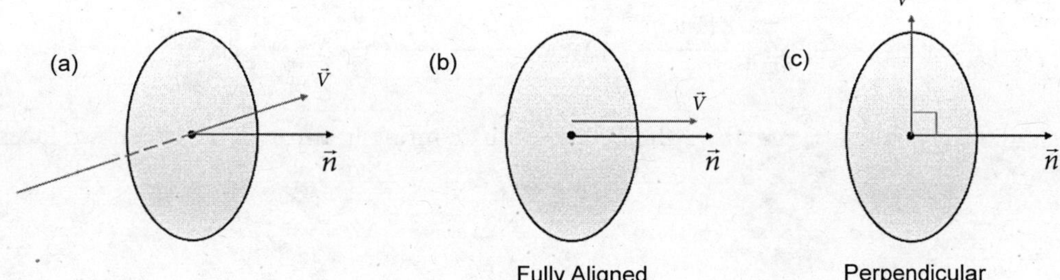

Figure 4.3: (a) Determination of mass flux through an area when the area vector and velocity vector are arbitrarily oriented. (b) When they are collinear, (c) When they are perpendicular to each other.

The thing left to do now is to figure out the rate of influx of mass into the control volume. To understand this better, we take a small detour and refer to Fig. 4.3 which shows an area A with a normal vector $\vec{n}$. If we now consider a uniform velocity field at the area equal to $\vec{V}$, then the volume flux brought by the flow through the area is equal to

$$Q = \vec{V} \cdot \vec{n}A = \text{Projection of } \vec{V} \text{ onto the direction of area times area} \quad (4.10)$$

The reason why we must consider the projection of the velocity onto the area is clear from the Fig. 4.3(b) which shows that if the area is along the surface of the area, i.e. the normal vector of the area is orthogonal to the velocity vector, then essentially no velocity vector cross the area and hence there is actually no flux crossing the area. Using this idea, we can write down the mass flux due to a uniform velocity field $\vec{V}$ through an area having a normal vector $\vec{n}$ as

$$\dot{m} = \rho Q = \rho \vec{V} \cdot \vec{n}A \quad (4.11)$$

In case the velocity field and density field vary as a function of space over the area A then we can rewrite the total mass flux through the area A as

General expression for mass flux: $\dot{m} = \int \rho \vec{V} \cdot \vec{n} dA$ (4.12)

when velocity is normal to area: $\dot{m} = \int \rho V dA$ (4.13)

When the density is constant over the area: $\dot{m} = \rho \int \vec{V} dA$ (4.14)

When the velocity is uniform over the area: $\dot{m} = \rho V \int dA = \rho VA$ (4.15)

In the expressions above, V represents the magnitude of the velocity vector $\vec{V}$.

With this information we are now able to write down the rate of influx into the control volume. We refer to Fig. 4.2. Let us first consider the flux through the area given by $\Delta y \Delta z$. The mass flux per unit area on the left face is given by , where represents the component of the velocity field. Using the Taylor series expansion, we can write down the mass flux per unit rea on the right face as

$$\rho u + \frac{\partial \rho u}{\partial x} \ddot{A} x + \ldots \quad (4.16)$$

Therefore, the net mass flux coming in through the two areas having area $\Delta y \Delta z$ is given by

$$\dot{m}_x = \rho u \Delta y \Delta z - \left(\rho u + \frac{\partial(\rho u)}{\partial x} \Delta x \right) \Delta y \Delta z = -\frac{\partial(\rho u)}{\partial x} \Delta x \Delta y \Delta z \quad (4.17)$$

On similar lines we can write down the net mass flux coming in through the other two faces as

$$\dot{m}_y = -\frac{\partial(\rho v)}{\partial y} \quad (4.18)$$

$$\dot{m}_z = -\frac{\partial(\rho w)}{\partial z} \quad (4.19)$$

We can now combine the Eq. (4.9), Eq. (4.17), and Eq. (4.19) to obtain

$$\frac{d}{dt}(\rho \Delta x \Delta y \Delta z) = -\frac{\partial(\rho u)}{\partial x} \Delta x \Delta y \Delta z - \frac{\partial(\rho v)}{\partial y} \Delta x \Delta y \Delta z - \frac{\partial(\rho w)}{\partial z} \Delta x \Delta y \Delta z \quad (4.20)$$

If we now consider a control volume that does not change in time, we can reduce the above equation by cancelling the terms $\Delta x \Delta y \Delta z$ and obtain

$$\frac{\partial \rho}{\partial t}+\frac{\partial(\rho u)}{\partial x}+\frac{\partial(\rho v)}{\partial y}+\frac{\partial(\rho w)}{\partial z} \tag{4.21}$$

which eventually yields the same result that we have obtained earlier from a Lagrangian approach of a control volume moving with the fluid.

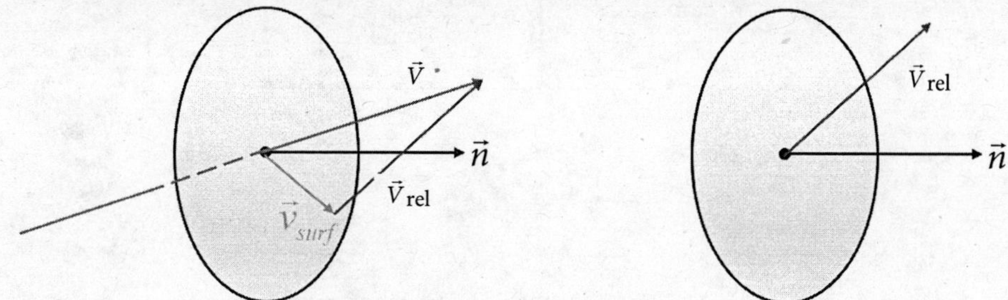

Figure 4.4: Flux through a surface when the surface is moving with a velocity $\vec{V}_{\text{surf}}$.

We now address another important aspect of mass conservation, in particular, the flux passing through a surface. If we consider Fig. 4.4, where we have a surface moving with a velocity $\vec{V}_{\text{surf}}$ and if the flow velocity is now $\vec{V}$, then the expression for the mass flux has to be modified to

$$\dot{m}=\int \rho\left(\vec{V}-\vec{V}_{surf}\right)\cdot \vec{n}dA=\int \rho \vec{V}_{rel}\cdot \vec{n}dA \tag{4.22}$$

where $\vec{V}_{\text{ref}}$ represents the relative velocity of the flow relative to the velocity of the surface. This is important to note as the flow velocity relative to the control surface determines the flux, not the velocity only. In fact if we have the special situation where the velocity of the control surface is exactly equal to the velocity of the flow, then there is no flux that crosses the surface.

Therefore, we can write down the conservation of mass as

$$0=\frac{\partial}{\partial t}\left(\int \rho d\forall\right)+\int \rho \vec{V}_{\text{rel}}.\vec{n}dA \tag{4.23}$$

Conservation of mass for control volumes therefore involves understanding the fluxes entering the control volume through various control surfaces and how it leads to a net change in the mass of the system. A better understanding of this may be obtained through the solved examples below.

4.1.1 Examples of conservation of mass

Let us now consider some examples where we can apply the ideas learnt in the previous section.

Example 4.1

Consider the tank shown in the figure. It has two pipes, one inlet and another outlet. Fluid enters the tank at a constant and uniform velocity $\vec{V}_i$ into the tank and exits the tank at a constant and uniform velocity $\vec{V}_e$. The diameter of the inlet and outlet are a and b respectively. Determine a relationship between the magnitudes of the inlet and outlet velocity.

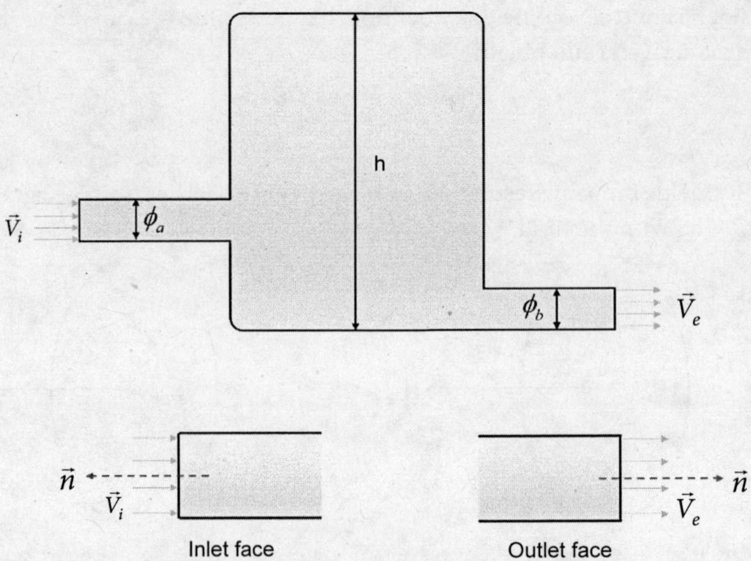

Solution:

Let us make use of the conservation of mass obtained in the section above. Here we note that the control volume is chosen as the entire tank along with the inlet and outlet pipes. In this case, we can make the following observations:

1. The chosen control volume is stationary and hence, $\vec{V}_{surf} = 0$
2. The problem is a steady one since there is no accumulation of mass in the control volume

In that case, we can reduce Eq. (4.23) to $0 = \int \rho \vec{V} \cdot \vec{n} dA$, where the integral can now be expressed in terms of the inlet and outlet solely. We can therefore write the equation by eliminating ρ from both the inlet and outlet fluxes to write:

$$0 = \int_{inlet} \vec{V} \cdot \vec{n} dA + \int_{outlet} \vec{V} \cdot \vec{n} dA \Rightarrow 0 = -V_i \pi \frac{a^2}{4} + V_e \pi \frac{b^2}{4} \Rightarrow \frac{V_i}{V_e} = \frac{b^2}{a^2}.$$

Let us spend some time to discuss the way the integrals for the inlet and outlet flux are evaluated. At the inlet, we refer to the subfigure (b). It is seen that the velocity vector and the normal vector are in opposite directions. In fact, in this situation, the angle between the two vectors is 180°. This means that the expression for the integrand can be expressed as $\vec{V}_i \cdot \vec{n} dA = V_i \cos 180^\circ dA = -V_i dA$. Therefore, the integral for the inlet flux can be finally expressed as $\int V_i dA = V_i A_i = V_i \pi \frac{a^2}{4}$, where we have made use of the fact that the magnitude of the inlet velocity, V_i remains constant over the entire face, i.e. the velocity is homogeneous on the entire inlet face.

On similar lines, we can find out the integral at the outlet face. If we refer to subfigure (c), we observe that the velocity at the normal area vector is aligned in the same direction. Therefore the integrand for the flux can be expressed as $\vec{V}_e \cdot n dA = V_e dA$ and therefore the integral can be written as $\int V_e dA = V_e \pi \frac{b^2}{4}$, where once again we have made use of the fact that the magnitude of the exit velocity, V_e remains constant over the entire face.

The above solution can be also interpreted in the simpler way learnt at the high school level as the continuity equation i.e. $V_1 A_1 = V_2 A_2$ where points 1 and 2 are at two different cross section of a pipe. One can infact easily derive this result with a more concrete theoretical framework as described in the above section.

Example 4.2

Let us now consider the same problem as above but instead of having a uniform velocity at the inlet and outlet we have a parabolic velocity profile, the magnitudes of which are given by:

$$V_i = A\left(1-\frac{r^2}{a^2}\right) \quad V_e = B\left(1-\frac{r^2}{b^2}\right)$$

The velocities at the inlet and exit faces are shown in the figure. The remaining figure is the same as that of the previous example.

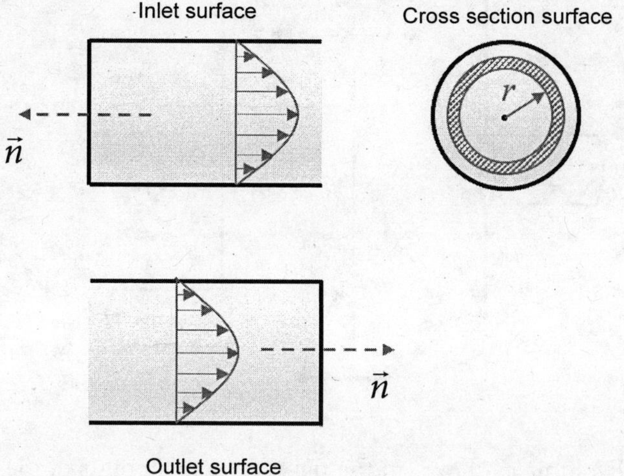

Solution:

Given that the problem has not fundamentally changed, i.e. the control volume is the same and it is stationary. We must only figure out how to determine the flux. Let us consider the inlet flux first. The direction of the velocity and the normal vector of the area are still 180 in directions. Therefore, we can write down the integral as

$$\int \vec{V}_i \cdot n dA = -\int V_i dA = -\int A\left(1-\frac{r^2}{a^2}\right) dA = -\int_0^a A\left(1-\frac{r^2}{a^2}\right) 2\pi r dr = -2\pi A\left(\frac{a^2}{2}-\frac{a^2}{4}\right) = -\pi A \frac{a^2}{2}$$

On similar lines, we can write down the flux at the exit as

$$\int \vec{V}_e \cdot ndA = \int V_e dA = \int B\left(1-\frac{r^2}{b^2}\right)dA = -\int_0^b B\left(1-\frac{r^2}{b^2}\right)2\pi r dr = -2\pi B\left(\frac{b^2}{2}-\frac{b^2}{4}\right) = -\pi B\frac{b^2}{2}.$$

We can therefore combine these two fluxes with the result of the conservation of mass and write

$$0 = -\pi A\frac{a^2}{2} + \pi B\frac{b^2}{2} \Rightarrow \frac{A}{B} = \frac{b^2}{a^2}.$$

Example 4.3

Let us now consider a problem where the control volume is chosen in such a way that it deforms. Such a problem is quite relevant for problems occurring in real life as we shall see.

A large tank with its top open has an inlet and outlet such that the velocities are $\vec{V}_i$ and $\vec{V}_e$ respectively. The base area of the tank is given by A_b is significantly larger than the area of the inlet and outlet, i.e. A_i and A_e. At a given instant of time, the height of water in the tank is h. Find out the rate of change of the water in the tank in the situation described above.

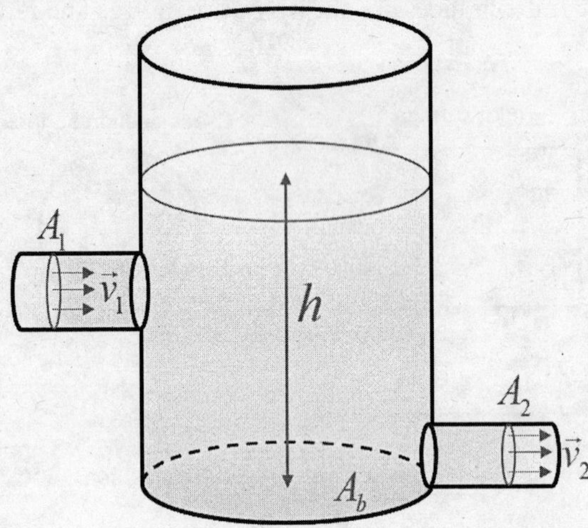

Solution

Let us consider the control volume as shown in the figure. Notable points of the control volume are as follows:

1. The control volume is such that the control surface contains the inlet and outlet surfaces.
2. Apart from the surface consisting of the inlet and outlet faces, the control volume is such that it is coincident with the top surface of water inside the tank. This implies that when the water level in the tank rises, the control volume also rises along with the tank level. Mathematically this means that the there is no flux which crosses the surface which is coincident with the top surface of water. Essentially, we can mathematically recast the statement as in the tank. This directly implies that the relative velocity at this surface in the tank has a zero relative velocity.

Let us first write down the statement of the conservation of mass for such a system

$$0=\frac{\partial}{\partial t}\int rhodV+\int\rho\vec{V}_r\cdot\vec{n}dA\Rightarrow\frac{\partial}{\partial t}(\rho V)+\int_{inlet}\vec{V}_i\cdot ndA+\int\vec{V}_e\cdot ndA+\int_{top}\vec{V}_{rel}\cdot\vec{n}dA$$

where we have now conveniently broken up the fluxes into the three different parts. Using the logic mentioned before, we can now simply the equation to yield

$$0=\rho\frac{d}{dt}V-\rho V_iA_i+\rho V_eA_e\Rightarrow\rho\frac{d}{dt}V=\rho V_iA_i-\rho V_eA_e$$

where we have made use of the fact that for the top surface, there is no flux owing to a net zero velocity. The above equation may be easily be physically understood. The rate of change of volume of the system is due to the net influx of the system minus the net outflux of the system. Upon simplifying the above equation by cancelling ρ from all the terms, we can now obtain

$$0\ =\frac{d}{dt}A_bh-\rho V_iA_i+V_eA_e\Rightarrow\frac{dh}{dt}=\frac{V_eA_e-V_iA_i}{A_b}$$

4.2 CONSERVATION OF MOMENTUM

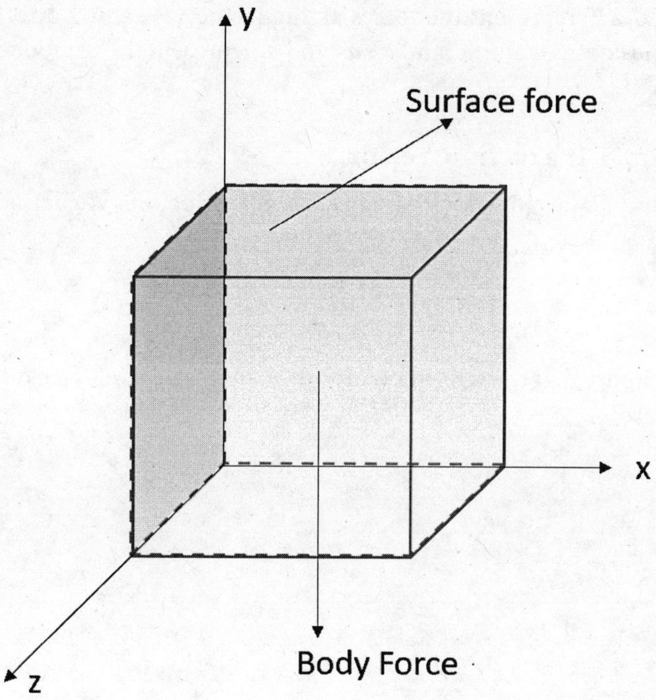

Figure 4.5: A rectangular parallelepiped is acted upon by body forces and surface forces thereby causing the mass inside the control volume to experience acceleration.

Let us first consider the derivation of the equation of motion for a flow in a Cartesian coordinate system. We consider a rectangular parallelepiped as shown in Fig. 4.5 where the length of the sides are represented by Δx, Δy and Δz respectively along the x, y, and z directions. The forces acting on the system are due to action of pressure, viscous stresses acting on the element due to the presence of the flow, and body forces acting on the entire volume of the body. The force due to pressure has already been discussed in Unit 2 where we dealt with hydrostatics. We will later see how this is responsible for the generation of flow, however, from common experience we know that a flow happens from a region having a relatively higher pressure to a region of lower pressure, for example, water flows naturally from a higher height such as an overhead tank, to ground level. The body force that is most encountered in such studies are due to the action of gravity. The total body force acting on a volume $d\forall$ is equal to $\rho\forall g$. The viscous forces act primarily due to the presence of velocity gradients. In the absence of any flow or where the flow is uniform throughout the domain, i.e. where the velocity gradients are also zero, there is no viscous stress.

Let us now consider the familiar Newton's second law of motion applied to the fluid element as shown above. In component form, we can write the sum of all forces is responsible for the fluid acceleration times the mass of the fluid element, as

$$\Sigma F_x = \Delta m a_x$$
$$\Sigma F_y = \Delta m a_y$$
$$\Sigma F_z = \Delta m a_z \tag{4.24}$$

where, ΣF_x, ΣF_y and ΣF_z represent the forces acting in the x, y, and z directions respectively, Δm represents the mass of the control volume, and a_x, a_y, and a_z represent the components of the acceleration vector $\vec{a}$.

Mathematics Concept 4.1: Traction Vector

Consider the rectangular parallelepiped in the same example as before. We define the traction vector as the force acting on the area dA oriented along the $\vec{n}$ direction.

$$\overline{T}^n = \lim_{\delta A \to 0} \frac{\delta \vec{F}}{\delta A} \tag{4.25}$$

which essentially implies that the traction vector, is the force per unit area outward at a point on the surface (refer to the figure).

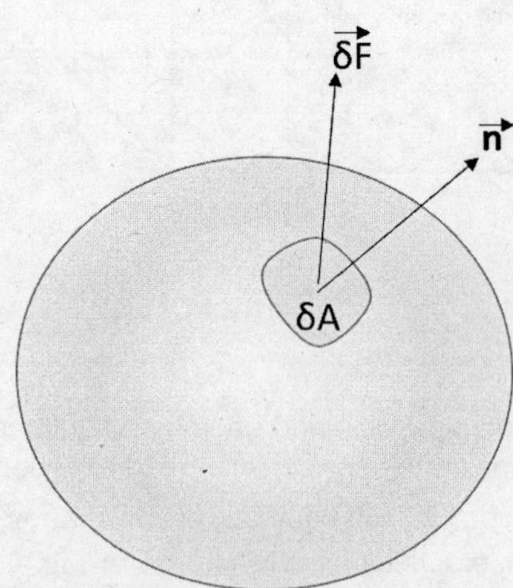

$\overline{T}^n$ has three components T_1^n, T_2^n, and T_3^n which represent the force per unit area acting on the x, y and z directions on a surface oriented along the $\vec{n}$ direction. When the control volume is consider as shown above, we can write down the traction vector in the more popular motion of

$$\tau_{ij} = \overline{T}_j^i \quad i = 1,2,3 \quad j = 1,2,3 \tag{4.26}$$

which simply implies that the area under consideration is oriented along the i direction and the force acts in the j direction. This is referred to as the Cartesian notation and the components τ_{ij} are referred to as the Cartesian stress tensor components.

Schematically, the various components of the stress are shown in the figure below. In the figure, we have not shown the pressure acting on the faces as this has been already discussed in Unit 2. The stresses on the faces visible to us on the figure are marked in black while the stresses on the faces not visible to us (i.e. the faces on the left, back, and bottom) are marked in red.

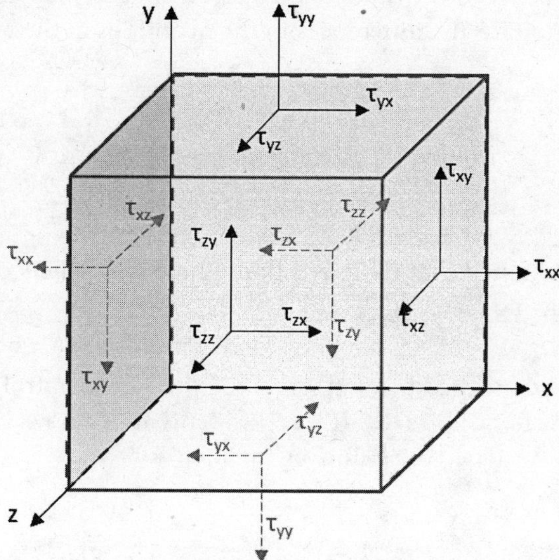

We can write down the force acting on any arbitrary surface oriented along the direction in terms of the various Cartesian components τ_{ij} as

$$T_i^n = \tau_{ji} n_j \quad i = 1,2,3 \quad j = 1,2,3 \tag{4.27}$$

Now, the force acting on the element may be written down in terms of the definition of the stress tensor:

$$d\vec{F} = T_i^n dA = \tau_{ji} n_j dA \tag{4.28}$$

which implies that the total force can be written as

$$\vec{F} = \int d\vec{F} = \int \tau_{ji} n_j dA = \int \frac{\partial}{\partial x_j} \tau_{ji} d\forall = \int \nabla \cdot \overline{\overline{\tau}} d\forall \tag{4.29}$$

where we have made use of the Gauss divergence theorem to convert from the surface integral to the volume integral. The last term appearing in the equation above is read as the divergence of the stress tensor.

We can now proceed with Eq. (4.24) and we can write down the left hand side of the above equation as

$$\Sigma F_x = \text{Surface forces} + \text{Body forces} \tag{4.30}$$

where, the surface forces may be obtained by summing all the forces acting on the surface of the body, i.e. the force due to action of the pressure and the traction vector:

$$p_x A_{yz} - p_{x+\Delta x} A_{yz} - \tau_{xx} A_{yz} + \tau_{xx}\Big|_{x+\Delta x} A_{yz} - \tau_{xy} A_{xz} + \tau_{xy}\Big|_{y+\Delta y} A_{xz} - \tau_{xz} A_{yx} + \tau_{xz}\Big|_{z+\Delta z} A_{xz} \tag{4.31}$$

where the area of the face is represented by A_{yz} and is represented by

$$A_{yz} = \Delta y \times \Delta z, \quad A_{xy} = \Delta x \times \Delta y, \quad A_{xz} = \Delta x \times \Delta z$$

where p_x and $p_{(x+\Delta x)}$ represent the pressures acting on the faces at location x and $x + \Delta x$. Similarly, the Cartesian stress tensors are shown in the mathematical concept above. We can now make use of the Taylor series approximation for the differences in the quantities in the above expression and finally obtain:

$$F_{x,surf} = -\frac{\partial p}{\partial x}\Delta x\Delta y\Delta z + \frac{\partial \tau_{xx}}{\partial x}\Delta x\Delta y\Delta z + \frac{\partial \tau_{xy}}{\partial y}\Delta y\Delta x\Delta z + \frac{\partial \tau_{xz}}{\partial z}\Delta z\Delta x\Delta y \tag{4.32}$$

$$\text{And the body force} = \rho b_x \Delta V \tag{4.33}$$

Additionally, we have assumed that there is a body force which acts on the control volume. The body force is defined as following

$$F_{\text{body force}, x} = \text{mass} \times \text{body force per unit mass} = \rho \times \Delta V \times b_x$$

where b_x presents the body force per unit mass of the fluid control volume. An example of a commonly encountered body force is gravity. If we consider that the gravity acts in the $-z$ direction, we can then write down the body force in the following manner

$$F_{\text{body force}, z} = \rho \times \Delta V \times (-g)$$

$$-\frac{\partial p}{\partial x}\Delta x\Delta y\Delta z + \frac{\partial \tau_{xx}}{\partial x}\Delta x\Delta y\Delta z + \frac{\partial \tau_{xy}}{\partial y}\Delta y\Delta x\Delta z + \frac{\partial \tau_{xz}}{\partial z}\Delta z\Delta x\Delta y + \rho b_x \Delta V = \ddot{A} m a_x \tag{4.34}$$

$$-\frac{\partial p}{\partial x}\Delta x\Delta y\Delta z + \frac{\partial \tau_{xx}}{\partial x}\Delta x\Delta y\Delta z + \frac{\partial \tau_{xy}}{\partial y}\Delta y\Delta x\Delta z + \frac{\partial \tau_{xz}}{\partial z}\Delta z\Delta x\Delta y + \rho b_x \Delta V = \rho\Delta x\Delta y\Delta z\left[\frac{\partial u}{\partial t} + u\frac{\partial u}{\partial x} + v\frac{\partial u}{\partial y} + w\frac{\partial u}{\partial z}\right] \tag{4.35}$$

where we have explicitly written down the expression for the mass of the control volume and the component of acceleration in the direction as

Mass of the control volume, $\Delta m = \rho\Delta x\Delta y\Delta z$

Acceleration of the fluid element in x direction, $a_x = \dfrac{\partial u}{\partial t} + u\dfrac{\partial u}{\partial x} + v\dfrac{\partial u}{\partial y} + w\dfrac{\partial u}{\partial z}$

We can then rearrange this equation by cancelling the term $\Delta x \Delta y \Delta z$ from both sides of the equation to obtain

$$\rho\left[\frac{\partial u}{\partial t}+u\frac{\partial u}{\partial x}+v\frac{\partial u}{\partial y}+w\frac{\partial u}{\partial z}\right]=\rho b_x-\frac{\partial p}{\partial x}+\frac{\partial \tau_{xx}}{\partial x}+\frac{\partial \tau_{xy}}{\partial y}+\frac{\partial \tau_{xz}}{\partial z} \tag{4.36}$$

Similar equations may be obtained for the momentum balance in the y and z directions. We do not mention them here for the sake of brevity. However, the three equations may be combined in the form of a single vector equation for the case where $\vec{b}$ is the body force in the system as

$$\rho\frac{D\vec{V}}{Dt}=-\nabla p+\nabla\cdot\overline{\overline{\tau}}+\rho\vec{b} \tag{4.37}$$

The above equation describes the differential form of the conservation of momentum and is known as the Navier's equation.

4.2.1 Inviscid Flow: Euler's Equation

The origin of the stress, as described above through the stress tensor and its divergence $\nabla\cdot\overline{\overline{\tau}}$, lies in the fluid property of viscosity. In many cases, however, the influence of viscosity remains closely bound near the solid surface. This is particularly true for high-speed flows such as the flow over an aero plane wing. For such flows, a first approximation that may be done is to assume that the flow is inviscid. This means that the influence of viscosity is neglected, and that directly implies that we can drop the influence of the stress tensor.

When this is done, we obtain the *Euler's equation*. This equation was originally analyzed by Euler where the ideas behind boundary layer theory were not quite developed, despite there being a broad idea about the fluid property of viscosity. The analysis of inviscid flows has played a major role in the development of fluid mechanics on a whole and it is far from being irrelevant. We can therefore drop the viscous stress and write down the Euler's equation as We can write down the three equations, i.e. the Euler equations of motion along the x, y, and z directions as

$$\rho\left[\frac{\partial u}{\partial t}+u\frac{\partial u}{\partial x}+v\frac{\partial u}{\partial y}+w\frac{\partial u}{\partial z}\right]=\rho b_x-\frac{\partial p}{\partial x} \tag{4.38}$$

$$\rho\left[\frac{\partial v}{\partial t}+u\frac{\partial v}{\partial x}+v\frac{\partial v}{\partial y}+w\frac{\partial v}{\partial z}\right]=\rho b_y-\frac{\partial p}{\partial y} \tag{4.39}$$

$$\rho\left[\frac{\partial w}{\partial t}+u\frac{\partial w}{\partial x}+v\frac{\partial w}{\partial y}+w\frac{\partial w}{\partial z}\right]=\rho b_z-\frac{\partial p}{\partial z} \tag{4.40}$$

It is not always convenient to write down the above three equations in component-wise form. Often it is reported as a single vector equation. This may be useful for some of the readers and we mention it below for ready reference:

$$\rho\frac{D\vec{V}}{Dt}=\rho\vec{b}-\nabla p \tag{4.41}$$

$$\rho\left[\frac{\partial \vec{u}}{\partial t}+\left(\vec{u}\cdot\nabla\right)\vec{u}\right]=\rho\vec{b}-\nabla p \tag{4.42}$$

When gravity points in $-z$ direction

When the body force is such that, it only acts in the $-\hat{k}$ direction, i.e. along , such that we can write down the body force as $\vec{b} = -g\hat{k}$, we can write down the Euler equations of motion as

$$\rho\left[\frac{\partial u}{\partial t} + u\frac{\partial u}{\partial x} + v\frac{\partial u}{\partial y} + w\frac{\partial u}{\partial z}\right] = -\frac{\partial p}{\partial x} \tag{4.43}$$

$$\rho\left[\frac{\partial v}{\partial t} + u\frac{\partial v}{\partial x} + v\frac{\partial v}{\partial y} + w\frac{\partial v}{\partial z}\right] = -\frac{\partial p}{\partial y} \tag{4.44}$$

$$\rho\left[\frac{\partial w}{\partial t} + u\frac{\partial w}{\partial x} + v\frac{\partial w}{\partial y} + w\frac{\partial w}{\partial z}\right] = -\rho g - \frac{\partial p}{\partial z} \tag{4.45}$$

Reduction to hydrostatics equation in the absence of motion

Interestingly, we observe that when there is no motion, i.e. $\vec{u} = 0$, i.e. $(u, v, w) = (0, 0, 0)$, we can simplify Eq. (4.45) as

$$0 = -\frac{\partial p}{\partial x} \tag{4.46}$$

$$0 = -\frac{\partial p}{\partial y} \tag{4.47}$$

$$0 = -gz - \frac{\partial p}{\partial z} \tag{4.48}$$

4.2.2 Euler's Equation in Streamline Coordinate System

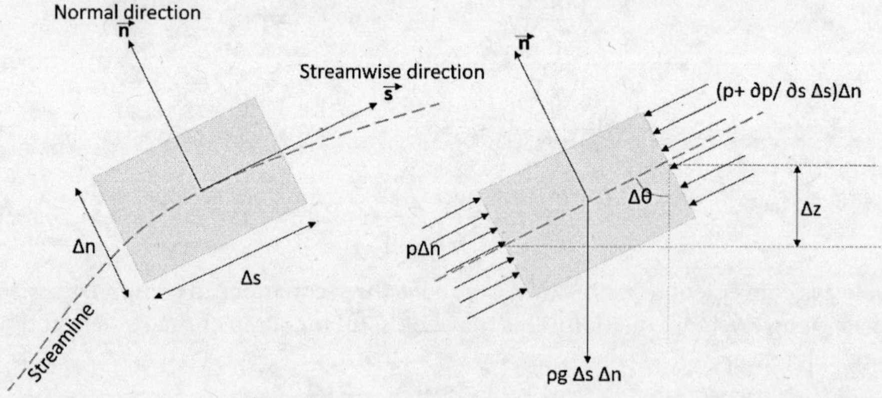

Figure 4.6: The streamline coordinate system along a streamline is shown. The two mutually orthogonal directions are $\hat{n}$ and $\hat{s}$ respectively. By the very definition of being a streamline, there is no flow in the $\hat{n}$ direction and hence the component of velocity is only V in the $\hat{s}$ direction.

While the above section dealt primarily with the Cartesian coordinate system, it is often desirable to describe the equation of motion in a *streamline coordinate system*. A streamline coordinate system is shown in Fig. 4.6. The streamline represents the direction of the flow by its very definition and hence an imaginary particle is bound to follow the trajectory of the streamline, at least instantaneously. In such a case, the flow of motion can therefore be decomposed into two directions, one along the streamline and the other orthogonal to the streamline. The first direction is represented by the unit vector $\hat{s}$ while the other is $\hat{n}$.

Euler's equation in the streamwise direction

We can begin by applying Newton's law along the streamline and obtain

$$\Sigma F_x = \Delta m a_s \tag{4.49}$$

"where,pressure on lower face is p, and on the upper face is $p+\frac{\partial p}{\partial s}\Delta s$ (4.50)

$$p\Delta n-\left(p+\frac{\partial p}{\partial s}\Delta s\right)\Delta n-\rho g\Delta s\Delta n\cos\theta=\rho\Delta s\Delta n a_s \tag{4.51}$$

But, $$\cos\theta=\frac{\Delta z}{\Delta s} \tag{4.52}$$

$\therefore$ $$-\frac{\partial p}{\partial s}-\rho g\frac{\Delta z}{\Delta s}=\rho a_s \tag{4.53}$$

In the expression above, it may be noted that the acceleration along the streamwise direction is obtained as

$$a_s=\frac{DV}{Dt}=\frac{\partial V}{\partial t}+V\frac{\partial V}{\partial s} \tag{4.54}$$

where V represents the velocity along the streamline. We reiterate here that the velocity vector is only along the $\hat{s}$ direction and hence we only have one component of the velocity vector. There is no component of velocity in the $\hat{n}$ direction.

We may now substitute the expression of the acceleration in the Newton's law to obtain

$$-\frac{\partial p}{\partial s}-\rho g\frac{\Delta z}{\Delta s}=\rho\left(\frac{\partial V}{\partial t}+V\frac{\partial V}{\partial s}\right) \tag{4.55}$$

We can now choose an infinitesimally small element to write

$$\frac{\Delta z}{\Delta s}=\frac{\partial z}{\partial s}$$

and obtain the Euler's equation of motion along a streamline as

$$\rho\left(\frac{\partial V}{\partial t}+V\frac{\partial V}{\partial s}\right)=-\frac{\partial p}{\partial s}-\rho g\frac{\partial z}{\partial s} \tag{4.56}$$

Euler's equation in the direction normal to the streamline

We can now consider the Newton's laws in the normal direction. We can write down

$$\Sigma F_n = \Delta m a_n \tag{4.57}$$

$$p\Delta s - \left(p + \frac{\partial p}{\partial n}\Delta n\right) - \rho g \Delta s \Delta n \cos\theta = \rho \Delta s \Delta n a_n \tag{4.58}$$

$$\Rightarrow \quad -\frac{\partial p}{\partial n} - \rho g \frac{\Delta z}{\Delta n} = \rho a_n \tag{4.59}$$

Again, considering the limit of an infinitesimally small element, we can write down

$$-\frac{\partial p}{\partial n} - \rho g \frac{\partial z}{\partial n} = \rho a_n \tag{4.60}$$

We note that the acceleration of a fluid element going about a streamline which has an instantaneous radius of curvature r is the centripetal acceleration which is given by

$$a_n = -\frac{V^2}{r} \tag{4.61}$$

Combining the expression for a_n in the Euler's equation, we can finally obtain the equation of motion as

$$\frac{\partial p}{\partial n} + \rho g \frac{\partial z}{\partial n} = \rho \frac{V^2}{r} \tag{4.62}$$

Let us now consider two cases of the above equation. First, if the motion of the fluid is occurring in a single plane, then we have $\frac{\partial z}{\partial n} = 0$ and therefore Eq. (4.60) can be simplified to

$$\frac{\partial p}{\partial n} = \rho \frac{V^2}{r} \tag{4.63}$$

Which implies that the pressure gradient in the normal direction is what balances the centripetal acceleration of the fluid element if it is going along a streamline which has an instantaneous radius of curvature r.

The second case that we can study is a further simplification of Eq. (4.63) for the case where the motion is along a straight streamline. In that case, the curvature is zero and essentially that means that the radius of curvature is infinite. In that case we obtain

$$\frac{\partial p}{\partial n} = 0 \tag{4.64)}$$

This means that for flows along a straight path, there is no normal pressure gradient to the streamline.

4.2.3 Integral form of the Conservation of Momentum

As was the case for the conservation of mass, we can now write down the statement for the conservation of momentum for a control volume. Let us first mention the statement in plain language after which we will cast it in a more convenient mathematical form. The total rate of change of momentum within a control volume is due to the net influx of momentum into the control volume and the total force acting on the control volume. Let us now break down the different terms mentioned in the statement above.

Total rate of change of momentum inside CV:

Total influx of momentum into CV: $\int \rho \vec{V}\left(\vec{V}\cdot\vec{n}\right)dA$ (4.66)

When the CV is moving with respect to the fluid: $\int \rho \vec{V}\left(\vec{V}_{rel}\cdot\vec{n}\right)dA$ (4.67)

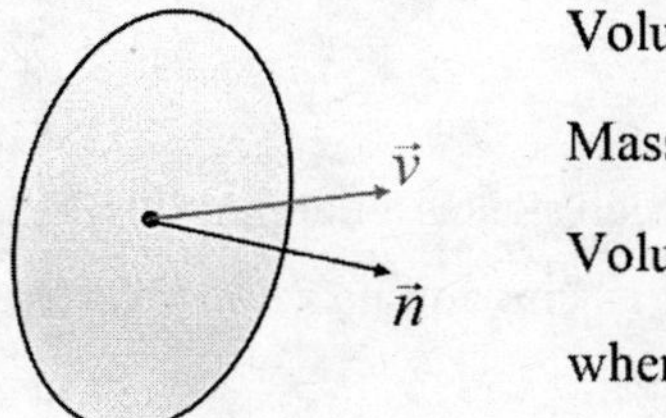

Volume flux: $\int(\vec{v}.\vec{n})dA$

Mass flux: $\int \rho(\vec{v}.\vec{n})dA$

Volume flux: $\int \rho\vec{v}(\vec{v}.\vec{n})dA$

when surface moves: $\int \rho\vec{v}_r(\vec{v}_r.\vec{n})dA$

Figure 4.7: Flux of momentum crossing an area.

Let us take a moment to elaborate on the term above. The flux crossing any surface has already been discussed in detail in the discussion of the conservation of mass and is written down as $\int \rho \vec{V} dA$. Now when this flux of mass is entering into the control volume, what momentum is it bringing along with itself? If we refer to the Fig. 4.7, we see that at the surface shown the momentum brought in is equal to $\int \rho \vec{V}\left(\vec{V}\cdot\vec{n}\right)dA$. The term inside the bracket is the volume flux that crosses the area and it is obviously the projection of the velocity in the direction of the area vector. However, the term outside the bracket, i.e. $\rho\vec{V}$ represents the momentum that this flux is bringing. Note that this is not any projection however it is the total momentum. In the same way, if the surface is now moving with respect to the flow, then we must account for this relative velocity through the term mentioned in equation Eq. (4.67).

This brings us to the last point, i.e. the total force acting on the control volume. This force could be due to the any of the points mentioned in the discussion of the stress tensor. This could be due to the pressure gradient or due to the presence of viscosity, or due to any other body force acting on the fluid.

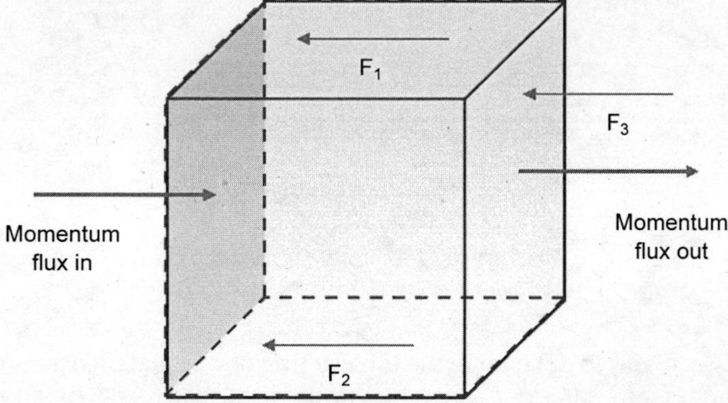

Figure 4.8: All the forces acting on a control volume causing change of momentum of the fluid.

In the control volume depicted in Fig. 4.8, we have shown the force acting on the control volume as $\vec{F}_{CV}$ and is responsible for slowing down the flow.

We can therefore compile all the terms into a single equation and write down the statement of the conservation of momentum for a control volume as

$$\frac{\partial}{\partial t}\int \rho \vec{V} d\forall + \int \rho \vec{V}\left(\vec{V}_{rel} \cdot \vec{n}\right) dA = \sum \vec{F}_{CV} \tag{4.68}$$

Let us now look at some examples where we can apply the conservation of momentum for practical problems.

Example 4.4

Let us consider the jet of water shown in the figure which strikes a flat plate. The plate is held in place by the application of a force $\vec{F}$. If the diameter of the jet is cm, and if the jet velocity is 5 m/s, find out the force required to hold the plate in place.

Solution

From common experience, we can first conclude that the jet which strikes the plate, also known as an impinging jet would lead to a force acting on the jet towards the right. For the plate to be held in place, one must therefore apply a force in the opposite direction, i.e. the direction which is shown in the figure.

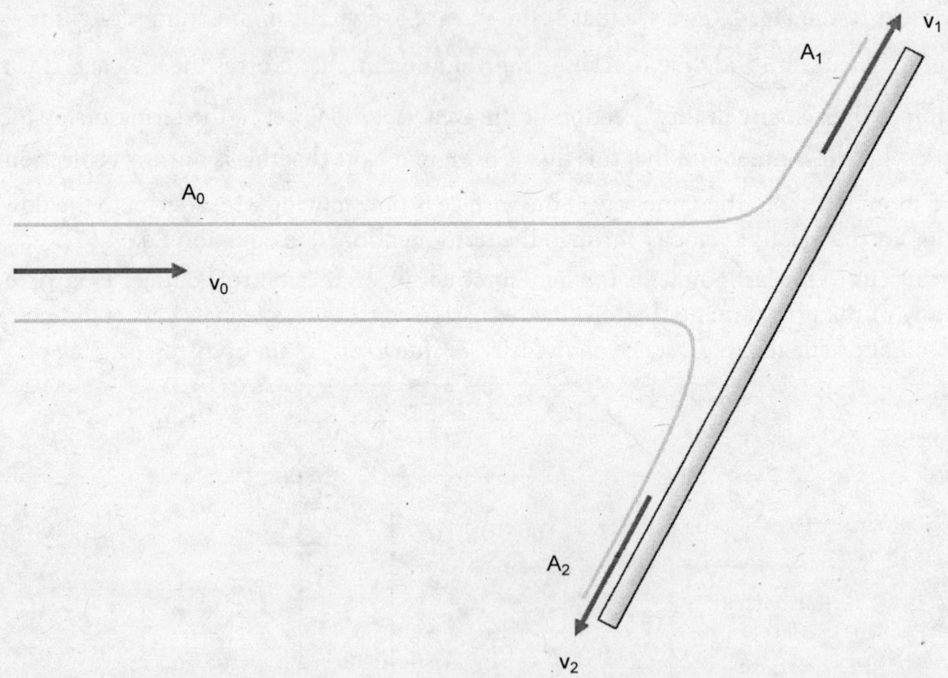

Let us now apply Eq. (4.68) to determine the force acting on the plate. We choose the control volume as shown in the figure. We note that the control volume in this case only encloses the water jet which strikes the plate. We can therefore write

$$\frac{\partial}{\partial t}\int \rho \vec{V} dV + \int \rho \vec{V}\left(\vec{V}\cdot\vec{n}\right)dA = \vec{F}_{CV,plate}$$

where we have made use of the following observations

1. The control volume is stationary, i.e. $\vec{V}_{surf} = 0$.
2. $\vec{F}_{CV,plate}$ represents the force acting on the control volume due to the plate.
3. The control volume is such that there is no accumulation of mass or momentum inside. We can therefore drop the first-time derivative in the equation above.

If we now focus only on the x momentum equation, we can write down

$$\int \rho u\left(\vec{V}\cdot\vec{n}\right)dA = F_{x,cv,plate}$$

If we focus on the flux that enters the left face, we can write down the first term in the equation above as

$$\int \rho u\left(\vec{V}\cdot\vec{n}\right)dA = \int \rho u(-u)dA$$

where we have made use of the fact that the area vector and the velocity vector are 180° apart.

Now, utilizing the fact that the velocity in the jet can be homogeneous over the entire cross section, we can write the above equation as

$$\int \rho u(-u)dA = -\rho u^2 A = -1000\times 25\times \pi\frac{0.1^2}{4} = -196.35N$$

Let us now focus on the two outfluxes of the control volume. If we now write down the x momentum at the two faces, we have

$$\text{Momentum flux exiting}: \int \rho u\left(\vec{V}\cdot\vec{n}\right)dA = 0$$

The above expression is zero because there is no u or x component of velocity the exit faces. The jet gets turned in the direction orthogonal or perpendicular to x direction and hence, it is devoid of any momentum in the x direction.

Therefore, we can now write down the statement of the conservation of momentum as

$$-196.35 = F_{x,\text{ CV, plate}}$$

This means that the plate exerts a force on the jet of water in the $-x$ direction with a magnitude of −196.35 N.

If we now consider a force balance of the plate, we have, for the equilibrium of the plate:

$$F_{x,\text{ plate, CV}} + \text{F} = 0$$

where now we can make use of Newton's third law and make use of

$$F_{x,\text{ plate, CV}} = -F_{x,\text{ CV, plate}} = 196.35\text{ N}$$

In that case, the equation yields

$$F = -F_{x,\text{ plate, CV}} = -196.35\text{ N}$$

This means that the force that must be exerted on the plate has a magnitude of 196.35N and points in the $-x$ direction, as we had logically reasoned in the beginning of the solution.

The same solution can be directly be obtained if we expand the control volume directly to include the plate as well. In this case, we will not have to apply Newton's third law and perform the equilibrium condition for the plate. This internal force balance is then directly built into the control volume.

The reader is instructed to go through this exercise once to consolidate this idea.

4.2.4 Bernoulli's Equation

The Euler equation in the streamwise direction serves as the basis for the derivation of the Bernoulli equation. For convenience, let us first write down the Euler equation in the $\hat{s}$ direction. We have

$$\rho\left(\frac{\partial V}{\partial t}+V\frac{\partial V}{\partial s}\right)=-\frac{\partial p}{\partial s}-\rho g\frac{\partial z}{\partial s}$$

Let us now assume that the flow is steady so that $\frac{\partial V}{\partial t}=0$. The equation for steady flow is therefore

$$\rho\left(V\frac{\partial V}{\partial s}\right)=-\frac{\partial p}{\partial s}-\rho g\frac{\partial z}{\partial s} \tag{4.69}$$

which can be simplified to

$$V\frac{\partial V}{\partial s}+\frac{1}{\rho}\frac{\partial p}{\partial s}+g\frac{\partial z}{\partial s} \tag{4.70}$$

Given that the above equation is valid for any arbitrary streamline *s*, we can simplify the above to

$$\frac{\partial p}{\rho}+V\partial V+g\partial z=0 \tag{4.71}$$

Let us now assume that the density is constant, i.e. the flow is incompressible. The above equation can now be integrated to obtain

$$\frac{p}{\rho}+gz+\frac{V^2}{2}=constant \tag{4.72}$$

or,

$$\frac{p}{\rho g}+\frac{V^2}{2g}+z=constant \tag{4.73}$$

Equation (4.73) represents the Bernoulli equation of a steady, inviscid, incompressible flow.

In the expression above, the first term, i.e. $p/\rho g$ represents the pressure head or the pressure energy per unit weight of the fluid. The second term, i.e. $V^2/2g$ represents the kinetic energy per unit weight of the fluid or the kinetic head while the third term, *z*, represents the potential energy per unid weight of the fluid, or the potential head of the system.

To this book, we mention here the important assumptions behind the Bernoulli's equation

1. The flow is steady
2. The flow is incompressible
3. The flow is inviscid
4. The equation is applied along a certain streamline; unless the flow is irrotational everywhere, we cannot apply Eq. (4.73) between any two points in the flow.

5. No heat transfer to and from the flow
6. No work is being done between the two points across which the Bernoulli equation is applied; i.e. no pump or turbine exists

The terms appearing in Eq. (4.73) have the units of length, and are often reported as meter. Therefore, each of the three terms in the equation are referred to as the potential head, kinetic energy head, and the potential head respectively.

Figure 4.9: Explanation of the work done due to pressure and its connection with the pressure head.

While the meanings of the second and third terms are clear, i.e. the kinetic energy per unit weight and the potential energy per unit weight, the physical meaning of the first term can be understood by looking at the work done to maintain the flow. Consider the schematic below where the pressure acting on the fluid element is p, the cross-section area is A. In that case, the work done to maintain the flow over a distance Δx is $pA\Delta x$. The work done per unit weight is then

$$\frac{pA\Delta x}{\rho A\Delta xg}=\frac{p}{\rho g}$$

This basically defines the flow work of the system.

To conclude, Bernoulli's equation describes that along a streamline, the total of the flow energy, kinetic energy, and potential energy per unit weight remains conserved, under the assumptions of an inviscid, steady, and constant density flow.

UNIT SUMMARY

- Continuity equation

$$\frac{\partial \rho}{\partial t}+\frac{\partial(\rho u)}{\partial x}+\frac{\partial(\rho v)}{\partial y}+\frac{\partial(\rho w)}{\partial z}=0$$

- Conservation of momentum

$$\rho\frac{D\vec{V}}{Dt}=-\nabla p+\nabla\cdot\bar{\bar{\tau}}+\rho\vec{b}$$

- Euler's equation

$$\rho\frac{D\vec{V}}{Dt}=\rho\vec{b}-\nabla p$$

- Integral form of conservation equation

$$\frac{\partial}{\partial t}\int \rho \vec{V} d\forall + \int \rho \vec{V}\left(\vec{V}_{rel}\cdot\vec{n}\right) dA = \sum \vec{F}_{CV}$$

- Bernoulli's equation

$$\frac{p}{\rho} + gz + \frac{V^2}{2} = constant$$

or,

$$\frac{p}{\rho g} + \frac{V^2}{2g} + z = constant$$

EXERCISES

Multiple Choice Questions

1. For an incompressible flow through a pipe of constant diameter, which of the following is/are same at the inlet and outlet of the pipe:

(a) Mass flow rate
(b) Density
(c) Velocity
(d) Cannot be determined from the above information

2. For a compressible fluid flowing through a pipe of constant diameter, which of the following is/are same at the inlet and outlet of the pipe:

(a) Mass flow rate
(b) Density
(c) Velocity
(d) Cannot be determined from the above information

3. For an inviscid fluid flow; pressure in the normal direction (neglecting gravity) to streamlines which are straight in nature:

(a) Decreases in the direction of the normal vector
(b) Increases in the direction of the normal vector
(c) Remains constant
(d) Cannot be determined

4. For a surface which is normal to the direction of flow:

(a) Mass flux is zero
(b) Mass flux is non-zero
(c) Momentum flux is zero
(d) Momentum flux is zero

5. For an incompressible inviscid fluid flow through a nozzle (a conduit whose area continuously decreases in the direction of flow), which of the following is/are true:

(a) Velocity is higher at the inlet than at the outlet
(b) Velocity is lower at the inlet
(c) Fluid pressure reduces in the direction of the flow
(d) Fluid pressure increases in the direction of the flow

6. For an incompressible inviscid fluid flow through a variable cross-sectional area conduit, how are the pressure at the inlet related to pressure at the outlet?

(a) Pressure at the inlet is always greater than at the outlet
(b) Pressure at the inlet is always equal to that at the outlet
(c) Pressure at the inlet can be lower or higher than the pressure at the outlet
(d) Cannot be evaluated from the given information

7. If the streamlines in the fluid flow are curving, then:

(a) Bernoulli's principle cannot be applied along the streamline
(b) Bernoulli's principle cannot be applied along the normal to the streamline
(c) Bernoulli's principle can be applied along the streamline
(d) Bernoulli's principle can be applied between any two points in the flow field

ANSWERS													
1.	(a), (b), (c)	2.	(a)	3.	(c)	4.	(a), (c)	5.	(a), (c)	6.	(c)	7.	(b), (c)

Unsolved Questions

Level - I

1. A 30 cm diameter pipe carries oil of specific gravity 0.8 at a velocity of 4 m/s. In another section, the diameter is 15 cm. Determine the mass flow rate of oil as well as the velocity of oil at a smaller section of the pipe.
2. Water flowing through an 8-cm-diameter pipe enters a porous section, as shown in the figure, which allows a uniform radial velocity v_w through the wall surfaces for 1.2 m. If the entrance average velocity V_1 is 10 m/s, find the exit velocity V_2 if (a) $v_w = 13$ cm/s out of the pipe walls; (b) $v_w = 9$ cm/s into the pipe. (c) What value of v_w will make $V_2 = 7$ m/s?

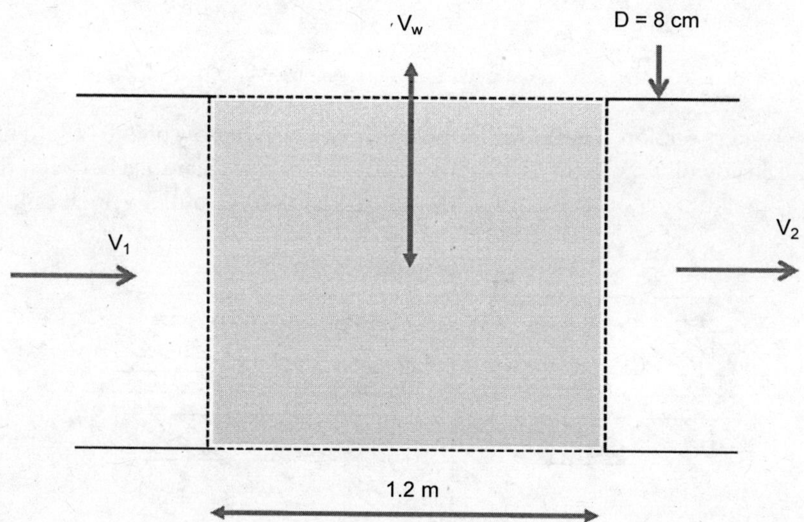

3. Water at 20 °C flows steadily through a closed tank, as shown in the figure. At section 1, $D_1 = 8$ cm and the volume flow 50 m³/h. At section 2, $D_2 = 7$ cm and the average velocity is 3 m/s. If $D_3 = 6$ cm, what is (a) Q_3 in m³/h and (b) average velocity V_3 in m/s.

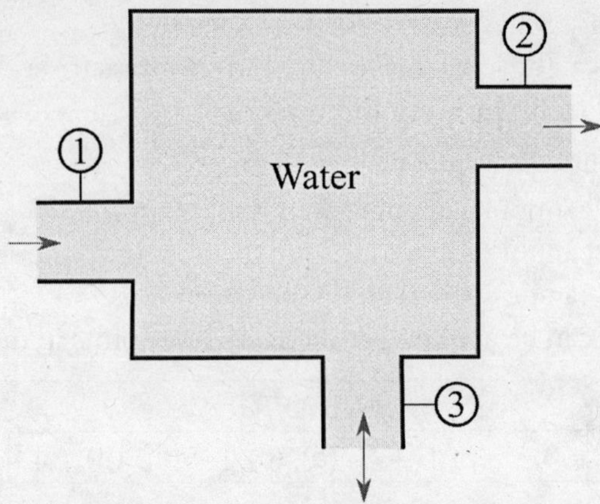

4. The converging-diverging nozzle shown in the figure expands and accelerates dry air to supersonic speeds at the exit, where $p_2 = 10$ kPa and $T_2 = 250$ K. At the throat, $p_1 = 300$ kPa, $T_1 = 670$ K, and $V_1 = 515$ m/s. For steady compressible flow for an ideal gas, estimate (a) the mass flow in kg/h, (b) the velocity V_2.

5.

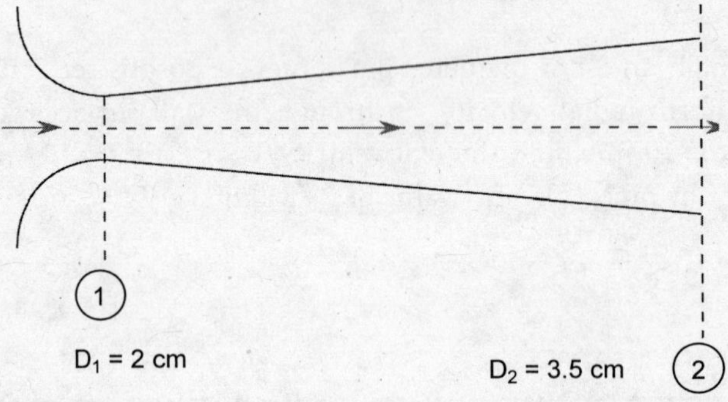

6. Oil (SG 0.88) enters section 1 as shown in the figure at a weight flow of 200 N/h to lubricate a thrust bearing. The steady oil flow exits radially through the narrow clearance between the thrust plates. Compute (a) the outlet volume flux in m³/s and (b) the average outlet velocity in m/s.

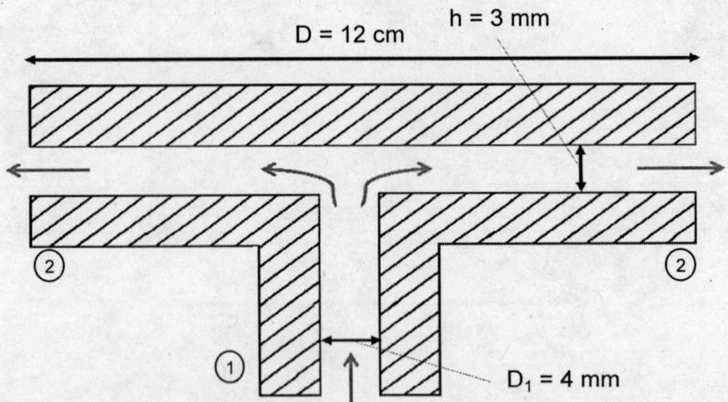

7. Water, considered incompressible, flows slowly through the round pipe depicted in the figure. The entrance velocity is constant, $u = U_{in}$, and exit velocity follows the hypothetical velocity profile i.e., $u = u_{max}(1-\frac{r^2}{R^2})^{\frac{1}{6}}$. Determine the ratio for this flow.

8. According to Torricelli's theorem, the velocity of a fluid draining from a hole in a tank is $V = (2gh)^{1/2}$, where h is the depth of water above the hole, as in Fig. Let the hole have area A_0 and the cylindrical tank has a bottom area A_b. Derive a formula for the time to drain the tank from an initial depth h_0.

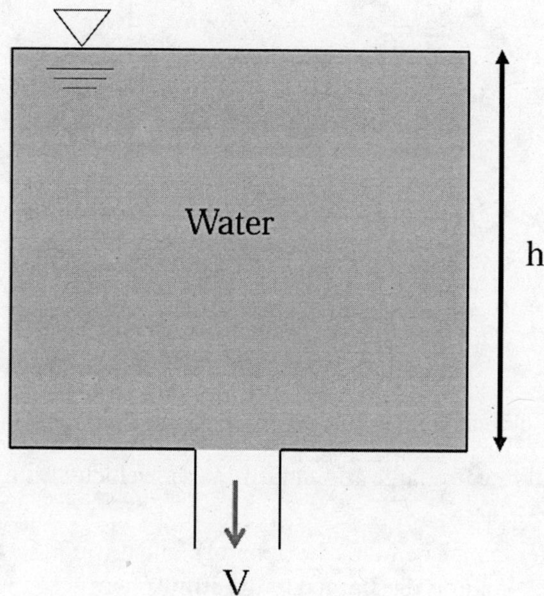

9. Three pipes steadily deliver water at 20°C to a large exit pipe as shown in the figure. The velocity $V_2 = 500$ cm/s, and the exit flow rate $Q_4 = 0.033$ m³/s. Find (a) V_1; (b) V_3; and (c) V_4 if it is known that increasing Q_3 by 20 percent would increase Q_4 by 10 percent.

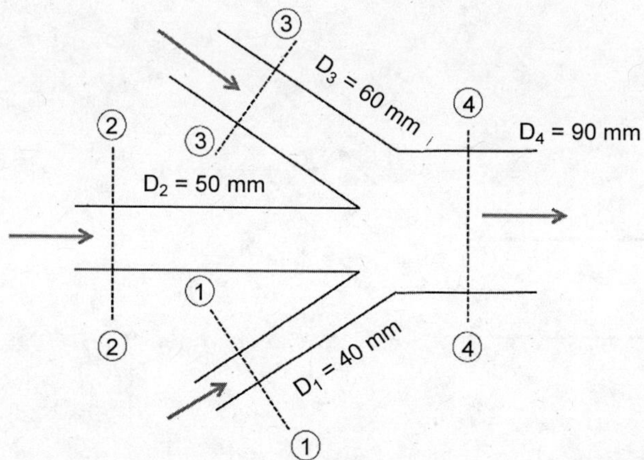

10. A laboratory test tank contains the seawater of salinity S and density ρ. Water enters the tank at conditions (S_1, ρ, A_1, V_1) and is assumed to mix immediately in the tank. Tank water leaves through an outlet A_2 at velocity V_2. If salt is a "conservative" property (neither created nor destroyed), use the conservation of mass principle to find an expression for the rate of change of salt mass M_{salt} within the tank.

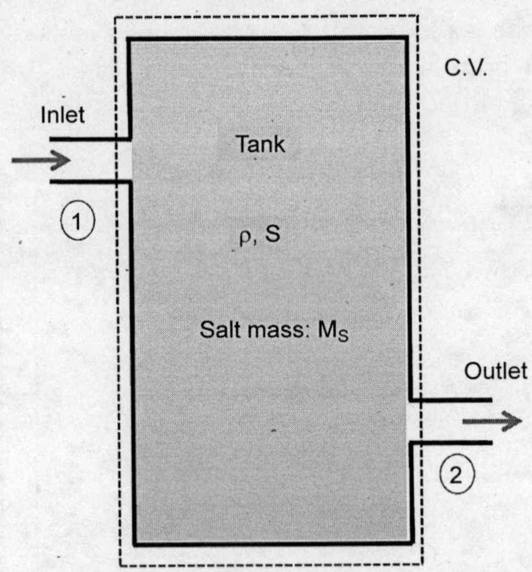

11. The diameters of the pipe at sections 1 and 2 are 25 cm and 30 cm respectively. If the velocity at section 1 is 5 m/s find the discharge and determine the velocity ratio i.e., $\frac{V_1}{V_2}$.

12. A stream of water enters a pipe bend in a horizontal plane through section 1 at the rate of 1800 m³/h as shown in the figure. Water is discharged to the atmosphere at section 2. The cross-sectional area at sections 1 and 2 are 1 m² and 0.25 m² respectively. The central axis of the pipe makes an angle 60 with *x*-direction. Neglect the friction, determine (a) Velocity at section 2, (b) Pressure at section 1, and (c) Force required to hold the pipe in the *x*-direction.

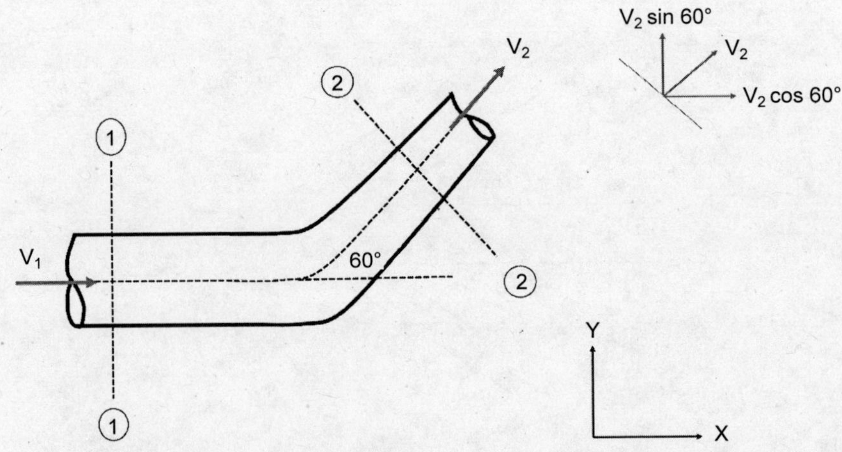

Level - II

1. Water assumed incompressible, flows steadily through the round pipe as shown in the figure. The entrance velocity is constant, $u = U_0$, and exit velocity approximates turbulent flow, $u = u_{max}(1 - \frac{r}{R})^{\frac{1}{7}}$. Determine the ratio U_0/u_{max} for this flow.

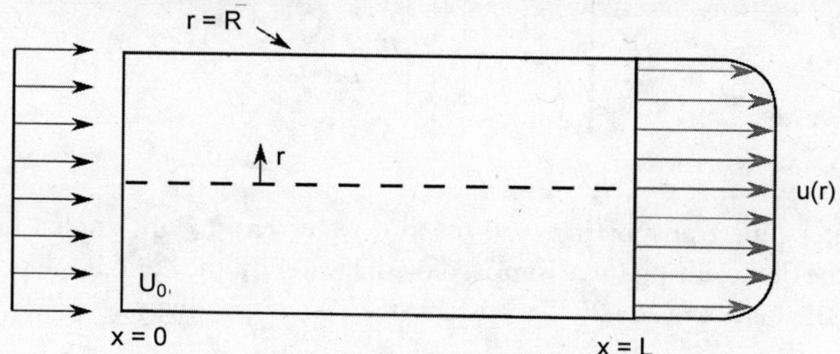

2. The open tank depicted in the image has water at 20°C and is being filled via section 1. Consider incompressible flow. First, calculate the water-level change $\frac{dh}{dt}$ in terms of arbitrary volume flows (Q_1, Q_2, Q_3) and tank diameter d. Determine the exit velocity v_2 for the provided data $v_1 = 5$ m/s and $Q_3 = 0.02$ m³/s if the water level h is constant.

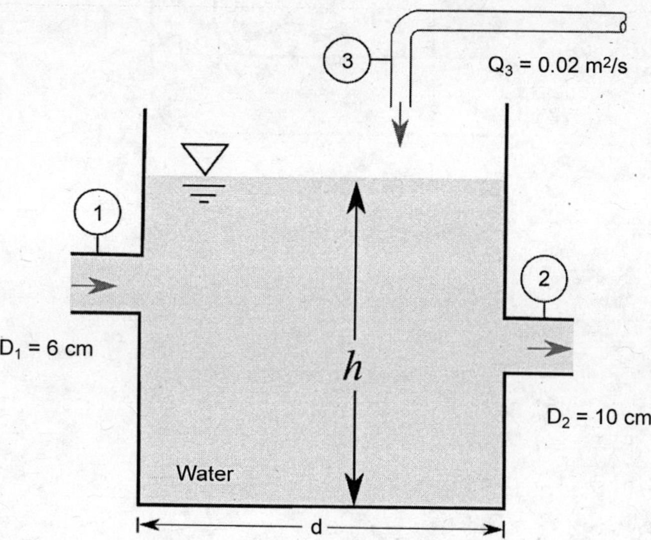

3. Water enters the bottom of the cone at uniformly increasing average velocity $V = kt$ as shown in the figure. If d is very small, derive an analytical formula for the water surface rise $h(t)$ for the condition $h = 0$ at $t = 0$. Assume the flow is incompressible.

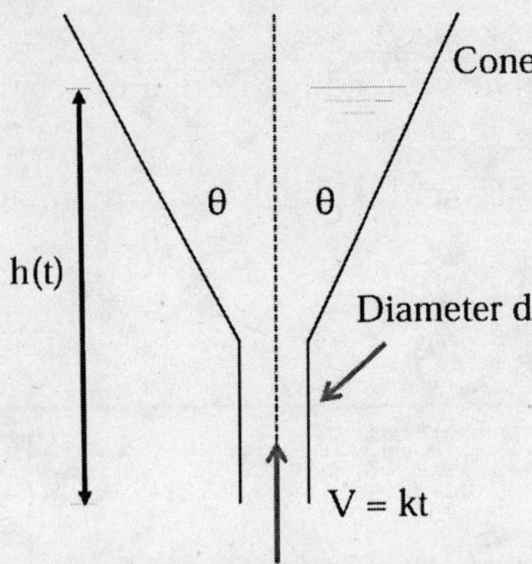

4. A 0.3 m diameter pipe transporting 0.30 m³/s of water has a right-angled bend in a horizontal plane. Determine the resulting force imposed by the water on the bend if the pressures at the intake and outflow of the bend are 245.25 kPa and 235.44 kPa, respectively.

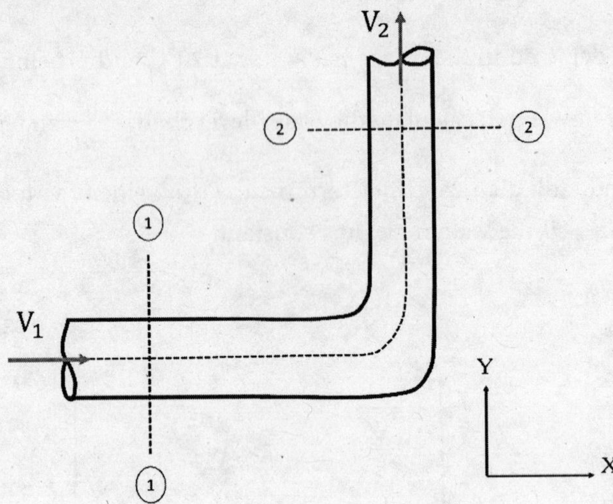

To know more about conservation lasw

ANSWER KEY

Level – I

1. $\dot{m}$= 225.92 kg/s, V_2 = 16.61 m/s
2. (a) V_2 = 2.2 m/s, (b) V_2 = 4.6 m/s, (c) V_w = 5.0 cm/s
3. (a) Q_3 = 8.44 m³/h, (b) V_3 = 0.83 m/s
4. (a) $\dot{m}$= 908.67 kg/h, (b) V_2 = 1883.26 m/s
5. (a) Q = 6.43 ×10⁻⁶ m³/s, (b) V = 5.68 × 10⁻³ m/s
6. U_{in} / u_{max} =6/7
7. $t = \frac{A_b}{A_0}\sqrt{\frac{h_0}{2g}}$

8. (a) $V_1 = 5.45$ m/s, (b) $V_3 = 5.89$ m/s, (c) $V_4 = 5.24$ m/s
9. $\left(\frac{dM_s}{dt}\right)_{CV} = S_1\rho_1 A_1 V_1 - S_2\rho_2 A_2 V_2$
10. $Q = 0.25\ m^3/s$, $\frac{V_1}{V_2} = 1.44$

Level – II

1. $U_0/u_{max} = 49/60$
2. $v_2 = 4.3464$m/s
3. $h(t) = [\frac{3}{8} kt^2 d^2 cot^2\theta]^{1/3}$
4. 25829.3 N

PRACTICAL

Aim: Understanding conservation of mass and momentum by means of PET bottle experiments

Apparatus: A 2-liter PET bottle (these kinds of bottles are the typical ones in which soft drinks such as Coca-Cola are sold), plastic straw, m-seal or araldite, binder clip.

Theory: The fundamental ideas behind mass conservation can be easily understood with the help of understanding the draining rates of the bottle mentioned above. The conservation of momentum can be deduced through a slightly indirect, but related, concept discussed in this Unit. By allowing water to exit through the PET bottle in certain directions, a net force and torque is seen to act on the bottle, thereby causing it to rotate. We will now discuss the specifics of the experiments.

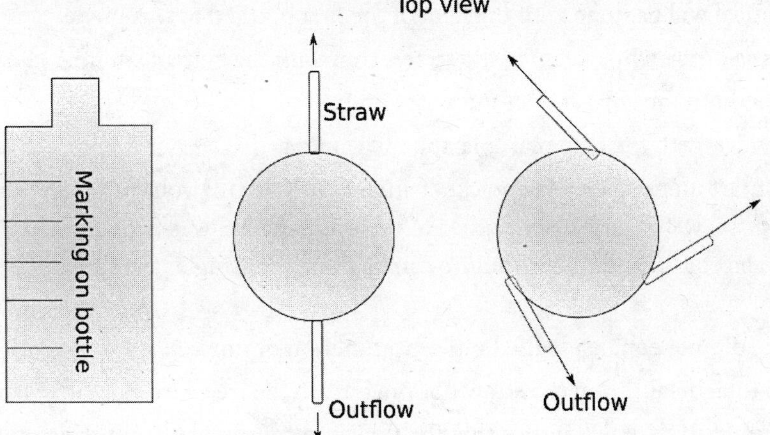

Figure 4.10: The bottle with markings, and the top view of the bottle with the straws inserted at the bottom for the 2 straw and 3 straw setups. Owing to the outflow of water in a relatively tangential direction for the 3-straw setup, the bottle will begin to rotate.

Preparation of the bottle: Take one such bottle mentioned above.

1. Remove the label and clean the bottle.
2. Use a measuring cylinder to pour 50 ml of water into the bottle.
3. With the help of a marker mark the location of this level.
4. Similarly, add 50 ml more and make a new mark of 100 ml.
5. In this manner, keep marking the bottle until the bottle is completely full. This particular step helps in the student to use odd shaped PET bottles.
6. Make a small vertical hole towards the bottom of the bottle so that the plastic straw goes into the hole. After this use some m-seal or araldite to make this assembly watertight.
7. In another bottle make the same volume markings. Apart from this create 4 equal angle markings on the bottle and clearly label it 1, 2, 3, 4. This label will help the student to measure the rotation rate of the bottle.
8. On this bottle make 3 holes in the bottom so that the straws can be inserted as shown in the figure.
9. On the cap of this bottle, make a hole so that some fine nylon string (fishing wire) can be passed. Connect this string to the cap by passing it through the hole and then making a broader knot inside. Connect the other side of the string to a fixed stand so that the bottle can freely rotate.
10. In both the bottles, it is also useful to have the height of the free surface of water in the bottle from the centerline of the straws. You can do so with the help of a ruler and marking different spots on the bottle surface with the help of a different colour.

Conservation of Mass

Procedure:

1. The first practical will be done with the help of the first bottle that was made.
2. Fill water in the bottle while keeping the straws shut with the help of a binder clip.
3. Carefully open both the clips and let the water drain.
4. Capture the video with the help of a smartphone camera.
5. If without a smartphone, then a friend can help to loudly say the volume that he sees and someone else can use a stopwatch with a split timer to record that what was the volume in the bottle at the certain time elapsed. This same procedure can be done by a single person with the help of a video as well.
6. Plot the total volume seen inside the bottle as a function of time.
7. Also plot the total height from the straw bottom part to the free surface of the water.

Theory: The observant student will notice that the flow rate coming out from the straws reduces as the level of water reduces in the bottle. This aspect will be solved as an example in the next Unit. For now, we recall the famous result studied in high-school due to Torricelli, which says that the average velocity at the output of the straw will be $V = \sqrt{2gh}$, where h is the height of the water above the straw.

We can now apply the conservation of mass and say that

$$\text{Rate of ch.of mass} = 2 \times \text{mass outflux} \quad \Rightarrow \frac{dM}{dt} = -2\times\rho AV = -2\times\rho A\sqrt{2gh}$$

The area of the straw can be figured out before the experiment itself. From the plot that the student has draw, they will have an estimate of the rate of change of the mass in the control volume as the slope of the curve.

By substituting the value the area of the straw, gravity, density of water, and the height vs. time, the student can find another theoretical estimate of the rate of change of mass.

Ideally these two should be equal. However, they will appear to be different slightly. The reason will be the deviations in the exit velocity through the straw as $\sqrt{2gh}$, which was derived by Toricelli with appropriate assumptions.

Conservation of Momentum (Optional)

Procedure

1. The practical of the conservation of momentum will make use of the second bottle that was made.
2. Hang the bottle so that the bottle is now steady with the mouths of the three straws closed with the help of a clip.
3. Now fill water in the bottle slowly so that no motion is created in the bottle.
4. Allow for the bottle to be as motionless as possible.
5. Remove the clips from the three straws. This particular step may require some friend's assistance.
6. Now capture the video of the bottle as it begins to let water out through the three straws.
7. The important things to capture are the angles as seen by the user, the height of water in the bottle and the volume markings.
8. Plot the volume of the water in the bottle as a function of time, plot the height of the free water surface from the center of the straw as a function of time, and plot the angles seen as a function of time. For plotting the angles, remember that once 360 degree is passed, we have to to keep adding the angles to this. The angles can be estimated from the video by looking at the angle markings that you had made on this bottle.

Theory: The force exerted by the water as it exits the jet is responsible for the bottle to start rotating. If the bottle were to be held still, then the total force would be equal to ρAV^2, where $V = \sqrt{2gh}$. At this very instant, if the bottle is now let to rotate freely, then the angular acceleration would be equal to $I\dot{\omega} = 3\times\rho AV^2 \times r$, where r is the radius of the bottle, I represents the moment of inertia of the filled bottle, and $\dot{\omega}$ represents the rate of angular acceleration.

However, when the bottle starts to rotate, then the relative velocity of the water at the exit of the bottle must be accounted for when trying to find out the total torque acting on the bottle. The observations and calculations are left to the reader as an exercise.

KNOW MORE

Daniel Bernoulli (1700-1782), son of Johann Bernoulli, spent several years teaching mathematics in St. Petersburg. During this time, he began writing *Hydrodynamics* in 1729, leaving an incomplete manuscript in Russia upon returning to Basel four years later. The book was published in Germany in 1738, but Bernoulli requested the destruction of the Russian manuscript, which still exists at the Soviet Academy of Science.

After *Hydrodynamics* was released, Bernoulli faced competition from his father, whose *Hydraulics* (1743) and Jean le Rond d'Alembert's *Traité de l'équilibre et du mouvement des fluides* (1744) were published around the same time. This led to a distinction between hydrodynamics, which focuses on the theory of fluid motion, and hydraulics, which deals with practical applications. In the Preface to *Hydraulics*, Johann Bernoulli argued that the field needed a stronger foundation in Mechanics rather than relying on uncertain empirical theories, advocating for a basis in Newtonian principles.

[Source: https://www.researchgate.net/publication/333766804_History_of_the_Bernoulli_Principle]

REFERENCES AND SUGGESTED READINGS

1. **Introduction to Fluid Mechanics and Fluid Machines** (3rd Edition), S K Som, Gautam Biswas, Suman Chakraborty, McGraw Hill
2. **Fluid Mechanics** (9th Edition), Frank M. White, Henry Xue, McGraw Hill
3. **Fluid Mechanics** (6th Edition), Piyush K. Kundu, Ira M Cohen, David R. Dowling, Academic Press

5 Engineering Applications 1: Flow through Pipes

UNIT SPECIFICS

In this unit we will discuss about the following topics:

1. Applications of Bernoulli's theorem
2. Concept of laminar and turbulent flow
3. Losses in flow through pipes
4. Concept of friction factor

RATIONALE

In the previous Unit, we have discussed various aspects of conservation laws in fluid mechanics. We concluded the previous Unit with the discussion of Bernoulli's theorem. In this Unit we aim to discuss various applications of the concepts and ideas laid down in the previous Unit. Bernoulli's theorem leads the way for a broad variety of applications which we first discuss in details. We then discuss about the various losses and their quantification in flow through pipes. We delineate between laminar and turbulent flow after which we quantify the pressure loss in terms of the friction factor. We discuss the flows through pipes in a network, the hydraulic grade line, and the total energy line. Finally we discuss about water hammer to conclude this Unit.

PRE-REQUISITES

Elementary calculus

UNIT OUTCOMES

U5-O1: Application of Bernoulli's principle to

- (a) Hydraulic siphon
- (b) Flow measuring instruments
- (c) Pressure measuring instruments

U5-O2: Concept of laminar and turbulent flow

U5-O3: Frictional losses in laminar and turbulent flow

U5-O4: Minor losses in flows

U5-O5: Pipes networks

U5-O6: Water hammer

Unit-5 Outcomes	EXPECTED MAPPING WITH PROGRAMME OUTCOMES (1 - Weak Correlation; 2 - Medium Correlation; 3 - Strong Correlation)						
	CO-1	CO-2	CO-3	CO-4	CO-5	CO-6	CO-7
U1-O1	2	1	1	3	3	2	3
U1-O2	2	-	2	2	-	2	3
U1-O3	1	-	-	3	2	2	3
U1-O4	-	-	1	2	2	-	2
U1-O5	-	-	-	3	-	-	2
U1-O6	-	-	-	3	1	1	2

5.1 PRINCIPLES OF HYDRAULIC SIPHON

From the previous unit, we have seen that Bernoulli's equation describes the nature of the variation of the pressure, kinetic energy, and potential energy in a flow, with appropriate assumptions. We start with the application of the Bernoulli's equation by analyzing a very common phenomenon, the hydraulic siphon. Let us consider a tube T_1, which is initially completely filled with the working liquid. Consider also a tank T_2 which contains stationary water. If one end of the tube, T_1 is then dipped into the tank, then we observe that the liquid will flow from the tank T_2 and get discharged from the other end of the tube T_1. This particular action is known as the siphonic action.

Several real-world applications of a siphon can be seen around us. A very commonly seen phenomenon is where one wants to withdraw liquid from a tank without any pump - say for example, you would like to extract some fuel from a fuel tank of your friend for your vehicle. In this case, a small suction on the exit side of the tube is maintained by sucking on the tube and closing it suddenly with a thumb and then inserting the other end into the tank. Another place it is seen is in emptying on a large sump or reservoir - a pump is used to start the flow after which the tube is immediately put into the reservoir, thereby using the siphon action to continue the drainage of the sump.

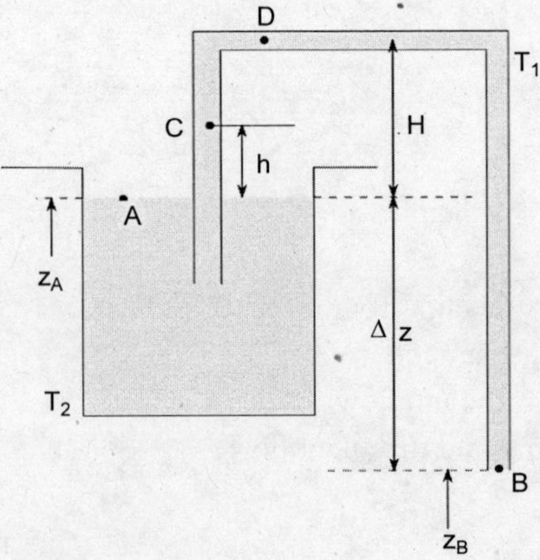

Figure 5.1: A hydraulic siphon. The point most prone to cavitation is the highest point .

Let us refer to the schematic. Let us label different points in the flow. Immediately important to us are points A and B. The difference in the height between the two points is Δz. It is immaterial to note the absolute locations of the points A and B. For the sake of the analysis, we may assume that the levels of the two points are Z_A and Z_B respectively, with respect to some arbitrary datum.

If we neglect the frictional effects, we can write down Bernoulli's theorem between points A and B as

$$\frac{p_A}{\rho g}+\frac{V_A^2}{2g}+z_A=\frac{p_B}{\rho g}+\frac{V_B^2}{2g}+z_B \tag{5.1}$$

$$\Rightarrow \quad \frac{p_A}{\rho g}+0+z_A=\frac{p_B}{\rho g}+\frac{V_B^2}{2g}+z_B \tag{5.2}$$

Note that we have neglected the velocity at point because the area of the tank is assumed to be large enough so that any change in the liquid volume in the tank leads to a negligible rate of change of the height of the liquid level. Mathematically, the velocity of the liquid at point A is $\frac{dz_A}{dt}$ and the rate of change of the volume of liquid is $Q=\frac{dz_A}{dt}A_{surf}$ which automatically implies that if the A_{surf} is very large, then for finite Q, $\frac{dz_A}{dt}\rightarrow 1$.

Now, we can further simplify Eq. (5.2) by noting that the pressures at both points and are equal to the atmospheric pressure. In that case we can simplify equation Eq. (5.2) to isolate an expression for the velocity as

$$V_B=\sqrt{2g\left(z_A-z_B\right)}=\sqrt{2g\Delta z} \tag{5.3}$$

We observe from equation Eq. (5.3) that the flow and velocity head at point is obtained solely through the difference in the potential head between points A and B.

While the flow being inviscid is a very fundamental assumption behind the Bernoulli's equation, we can account for the losses due to friction (which are essentially always present due to the presence of viscosity) through a *head loss*, h_L. We can modify equation Eq. (5.2) as

$$\frac{p_A}{\rho g}+z_A=\frac{p_B}{\rho g}+\frac{V_B^2}{2g}+z_B+h_L \tag{5.4}$$

$$\Rightarrow \quad V_B=\sqrt{2g\left(\Delta z-h_L\right)} \tag{5.5}$$

It is clear that the presence of head loss leads to a reduction of the velocity V_B at the exit as compared to the case where there is no head loss.

Let us now consider a point inside the tube, C. We can then write down Bernoulli's equation between points A and B.

$$\frac{p_A}{\rho g}+z_A=\frac{p_C}{\rho g}+\frac{V_C^2}{2g}+z_C \tag{5.6}$$

realizing that $V_B = V_C$ owing to the fact that the tube cross section does not change and the density as well does not change, we can write down the above equation as

$$\frac{p_C}{\rho g} = \frac{p_{atm}}{\rho g} - \frac{V_B^2}{2g} - h \tag{5.7}$$

where we have made use of the fact that $pA = p_{\text{atm}}$ and $z_C - z_A = h$. The above equation can be interpreted as the atmospheric pressure (at point A) being responsible for the velocity and potential head at point C and the drop in the pressure head at C. We can easily extend this analysis to the highest point in the tube, i.e. point D where we can write down the pressure as

$$\frac{p_D}{\rho g} = \frac{p_{atm}}{\rho g} - \frac{V_B^2}{2g} - H \tag{5.8}$$

We can now utilize the fact that

$$\frac{V_B^2}{2g} = \Delta z$$

and rewrite the expression for p_D as

$$\frac{p_D}{\rho g} = \frac{p_{atm}}{\rho g} - (\Delta z + H) \tag{5.9}$$

A careful reader may notice that the above expression can be equally easily be obtained by writing Bernoulli's equation between points D and B.

An important conclusion of the above expression for the pressure at the highest point is that it is lower than the atmospheric pressure. If this reduced pressure becomes equal to the vapour pressure of the liquid at the given temperature, then the liquid begins to boil. The boiling causes pockets of vapour to form in the tube and create a zone where there is more compressible vapour existing inside the tube. This causes the incoming flow to not be able to drive the fluid because the pressure is used to compress the vapour bubble. If the vapour pocket gets further carried downstream into the flow, it may so happen that it encounters a region of a relatively higher pressure. If the pressure again becomes larger than the vapour pressure at the given temperature, then the vapour pocket collapses, often violently. The collapse of the vapour pocket leads to a sudden rush of liquid and this may cause damage to the nearby solid surface (imagine the liquid hitting the former vapour pocket like a hammer). This particular phenomenon is called as *cavitation*. In order to avoid this matter, the height of the tube is kept such that the pressure does not fall below its vapour pressure. For the case of water, the minimum pressure is approximately 20 kPa, which corresponds to a head of 2 m. Therefore, in the context of the analysis done above, we must ensure that $p_D > p_{vap}$, where p_{vap} represents the vapour pressure of the liquid at the given temperature.

Example 5.1

Consider a tube used as a siphon to discharge water from a large vessel to drain it to atmospheric pressure. Calculate

1. The pressure at point
2. The pressure at point .
3. Velocity of water through the tube

4. Maximum height of the tube above the liquid surface, assuming a vapour pressure of kPa. Take the atmospheric pressure to be 100 kPa and gravity as 9.81 .

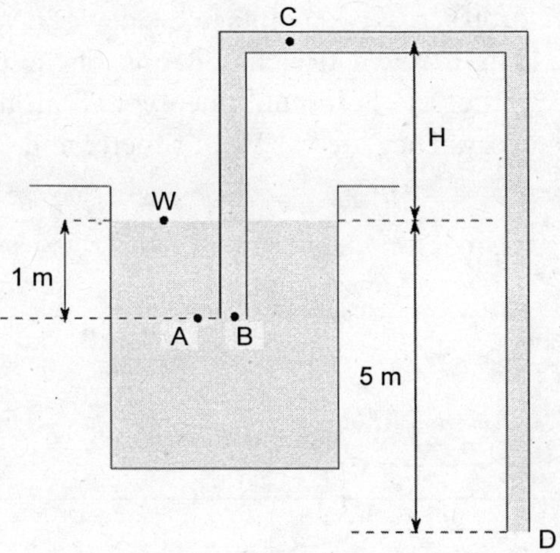

Solution:

Given that the velocity of the liquid in the tank is quite low, we can identify the pressure at point through the hydrostatic equation. Thus,

$$p_A = p_{atm} + \rho g z = 100 \times 10^3 + 1000 \times 9.81 \times 1.5 = 114.71 \text{ kPa}.$$

Before proceeding to determine the pressure at point *B*, we first find out the velocity of water in the tube. For this, let us apply Bernoulli's theorem between a point *W* on the surface of the tank and point *D*. We can therefore write

$$\frac{p_{atm}}{\rho g} + z_1 = \frac{p_D}{\rho g} + \frac{V_D^2}{2g} \Rightarrow \frac{100 \times 10^3}{1000 \times 9.81} + 5 = \frac{100 \times 10^3}{1000 \times 9.81} + \frac{V_D^2}{2g} + 0 \Rightarrow V_D = \sqrt{2g \times 5} = 9.9 m/s.$$

where, we have used the point *D* as the reference for the *z* displacement. We also note that the pressures at points *W* and *D* are equal to the atmospheric pressure.

Thus, we are now in a position to find out the pressure at point *B*. Writing Bernoulli's equation between points *A* and *B* by noting that the height of points *A* and *B* with respect to the datum *D* is the same, we have

$$\frac{p_A}{\rho g} = \frac{p_B}{\rho g} + \frac{V_B^2}{2g} = \frac{p_B}{\rho g} + \frac{V_D^2}{2g} \qquad \because V_B = V_D \Rightarrow p_B = p_A - \frac{\rho}{2} V_B^2 = 65.71 kPa.$$

Finally, we can figure out the maximum height of the tube above the surface of the water by writing Bernoulli's equation between the free surface and the highest point, which we assume to be at a distance *H* above the free surface. In order to do so, we will assume that the pressure at the top-most point will be equal to the vapour pressure. Thus, we have

$$\frac{p_{atm}}{\rho g} = \frac{p_{vap}}{\rho g} + \frac{V_C^2}{2g} + H \Rightarrow H = \frac{\left(p_{atm} - p_{vap}\right)}{rhog} - \frac{V_B^2}{2g} \Rightarrow H = 3.16 m$$

5.2 FLOW MEASUREMENT INSTRUMENTS

Measurement of flow through pipes can be achieved with the help of coaxial area contractions in the same pipe. Typically the pressure drop across two points in the device described above helps in assessing the flow rate through the pipe. The mathematical foundation of the principle behind the flow rate measurement lies in the application of the Bernoulli's theorem. Typically there are two primary flow meters which fall under this category. They are (a) Venturimeter and (b) Orificemeter. Next, we will discuss each of these devices.

5.2.1 Venturimeter

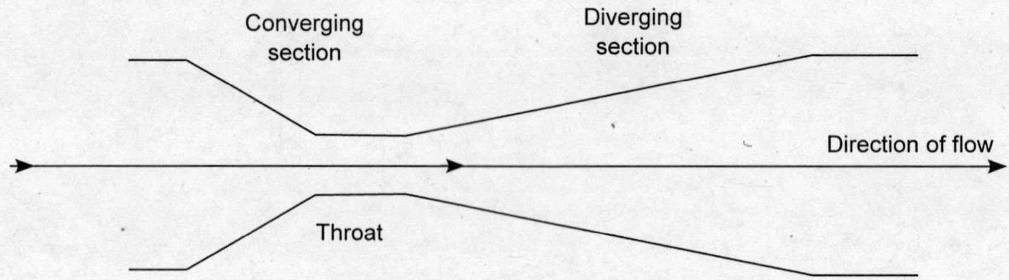

Figure 5.2: Schematic of a typical venturimeter.

A venturimeter, named after the Italian scientist Giovanni Venturi, is fundamentally a pipe which consists of three sections - a conical converging section, a constriction or throat, and a conical diverging section. It is obvious from the figure that the throat has the minimum cross section area. As is also observed from the figure, the converging section cone angle is higher than the diverging section cone angle. As a consequence, the inlet section is shorter than the outlet section, thereby causing a rapid convergence of the flow into the throat followed by a gradual divergence. This is done so that the flow does not *separate* after the throat. Any incidence of flow separation after the throat will cause undesirable energy loss and inaccurate measurements of the flow rate. It is clear that the geometry described would cause an increase in the velocity at the throat along with a reduction in the pressure.

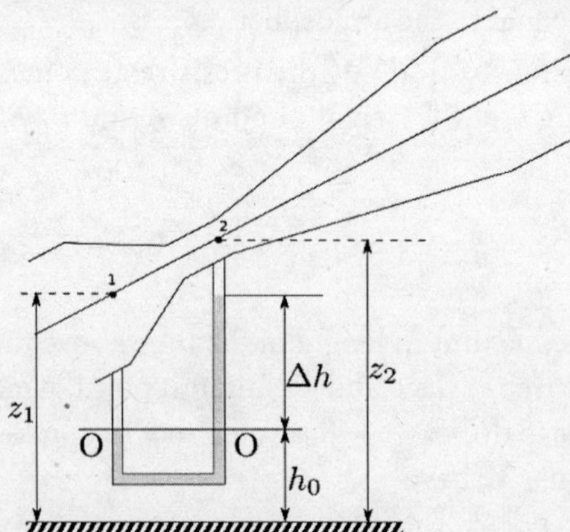

Figure 5.3: Schematic of an inclined venturimeter.

In order to mathematically assess a venturimeter, we consider the case of an inclined tube as shown. We consider a steady, inviscid, and one-dimensional flow along the axis of the venturimeter, thereby ensuring that the velocity and pressure profile at any cross section remain uniform. We consider two points, as shown in the figure. Apply Bernoulli's theorem between points 1 and 2, we obtain

$$\frac{p_1}{\rho g}+\frac{V_1^2}{2g}+z_1=\frac{p_2}{\rho g}+\frac{V_2^2}{2g}+z_2 \tag{5.10}$$

$$\Rightarrow \quad \frac{V_2^2-V_1^2}{2g}=\frac{p_1-p_2}{\rho g}+(z_1-z_2) \tag{5.11}$$

By utilizing the fact that from the continuity equation we have

$$V_2A_2 = V_1A_1$$

under the assumption that the density remains constant, we can simplify the expression above to

$$\frac{V_2^2}{2g}\left(1-\frac{A_1^2}{A_2^2}\right)=\left(\frac{p_1}{\rho g}+z_1\right)-\left(\frac{p_2}{\rho g}+z_2\right) \tag{5.12}$$

where the term inside the brackets is known as the *piezometric head*, i.e.

$$h^*=\frac{p}{\rho g}+z,\Rightarrow h_1^*=\frac{p_1}{\rho g}+z_1 \quad h_2^*=\frac{p_2}{\rho g}+z_2$$

Therefore, we can finally obtain an expression for V_2 and the volume flow rate, Q, as

$$V_2=\frac{1}{\sqrt{1-\frac{A_2^2}{A_1^2}}}\sqrt{2g\left(h_1^*-h_2^*\right)}\Rightarrow Q=A_2V_2=\frac{A_2}{\sqrt{1-\frac{A_2^2}{A_1^2}}}\sqrt{2g\left(h_1^*-h_2^*\right)} \tag{5.13}$$

We can now apply the principles of hydrostatics between points 1 and 2 for the u-tube manometer shown in the Fig. 5.3 and write down the pressure at the origin points O in the figure to obtain

$$p_1 + \rho g(z_1 - h_0) = p_2 + \rho g(z_2 - h_0 - \Delta h) + \Delta h \rho_m g \tag{5.14}$$

$$\rho_m = \text{Density of manometric fluid}$$

which may be simplified to

$$\left(\frac{p_1}{\rho g}+z_1\right)-\left(\frac{p_2}{\rho g}+z_2\right)=\left(\frac{\rho_m}{\rho}-1\right)\Delta h\Rightarrow h_1^*-h_2^*=\left(\frac{\rho_m}{\rho}-1\right)\Delta h \tag{5.15}$$

We can now substitute Eq. (5.15) in Eq (5.13) and write down an expression of the flow rate in terms of the height difference observed in the manometer as

$$Q=\frac{A_1A_2}{\sqrt{A_1^2-A_2^2}}\sqrt{2g\left(\frac{\rho_m}{\rho}-1\right)\Delta h} \tag{5.16}$$

As a simplification of the above equation, we have the situation where the venturimeter is not inclined but horizontal in which case, $z_1 = z_2$ which implies that the piezometric head is simply equal to the static pressure head, i.e.

$$h_1=\frac{p_1}{\rho g},h_2=\frac{p_2}{\rho g}\Rightarrow h_1-h_2=\left(\frac{\rho_m}{\rho}-1\right)\Delta h$$

A very important observation in the derivation is that regardless of the fact that the venturimeter is inclined or not, the final expression for the flow rate in the pipe remains unchanged.

Coefficient of discharge

In reality, the measured values of Δh are always larger than the ideal case of an inviscid fluid because of the additional pressure loss that happens due to the presence of friction. As a consequence, Eq. (5.16) would always overpredict the actual flow rate in practical cases. In order to correct this overprediction, we typically multiply the flow rate with a factor, C_d, the *coefficient of discharge* and write down

$$Q_{actual} = C_d Q = C_d \frac{A_1 A_2}{\sqrt{A_1{}^2 - A_2{}^2}} \left(\sqrt{2g(\frac{\rho_m}{\rho} - 1)\Delta h} \right) \tag{5.17}$$

In other words, the coefficient of discharge, C_d, is

$$C_d = \frac{\textit{Actual flow rate}}{\textit{Ideal flow rate}}$$

In practice, the value of C_d is somewhere between 0.95 and 0.98, thereby implying that the viscous effects do not signficantly cause us to overpredict the flow rate for flows through pipes as measured through venturimeters.

Example 5.2

For a given condition, water flows through a 400 mm × 200 mm venturimeter (which means that the pipe dimater is 400 mm and the throat diamter is 200 mm) at a rate of 0.068 m³/s and the differential gauge is deflected 1 m as shown in figure. Considering the specific gravity of the gauge liquid to be 1.27, determine the coefficient of discharge of the venturimeter.

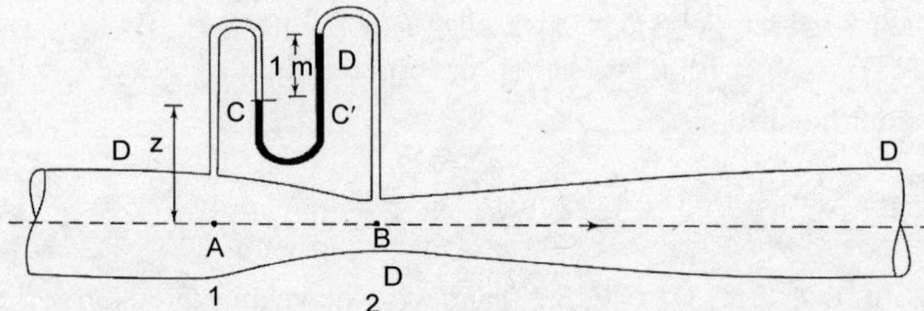

Solution:

The coefficient of discharge is defined as the ratio of actual rate of discharge to the theoretical rate of discharge. The theoretical discharge rate is predicted from the relation

$$Q = \frac{A_1 A_2}{\sqrt{A_1^2 - A_2^2}} \sqrt{2g\left(\frac{\rho_m}{\rho} - 1\right)\Delta h}$$

where, the value Δh is measured from the gauge height difference and the actual rate of discharge is the discharge rate measured in practice.

To find out the coefficient of discharge, first let's apply the Bernoulli's equation between A and B considering the fluid to be inviscid. Considering the axis of the venturimeter to be horizontal, the Bernoulli's equation between A and B becomes

$$\frac{p_1}{\rho g}+\frac{V_1^2}{2g}+0=\frac{p_2}{\rho g}+\frac{V_2^2}{2g}+0 \qquad (5.18)$$

Again from the continuity relation we get

$$V_A^{\ 2}=\left(\frac{A_B}{A_A}\right)^2 V_B^{\ 2} \qquad (5.19)$$

We proceed to solve for V_B from the Bernoulli's equation (Eq. 5.18) with the help of the continuity relation (Eq. 19), we obtain

$$V_B=\sqrt{\frac{2(p_A-p_B)/\rho}{1-(A_B/A_A)^2}}$$

We can write the actual rate of discharge (Q) to be a function of coefficient of discharge (C_D) and the theoretical discharge ($Q_{theo}=A_BV_B$), which is

$$Q=C_D\,A_B\,V_B$$

$$\Rightarrow \qquad Q=C_DA_B\sqrt{\frac{2(p_A-p_B)/\rho}{1-(A_B/A_A)^2}}$$

Again, from the principle of hydrostatics applied to the differential gauge, we get

$$p_A/\rho g-z)=p_B/\rho g-(z+1)+1.27\times 1$$

$$\Rightarrow \qquad \frac{p_A-p_B}{\rho g}=0.27\,m$$

Let's put the obtained values in the equation $Q=C_DA_B\sqrt{\frac{2(p_A-p_B)/\rho}{1-(A_B/A_A)^2}}$ to find the value of C_D.

After substituting the values the equation becomes

$$0.068=C_D\frac{\pi}{4}(0.2)^2\sqrt{2\times 0.27\times 9.81/(1-1/16)}$$

After solving the above equation we obtain the value of the coefficient of discharge for the present flow condition to be $C_D=0.91$.

5.2.2 Orificemeter

An orificemeter can be thought of a geometrically simpler and cheaper means for the measurement of flow through a pipe. An orificemeter does not have the requirement of having double conical geometry but rather it is simply a concentric circular blockage in the pipe as shown in the figure. The concentric hole is referred to as the orifice plate and is responsible for creating the obstruction to the flow. The orifice area is typically much smaller than the cross-section area of the pipe. The flow, after crossing the orifice, continues on due to its momentum where it contracts a point downstream of the orifice. This is the region where the streamlines converge to the smallest area. This is called as the vena contracta. After this, the flow then expands to the entire pipe again. Pressure tappings are provided at a location upstream and just after the orifice. The presence of the sudden orifice leads to the formation of the vortices in the corners of the orifice downstream of the flow, thereby leading to a loss of energy and therefore we expect that such a flow measurement device would significantly overpredict the flow rate - as aspect which we shall see in the derivations.

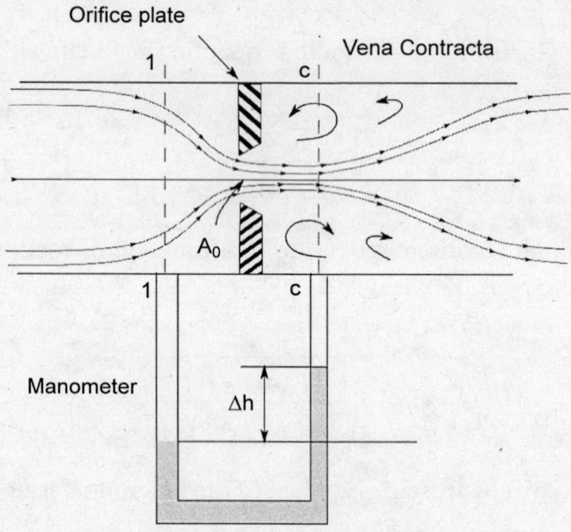

Figure 5.4: An orificemeter used to measure flow through a pipe.

The various parts can be seen in Fig. 5.4. We can apply Bernoulli's theorem at the centerline streamline between sections 1-1 and section c-c (which is at the vena contracta). We can therefore directly write

$$\frac{p_1^*}{\rho g}+\frac{V_1^2}{2g}=\frac{p_2^*}{\rho g}+\frac{V_c^2}{2g} \tag{5.20}$$

where $p^* = p + \rho g z$ represents the piezometric pressure. Proceeding further, we can make use of the continuity equation wherein for a constant density flow we have

$$V_1 A_1 = V_c A_c \tag{5.21}$$

where A_c represents the cross section area of the vena contracta. Therefore, we can substitute the continuity equation in the Bernoulli's equation to obtain an expression for the velocity at the vena contracta as

$$V_c = \sqrt{\frac{2\left(p_1^* - p_2^*\right)}{\rho\left(1-\frac{A_c^2}{A_1^2}\right)}} \tag{5.22}$$

We can further make refinements to the expression for the velocity by accounting for the viscous nature of the flow and introduce the *coefficient of viscosity*, C_v. This will help us to assess the actual velocity as measured through the height difference between the two limbs of the manometer and obtain

$$V_c = C_v\sqrt{\frac{2\left(p_1^* - p_2^*\right)}{\rho\left(1-\frac{A_c^2}{A_1^2}\right)}} = C_v\sqrt{\frac{2g\left(\frac{\rho_m}{\rho}-1\right)\Delta h}{\left(1-\frac{A_c^2}{A_1^2}\right)}} \tag{5.23}$$

where we have made use of the fact that $p_1^* - p_c^* = (\rho_m - \rho) g\Delta h$ as obtained from hydrostatics for the manometer.

We also define the coefficient of contraction as

$$C_c = \frac{A_c}{A_0} = \frac{\textit{Area of vena contracta}}{\textit{Area of the orifice}}$$

Consequently, we have the volumetric flow rate

$$Q = A_c V_c = C_c A_0 C_v \sqrt{\frac{2g\left(\frac{\rho_m}{\rho}-1\right)\Delta h}{\left(1-\frac{A_c^2}{A_1^2}\right)}} = C_v C_c A_0 \sqrt{\frac{2g}{1-\frac{C_c^2 A_0^2}{A_1^2}}}\sqrt{\left(\frac{\rho_m}{\rho}-1\right)\Delta h} \tag{5.24}$$

We can now combine the geometric and various coefficients into a single coefficient as

$$C = C_v C_c A_0 \sqrt{\frac{2g}{1 - C_c^2 A_0^2 / A_1^2}} \tag{5.25}$$

and finally obtain an expression for the flow rate as

$$Q = C\sqrt{\left(\frac{\rho_m}{\rho}-1\right)\Delta h} \tag{5.26}$$

For practical uses, the value of C depends on the diameter of the pipe and orifice and the actual flow rate (through the Reynolds number). Typically, the value of C is determined for various flow rates for a given pipe and this process is known as the calibration of the orificemeter.

To know more about orificemeter

Example 5.3

For a given system, water flows at a rate of 0.028 m/s through a 150 mm diameter orifice used in a 300 mm pipe. Considering the coefficient of contraction $C_c = 0.7$ and the coefficient of velocity C_v to be 0.9, find out the difference in pressure head between the upstream section and the vena contracta section.

Solution:

Before solving the above problem, let's recall that the measured value of the piezometric pressure drop for a real fluid is always more due to friction than that assumed in case of an inviscid flow. Hence a coefficient of velocity (C_v) is introduced which will always be less than, to determine the actual velocity at the vena contracta of the orifice V_c when the pressure drop is measured in terms of the manometric deflection. It is also worth mentioning that, the coefficient of contraction is defined as the ratio of cross sectional areas at the vena contracta to the area of the orifice i.e. $C_c = A_c/A_0$.

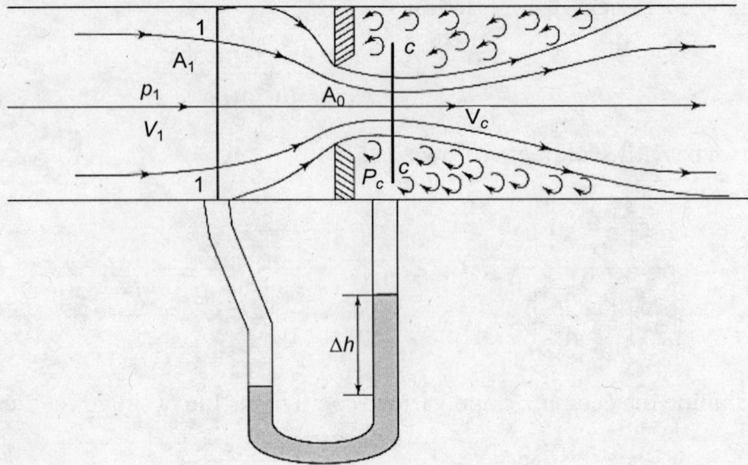

Solving the Bernoulli's equation, the actual velocity at the vena contracta of the orifice becomes

$$V_c = C_v\sqrt{\frac{2g(\rho_m/\rho - 1)\Delta h}{1-(A_c/A_1)^2}}$$

where, Δh is the difference in liquid levels in the manometer and ρ_m is the density of the manometric liquid. Then, the volumetric flow rate becomes

$$Q = A_c V_c = C_c A_0 C_v\sqrt{\frac{2g(\rho_m/\rho - 1)\Delta h}{1-(A_c/A_1)^2}}$$

$$\Rightarrow \quad Q = C_v C_c A_0\sqrt{\frac{2g}{(1-C_c^2 A_c^2/A_1^2)}}\sqrt{(\rho_m/\rho - 1)\Delta h}$$

for the differential pressure head in manometers, we can replace $(\rho_m/\rho - 1)\Delta h$ with since we want to write $\Delta p/\rho g$ down the difference in the pressure between the two limbs of the manometer in terms of the head of water.

$$\Rightarrow \qquad Q = C_v C_c A_0 \sqrt{\frac{2g}{\left(1 - C_c^2 A_c^2 / A_1^2\right)}} \sqrt{\frac{\Delta p}{\rho g}}$$

Hence, substituting all the known relevant values in the above equation, we can obtain the value of $\Delta p/\rho g$, which is the difference in pressure head between the upstream section and the vena contracta section.

$$\Rightarrow \qquad 0.028 = 0.9 \times 0.7 \times \frac{\pi}{4}(0.15)^2 \sqrt{\frac{2 \times 9.81}{\left(1 - 0.7^2 (1/2)^4\right)}} \sqrt{\frac{\Delta p}{\rho g}}$$

$$\Rightarrow \qquad \frac{\Delta p}{\rho g} = 0.559 \ m \ of \ water$$

5.3 PRESSURE MEASUREMENT INSTRUMENTS

5.3.1 Static Pressure

Static pressure is the manifestation of the thermodynamic collisions of molecules in a fluid. When the fluid is at rest, the pressure is known as the hydrostatic pressure. We have seen from Bernoulli's theorem that the static pressure is a means of quantifying the pressure head, which can also be used to estimate changes in the velocity in an ideal flow. Static pressure represents the pressure of the fluid which is measured while the fluid is in motion. In this section we will look at the means of measuring various pressures in a moving fluid.

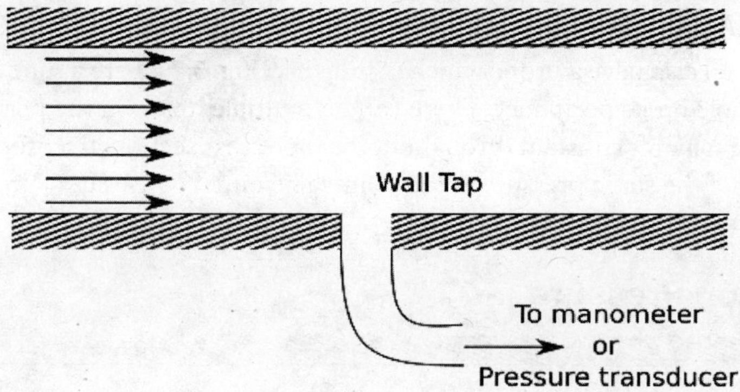

Figure 5.5: A wall tap used to measure the static pressure at a point in the pipe.

Consider the flow as shown in Fig. 5.5, which shows a pipe flow with a port, which is basically a hole which is flush with the internal surface. This is also known as a tap. The pressure measured by the tap will be the static pressure. An important aspect of the wall tap is that there is no flow into the tap, i.e. the flow remains along the axis. For an ideal flow, the pressure sensed at the location of the tap is constant throughout the cross section, and is the static pressure. Typically, the tap is connected with the help of a tube to a manometer or a pressure transducer. This helps us to measure the static pressure at the point in the fluid.

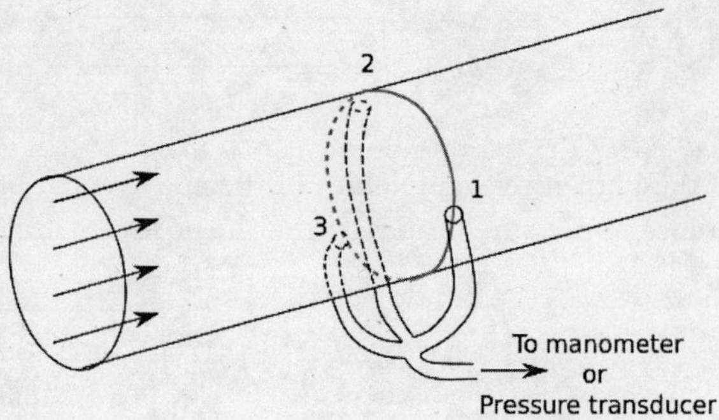

Figure 5.6: An arrangement of holes on the periphery of the tube to measure the static pressure at a point in the flowing fluid.

More generally, this can be extended to an arrangement of tubes all around the periphery of the tube so that the values of pressure can be averaged at a certain cross section. This helps to reduce the variability in the measurement of the static pressure. This arrangement is shown in Fig. 5.6.

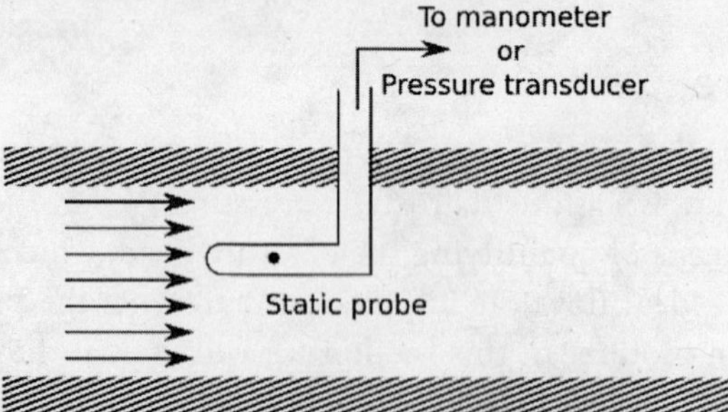

Figure 5.7: A static probe used to measure the static pressure at a point in the flow.

Besides this, we can also have a probe which is shown in Fig. 5.7 where a tube is introduced into the flow. The tube has a hole on its periphery. There can be multiple holes on the periphery. Given the fact that the static pressure remains constant throughout the pipe cross section, the probe, known as the *static probe* is able to measure the static pressure at the point. The outlet of the static pressure is connected to a manometer or a pressure transducer.

5.3.2 Stagnation Pressure

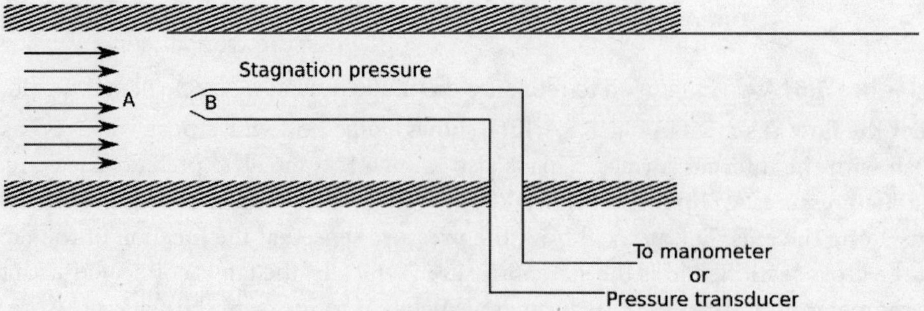

Figure 5.8: A stagnation pressure probe with a hole in the front as opposed to a hole on the side wall, which measures the static pressure.

The stagnation pressure concept is derived from the hypothetical process where in the flow is brought to rest in an isentropic manner, i.e. the process is reversible and adiabatic. The pressure measured at this point is known as the *stagnation pressure.* Consider the flow in Fig. 5.8 wherein the flow is thought to be brought to rest in isentropically between points A and B. At equilibrium, the velocity of the flow at point A will be equal to zero. Consequently, if we look at the Bernoulli's equation, we can conclude that the pressure at point B will be higher than the static pressure at point A. We can thus say that the pressure measured at point B where the fluid has zero velocity, i.e. the fluid is *stagnant* is the stagnation pressure.

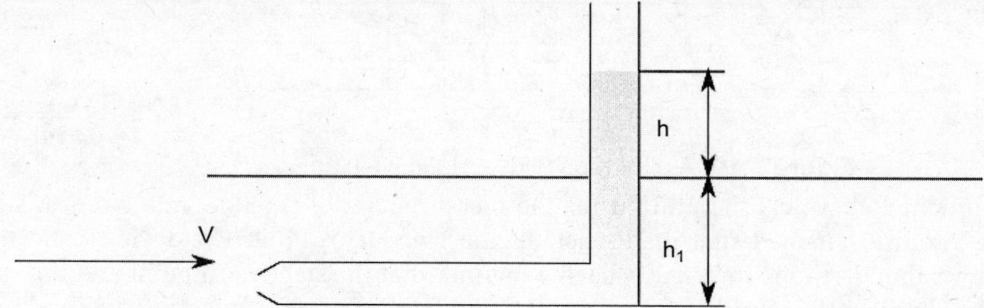

Figure 5.9: A stagnation pressure probe can be used to measure the stagnation pressure in an open flow as well.

Note, that the same arrangement can also be used to measure flow in an open channel where the probe, which is known as the Pitot tube, is simply inserted from the free surface. The level rise of the fluid in the tube is a measure of the stagnation pressure (see Fig. 5.9). This pressure, relative to the atmospheric pressure, can be easily determined with the help of the manometry pressure calculations between the stagnation point and the top surface of the liquid. The pressure can be related as $p_0 = p_{atm} + \rho g(h + h_1)$. This gives us a means to determine the stagnation pressure. However, the relation between the stagnation pressure and the flow velocity can be established through the Bernoulli's theorem between points A and point B.

$$p_0 = p + \frac{\rho V^2}{2} \tag{5.27}$$

$$\text{Stagnation pressure} = \text{Static pressure} + \text{Dynamic pressure} \tag{5.28}$$

$$V = \sqrt{\frac{2(p_0 - p)}{\rho}} \tag{5.29}$$

Therefore, it implies that the measurement of both the static and stagnation pressure of a flowing fluid would allow us the measurement of the flow velocity. We also note that in reality, the stagnation pressure that will be measured through experiments will be always be lower than the actual stagnation pressure because there will be some irreversibility in the flow. In that case some of the energy of the incoming flow will be converted into heat when it is being brought to a zero velocity. Therefore, we can account for this non-ideality by introducing a correction factor C which is multiplied to the expression of the velocity as

$$V \quad C\sqrt{(p \quad p)} \tag{5.30}$$

5.3.3 Pitot tube

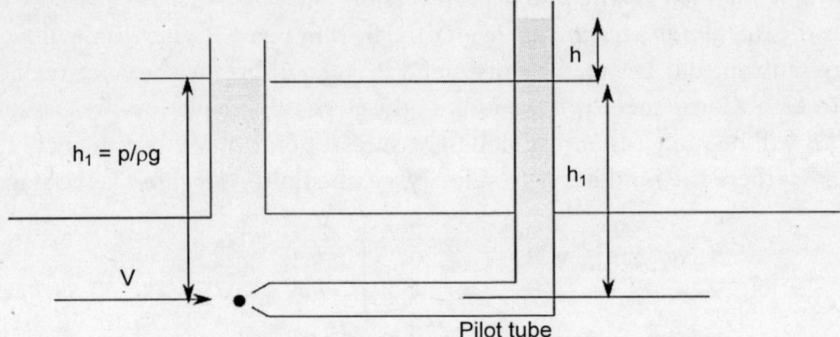

Figure 5.10: A pitot tube with a static wall tap.

The ideas mentioned above can be utilized for the measurement of the flow rate by means of a device known as a Pitot tube, named after the French engineering Henri Pitot. The device is shown in Fig. 5.10 which shows the Pitot tube inserted in such a manner that the stagnation point and the static pressure wall tap are almost at the same location. The same fluid moves up the stagnation tube and static wall tap tube as well. The difference in the height between the two limbs is the resultant flow velocity. We can write down the expression for the difference between the stagnation and static pressures in terms of the velocity as $p_0 - p = \frac{1}{2}\rho V^2$ and then relate the static pressure to the height of liquid in the tube as $p = \rho g h_1$, while the height of liquid in the other tube as $p_0 = \rho g h + \rho g h_1$. Upon substituting these values, we obtain that

$$V = \sqrt{2gh} \tag{5.31}$$

5.3.4 Pitot Static Tube

The above idea of simultaneously measuring the static and stagnation pressure is typically combined into a single device known as the Pitot static tube as shown in Fig. 5.11. In this there is one coaxial hollow section (colored blue) and a set of peripheral tubes (colored orange). The former measures the stagnation pressure while the peripheral tubes measure the static pressure. Therefore through this one probe, we are able to obtain static and stagnation pressure. The outlets of these tubes can be connected to individual manometers to get the absolute pressures. However, they are typically connected to two limbs of the same manometer, which are referred to as the *differential manometer.*

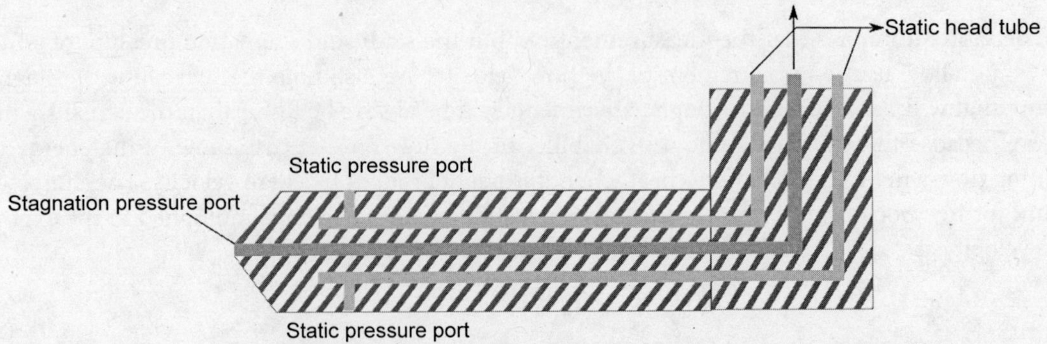

Figure 5.11: A Pitot static tube without the connection to the manometer.

5.4 CONCEPT OF LAMINAR AND TURBULENT FLOW

We will now proceed to discuss about the various aspects of internal flows. These typically include flows in pipes, ducts, and pipes of non-standard cross sections. Such systems are widely seen in almost all engineering applications which involve handling of fluids. While balances of mass and momentum can yield certain solutions of fluid flow, there exists a distinct demarcation of the nature of flow characterized through

$$\mathrm{Re} = \frac{\rho UL}{\mu} \tag{5.32}$$

where is termed as the *Reynolds number*, is the density of the fluid, is the dynamic viscosity of the fluid, represents the characteristic velocity of the flow - this could be something like the average velocity of flow in a flow through a pipe, and represents the characteristic length - this could be the diamter of the pipe through which the fluid is flowing. It is interesting to note that the Reynolds number is termed as a *dimensionless number*, a number which does not have any dimension attached to it.

The Reynolds number helps to delineate between *laminar* and *turbulent* flows. One of the fundamental characteristic of laminar flow is where the flow remains smooth and steady (if the inlet conditions of the fluid are steady. A very prominent example of such a flow is the flow of honey as it falls from a jar onto the plate. The column of honey which is falling down appears to do so in a very well-behaved fashion. In fact, this is also seen in the case of a tap which is opened ever so slightly so that the flow rate coming out from the tap is low. The flow appears to be almost still and there are no spontaneous changes in the geometry of the column falling down. However, as the flow rate is increased, which is tantamount to increasing the Reynolds number, we see that the flow of the column of water from the tap changes continuously with puffs and fluctuations present everywhere. This implies that the flow characteristics such as the shape of the interface or even the velocity field is having some components which are constantly changing even though the inlet conditions do not change, except for an increase in the flow rate. We then say that the flow has a Reynolds number which is higher than the critical Reynolds number, which is a fundamental parameter which helps in quantify the change in the fundamental behaviour. Besides having velocity fields are constant in a state of fluctuation, there is yet another hallmark of these two kinds of flows. In a laminar flow, there is hardly any mixing between two fluids. However, when the flow becomes turbulent, there is a drastic increase of the mixing between the two fluids.

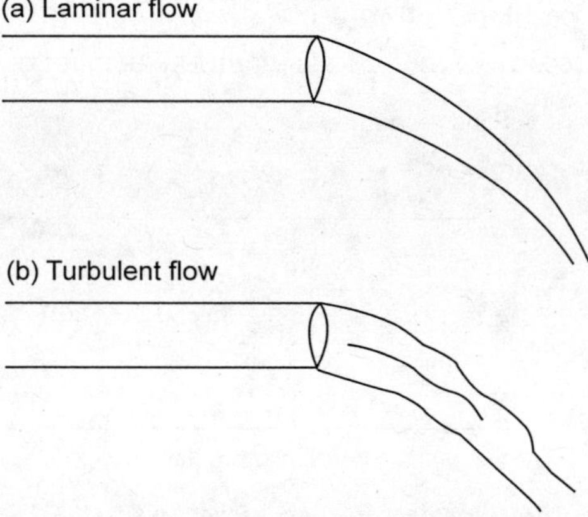

Figure 5.12: (a) Laminar flow and (b) Turbulent flow at the exit of a pipe through which water is flowing. An interested student may easily undertake this experiment with the help of a common tap. By increasing the flow rate coming out from a tap, one can observe the transition from laminar to turbulent flow.

To know more about laminar and turbulent flows

Hagen-Poiseuille flow

A very important result for internal flows was given by Hagen who was one of the first engineers to suggest the presence of two different regimes of flow. It was suggested that the pressure drop of the flow of water through a tube is proportional to the flow rate as

$$\Delta p = K\frac{QL}{D^4} + Const.,$$

where K is a constant, L is the length of the tube, D is the diameter of the tube, Q is the volume flow rate, and Δp represents the pressure drop. This suggested formula was then shown to be true for lower ranges of flow rates, i.e. when the flow was laminar. As the flow rate increased, there was a transition to the turbulent regime wherein the pressure drop increased at a faster rate as the flow rate was increased. Reynolds, after whom the Reynolds number is defined, definitely showed through experiments in flow of dye in a flow that the critical Reynolds number after which the flow becomes turbulent is $Re_{critical} = 2300$. Note that these transitions are not just restricted to internal flows, but is also valid for external flows, for example flow over aerofoils etc.

The rigorous solution involves the Navier-Stokes equation as the starting point. The Navier-Stokes equations represent the law of the conservation of momentum in a differential form and are beyond the scope of this particular text. However, we can still use the fundamental ideas developed so far to obtain the same result. This shows the power of utilizing such control volume approaches to derive very fundamental solutions.

Let us first consider the problem definition. We have a pipe through which a viscous fluid is being transported. The dynamic viscosity of the fluid is μ and the density of the fluid is ρ.

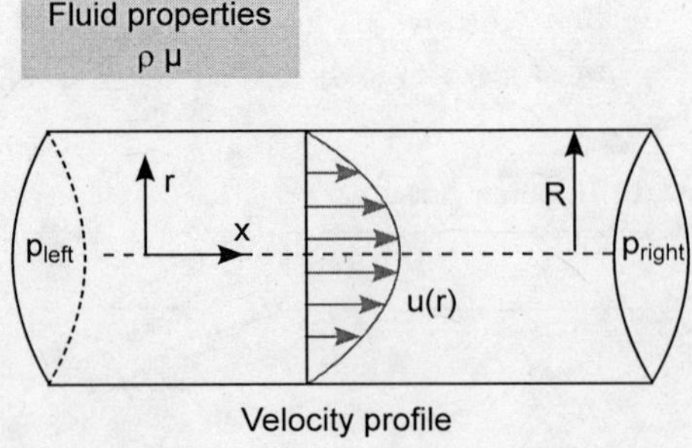

Figure 5.13: A fully developed flow through a pipe of a segment of length, L, and radius, R, is shown in the figure. Flow occurs from left to right and is driven by the pressure gradient. The pressure at the left end is higher than the pressure at the right end. The velocity profile is shown at the cross section and owing the fully developed flow, the velocity in the axial direction only depends on the radial direction and not on the axial direction.

The flow is driven by a pressure gradient. A pressure gradient in this context is the relatively higher pressure on the left side of the tube as compared to the right side which causes the fluid to be driven or flow from the left side to the right side. In such kinds of flows, the pressure gradient is achieved by means of a pump or by means of a reservoir which is at a larger height as compared to the other reservoir. Both of these mechanisms are commonly seen in water distribution systems in apartments and homes. In such typical systems, the fluid enters at the inlet where the fluid is now going from a state where it was almost stationary at the reservoir to the fully developed flow in the pipe. When the flow initially starts, the velocity does not immediately reach the same profile as shown in the figure. In fact, the fluid does not *feel* the influence of the wall immediately and hence the no-slip boundary condition is slowly felt as the fluid enters further into the pipe. In this region, the flow is accelerating and then it reaches stead y state at which point there is no more acceleration.

In this ensuing analysis, we will consider the fully developed region for simplicity. We must note however, that the same analysis can also be done for the so-called developing region as well. In the fully developed region, the pressure is also linearly reducing as we shall see soon. First, we look at what is happening physically. The flow is driven from left to right, i.e. along the positive direction by the larger pressure at the left face than the right. This is then counteracted by the influence of the viscosity which aims at slowing down the flow. Therefore at steady state, we may say that the pressure which drives the flow from left to right is equally retarded by the influence of the viscous stress from the right to left.

In that case, let us consider the control volume which encompasses the fluid inside the tube and write down this driving force. The net force due to the influence of the two pressure terms is

$$
\begin{aligned}
F_p &= F_{\text{left}} - F_{\text{right}} \\
&= p_{\text{left}} \times \pi r^2 - p_{\text{right}} \times \pi r^2 \\
&= p \times \pi r^2 - \left(p + \frac{dp}{dx}\Delta x \right) \times \pi r^2 \\
&= \left(-\frac{dp}{dx}\Delta x \right) \times \pi r^2 \\
&= -\frac{dp}{dx}\Delta x \pi r^2 \qquad (5.33)
\end{aligned}
$$

A very important conclusion from the above expression is that the flow is driven by the difference in pressure across the tube and *not* the absolute pressures given. Another important aspect in the expression obtained above is that the flow is driven in the positive direction when the pressure gradient is negative. What do we mean by a negative pressure gradient? It means that as the value of increases, the value of decreases, i.e. $\frac{dp}{dx} = \frac{p_{right} - p_{left}}{\Delta x} < 0$ because $p_{\text{left}} > p_{\text{right}}$.

Now, what would happen if there is nothing to stop this flow, the pressure gradient would keep acting on the fluid particles and keep them accelerating. However, owing to the flow, the lamella of the fluid gets retarted by the action of viscosity. We have seen an elementary discussion of viscosity where we have obtained Newton's law of viscosity. In a similar fashion, the velocity field which is developed in the pipe causes the presence of a shear stress. The genesis of this velocity profile also warrants a discussion. The condition at the solid surface of the pipe is known as the no-slip boundary condition, which implies that the velocity of the fluid particle at the contact of the pipe is zero. As we move towards the center of the pipe, the influence of the wall reduces and the velocity field increases at the center. The maximum velocity is seen at the centerline of the pipe.

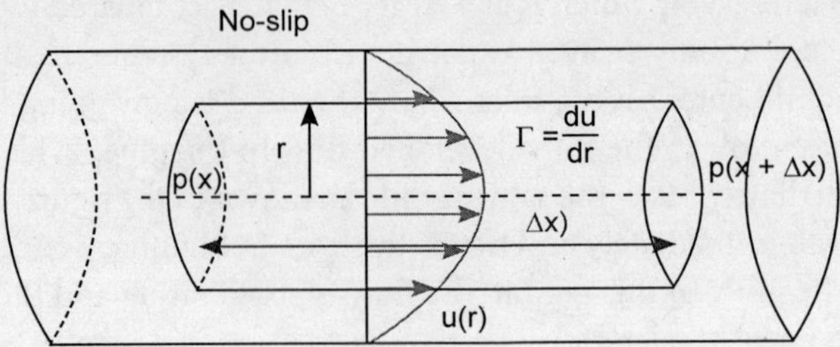

Figure 5.14: A representative volume which is used for the calculation of the shear stress.

We can write down the shear stress at any radial location as τ while the total force acting on the fluid element due to the shear stress is given by $\tau \times A_p$, where A_p represents the perimeter area, which in this case is $2\pi r\Delta x$

$$F_{viscous} = \tau A_p = \mu \frac{du}{dr} \times 2\pi r\Delta x \tag{5.34}$$

Accordingly we can now combine equations Eq. (5.33) and (5.34) to obtain

$$F_p + F_{viscous} = 0 because, fully\ developed\ flow\ does\ not\ accelerate - \frac{dp}{dx}\pi r^2 \Delta x + \mu \frac{du}{dr} \times 2\pi r\Delta x = 0$$

$$\frac{du}{dr} = \frac{1}{2\mu}\frac{dp}{dx} r \tag{5.35}$$

We can now rearrange the above expression as

$$\frac{1}{r}\frac{du}{dr} = \frac{1}{2\mu}\frac{dp}{dx} \tag{5.36}$$

where we can now make some more conclusions about the set of terms seen above. The left hand side represents a set of terms which depends only on the *r* coordinate while the term on the right hand side represents terms which only depend on the *x* coordinate. This is because of the assumption that the flow is a fuly developed one. In such a case where both such terms are equal, each must be a constant. In such a case, we can now integrate the terms above to obtain

$$u = \frac{1}{2\mu}\frac{dp}{dx}\frac{r^2}{2} + C \tag{5.37}$$

$$Because u(r = R) = 0 due\ to\ no - slip, C = -\frac{1}{2\mu}\frac{dp}{dx}\frac{R^2}{2} \tag{5.38}$$

$$\Rightarrow \quad u = \frac{1}{2\mu}\frac{dp}{dx}\left(\frac{r^2}{2} - \frac{R^2}{2}\right) \tag{5.39}$$

$$\Rightarrow \qquad u = -\frac{R^2}{4\mu}\frac{dp}{dx}\left(1 - \frac{r^2}{R^2}\right) \tag{5.40}$$

In the entire derivation above, we have only written down the static pressure, p, in the expressions. However, we note that the even if the pipe or tube is oriented along an arbitrary direction with respect to gravity, the expression of the velocity field only has to account for the piezometric pressure, i.e. $p + \rho gz$ instead of the static pressure. Everything else remains the same.

We can now find out the volume flow rate with the expression for the velocity field as

$$Q = \int_0^R u dA = \int_0^R \left(1 - \frac{r_2}{R_2}\right) 2\pi r dr = \pi \frac{R^4}{8\mu}\left(-\frac{dp}{dx}\right) \tag{5.41}$$

5.5 FRICTION LOSSES IN LAMINAR FLOW

From the above expression, it is clear that the flow rate and pressure drop in the pipe have a linear relationship between them. Infact, We have already shown that the pressure drop is constant, i.e. $\frac{dp}{dx} = const.$ (see Eq. (5.36) which means that we can write the pressure gradient as $\frac{dp}{dx} = \frac{\Delta p}{L}$. Thus, we can rearrange Eq. (5.41) as

$$\Delta p = \frac{8\mu LQ}{\pi R^4} \tag{5.42}$$

Furthermore, we can find out the wall shear stress as

$$\tau_w = \mu \frac{du}{dr}\Big|_{r=R} = 2\mu \frac{u_{max}}{R} = \frac{1}{2} R\left(-\frac{dp}{dx}\right) \tag{5.43}$$

The skin friction coefficient, C_f, is defined as

$$C_f = \frac{\tau_w}{\frac{1}{2}\rho V^2} = \frac{16}{\rho V d / \mu} = \frac{16}{\text{Re}_d} \tag{5.44}$$

We can write down the head loss, i.e. $h_f = \Delta \frac{p}{\rho g}$ by substituting the expression of pressure drop in Eq. (5.42) and obtain

$$h_f = \frac{128\mu QL}{\rho g \pi D^4} \tag{5.45}$$

The frictional head loss can also be expressed through the Darcy friction factor, f, named after the French engineering Darcy, through

$$h_f = f\frac{L}{D}\frac{V^2}{2g} = \frac{8\mu LQ}{\pi R^4} \Rightarrow f = \frac{64}{\text{Re}_d} \tag{5.46}$$

Alternately, we can also define the Darcy friction factor as

$$f = \frac{8\tau_w}{\rho V^2} = \frac{64\mu}{\rho V d} = \frac{64}{\text{Re}_d} \tag{5.47}$$

which shows that the friction factor is inverse proportional to the Reynolds number based on the diameter of the pipe, *d*.

To know more about fraction losses

5.6 MAJOR LOSSES IN TURBULENT FLOW

While the solution of the momentum equation for a laminar flow yields the friction factor as $f = \frac{64}{\text{Re}_d}$, we note that the case of turbulent flows is not so straight forward. In fact, the friction factor can be shown through dimensional analysis that it depends functionally on both the Reynolds number and the roughness of the pipe through which the flow is happening. The details of the dimensionless numbers is beyond the scope of this text. Pioneering works were undertaken by Standon, Nikuradse, and Moody towards a solution to obtain the variation of the friction factor as a function of the Reynolds number for different pipe roughnesses. Pipe roughness always exists in pipes due to the fundamental nature of the fabrication process it is made through. For laminar flows, the velocity is the lowest in the vicinity of the walls through the no-slip boundary condition due to the presence of viscosity. This low velocity leads to a poor influence of the roughness on the friction factor. However, for turbulent flows, which has a higher velocity and shear stress near the wall, the influence of the roughness, which plays an important role in causing more frictional losses, cannot be ignored. Although roughness is not constant for a pipe, the statistics can be predicted - for example the root mean square (rms) roughness of a pipe. Not only does this depend on the nature of fabrication, but is also strongly governed by the material of the pipe. For example, pipes of concrete inherently have a higher roughness than pipes made of aluminum or stainless steel.

5.6.1 Friction Factor: Moody Diagram

The combined results of the friction factor for both laminar and turbulent flows are summarized through a single plot known as Moody's diagram as seen in Fig. 5.15.

As we have already seen, for laminar fully developed flows, the friction factor is given by $f = 64/\text{Re}$. In the case of turbulent flows, however, the friction factor depends on two factors - the Reynolds number and the relative roughness of the pipe, $\frac{\in}{D}$, where $\in$ is the rms roughness of the pipe surface and *D* is the diameter of the pipe. The roughness of a pipe depends on the material from which the pipe is made, for example, cast iron, concrete, steel, etc. and the method by which the pipe is made, for example, machining, extrusion, casting etc. The nature of the variations is embodied in the form of a single plot called as the Moody diagram, as has been mentioned before. From the plot it can be seen that below Re < 2000, the friction factor is universal, i.e. it is independent of the roughness of the pipe. The value of Reynolds number at which the laminar flow transitions to turbulent flow is called as the critical Reynolds number. Beyond this Reynolds number, the flow becomes turbulent. Once we have turbulent flows, we see that the friction factor tails off as the Reynolds number goes beyond a certain value. Thereafter we say that the fully is fully turbulent; this is marked by the complete turbulence regime in the figure. For the fully developed flow, further increase in the Reynolds number does not yield any further reduction of the friction factor. Therefore, for a fixed relative roughness ratio we have a constant value of the friction factor. The dashed line represents the locus of all points beyond which the flow becomes fully developed for various values of relative pipe roughness. We also see that all the curves behave in a universal manner as the surface is made as smooth as possible. This curve is marked as the smooth pipe curve.

Example 5.4

Find the head loss due to friction when water flows through a 100 m long galvanized iron (GI) pipe, average surface roughness = 0.15 mm, with a diameter of 150 mm with a flow rate of 0.01 m^3/s. The kinematic viscosity of water is $\nu = 1.14 \times 10^{-6}$ m^2/s. Also determine the pumping power required to sustain the flow.

Solution:

Let us first determine the average velocity of the flow

$$V = \frac{Q}{\pi \frac{D^2}{4}} = \frac{0.01}{\frac{\pi}{4} \times 0.15^2} = 0.566 \frac{m}{s} \tag{5.48}$$

The Reynolds number corresponding to this velocity is

$$\text{Re} = \frac{VD}{\nu} = \frac{0.566 \times 0.15}{1.14 \times 10^{-6}} = 74.5 \times 10^3.$$

The relative roughness of the pipe is

$$\frac{\epsilon}{D} = \frac{0.15}{150} = 0.001$$

Looking at the chart for the relative roughness of 0.001 and the Reynolds number found out, we obtain

$$f = 0.025$$

Therefore the corresponding head loss is

$$h_F = f\frac{L}{D}\frac{V^2}{2g} = 0.025 \times \frac{100}{0.15} \times \frac{0.566^2}{2 \times 9.81} = 0.27m$$

The power required to overcome this head loss is

$$P = \rho g h_f \times Q = 10^3 \times 9.81 \times 0.27 \times 0.01 = 26.7W.$$

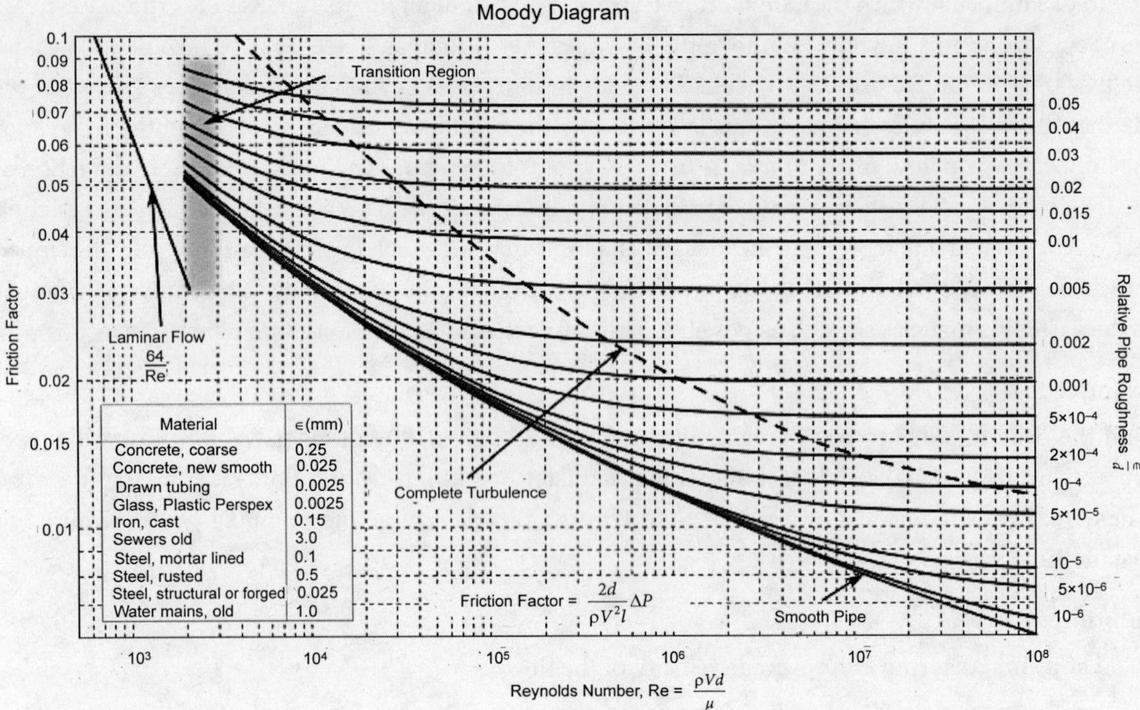

Material	∈(mm)
Concrete, coarse	0.25
Concrete, new smooth	0.025
Drawn tubing	0.0025
Glass, Plastic Perspex	0.0025
Iron, cast	0.15
Sewers old	3.0
Steel, mortar lined	0.1
Steel, rusted	0.5
Steel, structural or forged	0.025
Water mains, old	1.0

Figure 5.15: Moody diagram which shows the friction factor as a function of the Reynolds number for different relative pipe roughness ratios. By Original diagram: S Beck and R Collins, University of Sheffield (Done by the second law at English Wikipedia) Conversion to SVG: Marc.derumaux, CC BY-SA 4.0, https://commons.wikimedia.org/w/index.php?curid=52681200.

5.7 MINOR LOSSES IN PIPE FLOW

While we have considered the major loss in pipe flow due to the friction occurring through viscosity, there are other reasons why there might be a loss of energy of the flow. These losses are incurred due to geometric changes of the pipe encountered by the flow. In cases where the pipe is very long, these changes in geometry often cause a relatively smaller amount of energy loss in the flow as compared to the viscous losses. These losses are therefore termed as minor losses. In many cases, however, it may so happen that the so-called minor losses may be significant as compared to the frictional losses in the case of relatively short pipes. It is also important to note that such minor losses are usually over small regions of the duct, where geometric changes occur. Below, we consider the various causes for minor losses.

5.7.1 Minor Loss due to Sudden Enlargement

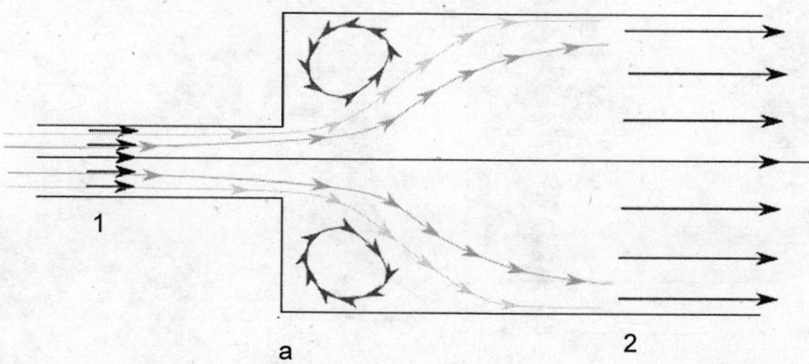

Figure 5.16: Expansion loss in a flow going into a pipe or duct of a larger diameter.

Consider a pipe with a uniform cross section such as that shown in Fig. 5.16. The pipe is then connected to another pipe section which has a larger diameter. The sudden deviation in the pipe cross section area leads to the creation of a region near the wall where turbulent eddies are formed. This is at cross section *a*. These vortical structures lead to a dissipation of the energy of the incoming flow and hence an overall loss in the mechanical energy. Such a kind of loss is fundamentally due to the separation of the flow near the corners and is something which we had seen was avoided in the case of the design of a venturimeter. The nature of the venturimeter was kept in such a way that the flow gradually reaches the throat and gradually again increases in area.

We can analyze this system by considering the transport of momentum between sections 1 and 2.

$$p_1A_1 + p'(A_2 - A_1) - p_2A_2 = \rho Q(V_2 - V_1) \tag{5.49}$$

where p' is the pressure at the turbulent face at the junction of the two pipes which acts over the annualar area, $A_2 - A_1$. The other terms have their usual meaning. From various experiments, it has been seen that the pressure at the recirculation zone is such that $p' = p_1$. This implies that the momentum equation can be simplified to

$$(p_1 - p_2)A_2 = \rho Q(V_2 - V_1) \tag{5.50}$$

The flow rate may be expressed as QV_2A_2. With the help of this we can further simplify the expression as

$$p_1 - p_2 = \rho V_2(V_1 - V_2) \tag{5.51}$$

Now, if we apply Bernoulli's equation between sections 1 and 2, we obtain

$$\frac{p_1}{\rho} + \frac{V_1^2}{2} = \frac{p_2}{\rho} + \frac{V_2^2}{2} + gh_e \tag{5.52}$$

where h_e represents the head loss due to the sudden expansion. Simplifying the expressions we can further simply the expression above using Eq. (5.51) to find out the head loss as

$$h_e = \frac{(V_1 - V_2)^2}{2g} = \frac{V_1^2}{2g}\left(1 - \frac{A_1}{A_2}\right)^2 \tag{5.53}$$

The above expression has been found to be accurate for several experimental and practical situations, and is therefore a good measure of the head loss due to a sudden expansion.

5.7.2 Minor Losses Due to Sudden Exit

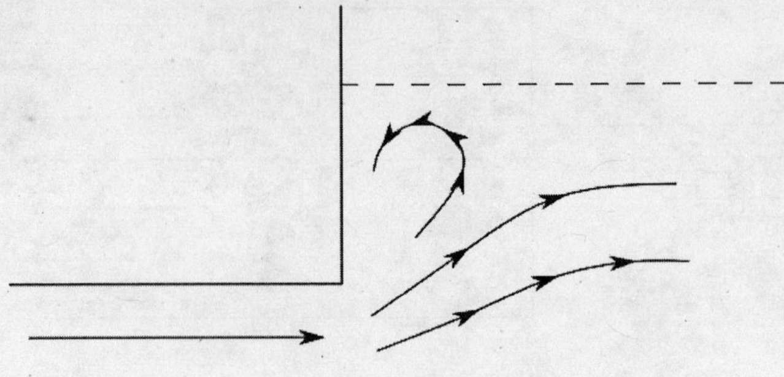

Figure 5.17: Losses due to fluid suddenly exiting into a large reservoir.

Much like the previous case, there may be a situation where the flow through a pipe discharges into a big reservoir so that the ratio $\frac{A_2}{A_1} \gg 1$. The eddy losses uccurs due to the recirculation zone just at the region of the exit. Such a kind of loss is known as exit loss. This is shown in Fig. 5.17. The expression for such a head loss can be easily obtained from the solution of the head loss in the previous section and is

$$h_o = \frac{V_1^2}{2g} \tag{5.54}$$

5.7.3 Minor Losses Due to Sudden Contraction

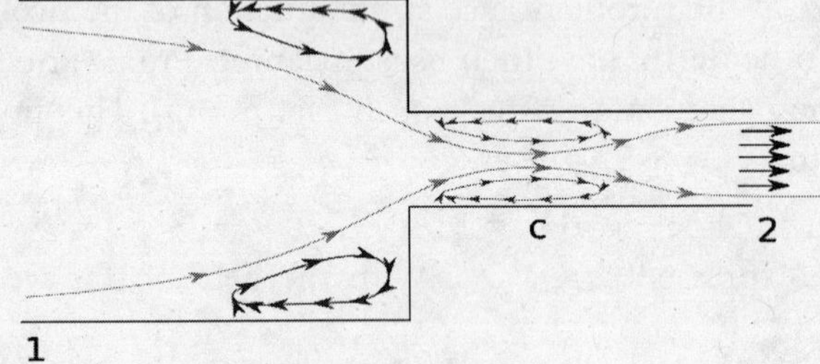

Figure 5.18: Losses due to fluid suddenly entering into a pipe or duct of a smaller diameter.

Similar to the minor loss due to the sudden expansion, there is also a similar situation where an abrupt contraction of the flow area also leads to a loss of energy. The streamlines in the flow are unable to converge enough so that the entire area is covered into the small area. Consequently, there is again a region of recirculation. The actual streamline convergence happens in such a manner that the stream tube, i.e. the bundle of streamlines, goes an area which is smaller than the pipe document. This section is called as the *vena contracta*. Beyond the vena contracta, the streamlines cover the entire area. This is shown in Fig. 5.18. We can proceed with a similar derivation as shown for the flow out to a sudden enlargement and write down the head loss as

$$h_c = \frac{V_2^2}{2g}\left(\frac{A_2}{A_c} - 1\right) \Rightarrow h_c = K\left(\frac{V_2^2}{2g}\right) \tag{5.55}$$

where A_c represents the area of the vena contracta. Typically, the head loss is represented in terms of a loss coefficient K as shown in the expression above.

5.7.4 Entry Loss

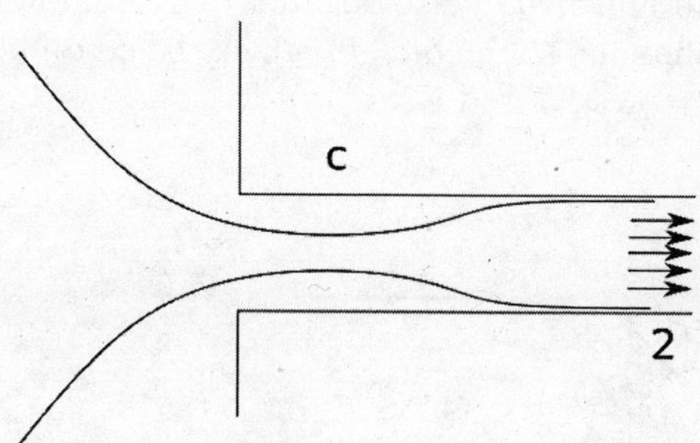

Figure 5.19: Loss due to a fluid entering into a pipe from a large reservoir.

In the above situation, we have a situation where a flow from a relatively large area is suddenly entering into a pipe. Mathematically, this situation is where $A_1 \rightarrow \infty$. This situation is done in Fig. 5.19.

The head loss due to sudden exit is given by

$$h_i = 0.5\frac{V^2}{2g} \tag{5.56}$$

Table 5.1: Loss coefficient for an elbow for various angles.

Elbow angle,	**Loss coefficient K for nominal diameter, D (inches)**			
	inch	**1 inch**	**2 inch**	**4 inch**
45°	0.39	0.32	0.30	0.29
90°	2.0	1.5	0.95	0.64
180°	2.0	1.5	0.95	0.64

Table 5.2: Loss coefficient for tees.

Tees	Loss coefficient K for nominal diameter, D (inches)			
	inch	1 inch	2 inch	4 inch
In-line flow	0.90	0.90	0.90	0.90
Branched flow	2.4	1.8	1.4	1.1

5.7.5 Flow at Elbows and Tees

In pipe networks several pipe connectors like elbows and tees are installed or flow regulating device like valves are also used. The minor head loss, h, is represented by means of the loss coefficient, K, as:

$$h = K\left(\frac{V^2}{2g}\right) \tag{5.57}$$

The numerical value of K is obtained through experiments and reported by the manufacturers in their component data sheet.

An elbow is installed in a pipe-network when fluid flow has to bend in a certain direction by a certain angle. As flow bends sharply through an elbow, there are losses due to the flow separation and eddy formation. The typical value of loss coefficient depends on the angle by which the fluid flow bends in an elbow. This is marked as in Fig. 5.20. It also depends on the elbow diameter. Typical values for 45, 90 and 180 degree elbows for diameter 1/2 inch to 4 inch are listed out in Fig. 5.20 (a). The data for the various losses are mentioned in tables 5.1 and 5.2.

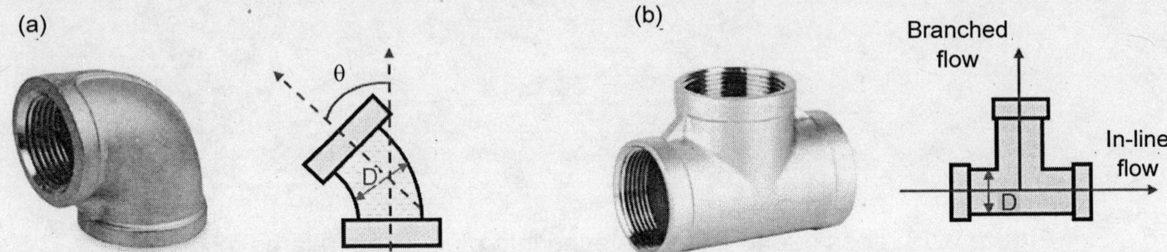

Figure 5.20: Minor loss coefficient for a) Elbows and b) Tee.

Tees are used to branch fluid flow in two parallel paths. There is pressure loss associated with this branching of flow. For tees, the value of depends on the flow direction inside the tee-joint. For straight "in-line" flow, the typical value of is 0.90 whereas for "branched" flow it depends on the pipe diameter. The values are tabulated in 1.20 (b). The numerical values of are significantly higher for branched flow, indicating larger pressure loss when compared to in-line flow. This is due to the fact that fluid bends through 90 degree inside the tee in case of branched flow. Example 5.7 will illustrate the mathematical treatment of flow through tees.

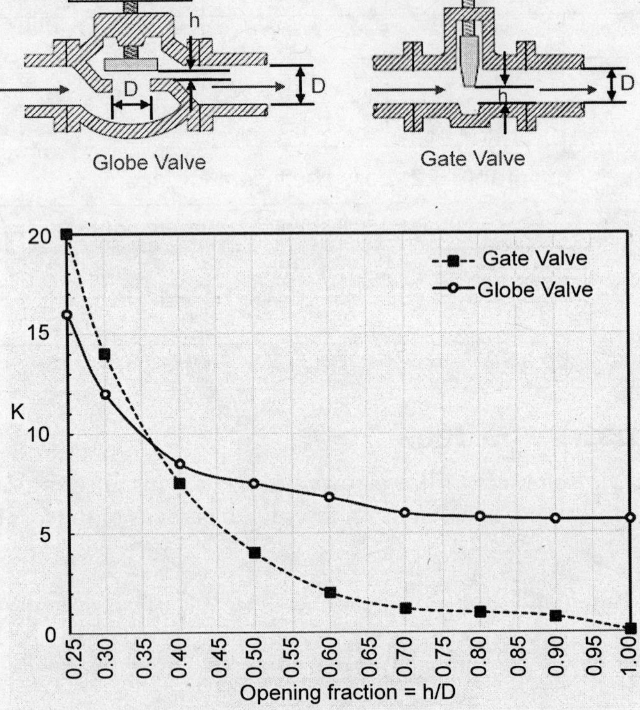

Figure 5.21: Plot of the minor loss coefficient, K, for globe and gate valves for various opening fractions, $\frac{h}{D}$.

Once a pipe-network along with the connectors like elbows and tees is sized and installed, the overall effective resistance offered by the pipe-network becomes constant. However, if a manual control over flow rate is desired in the pipe-network then usually a valve is installed. All valves control the flow rate by increasing or decreasing the opening ratio, $\frac{h}{D}$. Here h is the opening though which fluid flow occurs and D is the pipeline diameter (see Fig. 5.21). Therefore, the opening ratio for a valve can vary from 0 (completely closed) to 1 (completely open). Fig. 5.21 shows the cut-section of a Globe and Gate valve. The principle behind the flow control for both these valves are same; however, internal fluid flow path is different. Hence, it is expected that the loss coefficient K will be significantly different for different valves. A typical variation of K with valve opening ratio for Globe and Gate valve is provided which shows how the loss coefficient for both valves reduces as the valve opening ratio increases with the least value at $\frac{h}{D}=1$. There are many different types of valves available commercially, and the loss coefficient K for a same type of valve can vary from one manufacturer to another.

Example 5.5

Consider a flow of water through a horizontal pipe at a volume flow rate of 0.2 m^3/s. The diameter of the pipe is 200 mm. The pipe suddenly encounters an expansion to a pipe of diameter 400 mm. The pressure in the smaller pipe is 120 kPa. Find (i) The head loss due to the sudden enlargement, (ii) The pressure in the large pipe, (iii) The power lost due to the enlargement

Solution:

Let us first find out the velocity of the flow in the two pipes. As per the information about the flow rate, the velocity in the smaller pipe is

$$V_1 = \frac{Q}{A_1} = \frac{0.2}{\pi 0.2^2/4} = 6.36 m/s.$$

Using the continuity of mass, we obtain the velocity of water in the larger pipe as

$$V_2 = V_1 A_1/A_2 = V_1 (0.2/0.4)^2 = 6.36/4 = 1.59 \text{ m/s}.$$

Using this information we can find out the head loss due to the sudden enlargement as

$$h_e = \frac{(V_1 - V_2)^2}{2g} = (6.36-1.59)^2/(2\times 9.81) = 1.16m.$$

We can now find out the pressure in the larger pipe using Bernoulli's equation between points 1 and 2, which are before and after the expansion. We can therefore write

$$\frac{p_1}{\rho g} + \frac{V_1^2}{2g} + z_1 = \frac{p_2}{\rho g} + \frac{V_2^2}{2g} + z_2 + h_e$$

where we note that since the pipe is horizontal, $z_1 = z_2$. This allows us to find out an expression for p_2 using the values found so far. Substituting the values, we find

$$12.23 + 2.061 = \frac{p_2}{\rho g} + 0.1288 + 1.16 \Rightarrow p_2 = 127.55kPa.$$

The corresponding power lost due to the enlargement is

$$P = \rho g h_e Q = 2275.9 \text{ W} = 2.276 \text{ kW}$$

5.8 PIPES IN SERIES AND PARALLEL

The relationship between the flow rate and pressure drop can be mathematically represented as $\Delta p = Cf(Q)$, where C is some factor which depends on the geometry of the pipe. For a simple case of a two reservoirs connected by a pipe, we can represent the head loss H as

$$\text{H = Entry loss + friction loss in pipe + Exit loss} \tag{5.58}$$

$$H = \frac{1}{2}\frac{V^2}{2g} + f\frac{L}{D}\frac{V^2}{2g} + \frac{V^2}{g} = \left(\frac{3}{2} + f\frac{L}{D}\right)\frac{V^2}{2g} \tag{5.59}$$

The expression above can be easily be converted into a relationship between H and Q by making use of the fact that the flow rate and velocity can be represented by $Q = AV$ thereby allowing us to write

$$H = \left(\frac{3}{2} + f\frac{L}{D}\right)\frac{Q^2}{A^2} = RQ^2 \tag{5.60}$$

The factor represents the flow resistance and helps us to write down an *Ohm's law*-like relationship for the head loss and flow rate in a pipe as

$$H = RQ^2 \tag{5.62}$$

Therefore, this gives us an easy way of analyzing pipe networks with pipes in series and/or parallel. It is obvious to note that for pipes in parallel, the pressure drop across each pipe is the same, i.e. $H = equal$ for all pipes while the flow resistance encountered by the fluid for each pipe dictates how much flow happens in each of the parallel pipes. Similarly, for the case of pipes in series, the flow rate for each of the pipe is fixed while the head loss across each pipe depends on the flow resistance offered by the pipe. These ideas are better explained through the examples below.

Example 5.6

Two pipes of length L each. One has a diameter d and the other one is $\frac{d}{2}$. When they are arranged in series and water flow rate Q, the head loss is H. When they are arranged in parallel with the same flow rate Q the head loss is h. Determine the ratio of $\frac{H}{h}$ neglecting all minor losses.

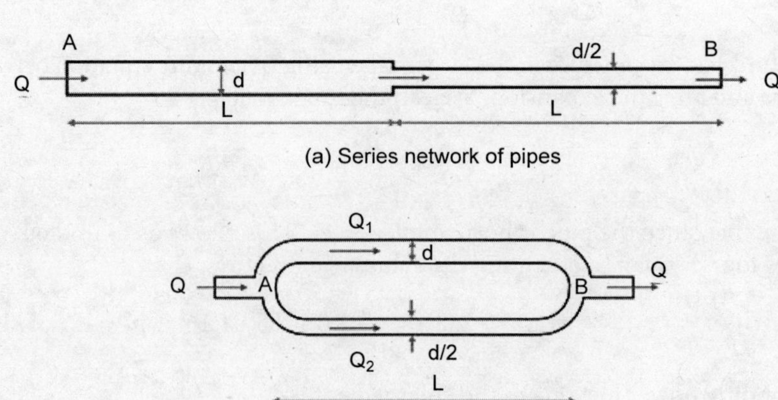

(a) Series network of pipes

(b) Parallel network of pipes

Solution 1:

(a) When pipes are in series the major loss is summation of the individual head loss in first and second pipes. Hence, we can write the expression for head loss between A and B as:

$$\Delta H_{A\rightarrow B,series} = f\frac{L}{d}\left(\frac{V_1^2}{2g}\right) + f\frac{L}{\left(d/_2\right)}\left(\frac{V_2^2}{2g}\right)$$

Where the flow rate is

$$Q = \frac{\pi}{4}d^2V_1 = \frac{\pi}{4}\left(\frac{d}{2}\right)^2 V_2$$

$$\Rightarrow \qquad V_1 = \left(\frac{Q}{\frac{\pi}{4}d^2}\right) and \Rightarrow V_2 = \left(\frac{4Q}{\frac{\pi}{4}d^2}\right)$$

Substituting the above expressions for velocities, we obtain the head loss H for the series arrangement as:

$$H = f\left(\frac{L}{d}\right)\frac{Q^2}{2g\left(\frac{\pi}{4}\right)^2 d^4}\left(1+2\times4^2\right) = 33f\left(\frac{L}{d}\right)\frac{Q^2}{2g\left(\frac{\pi}{4}\right)^2 d^4}$$

(b) When pipes are in parallel arrangement, the head loss is identical across each branch of pipe. Hence, we can write the expression for head loss between A and B as:

$$\Delta H_{A\rightarrow B,parallel} = f\frac{L}{d}\left(\frac{V_1^2}{2g}\right) = f\frac{L}{\left(d/_2\right)}\left(\frac{V_2^2}{2g}\right)$$

$$\Rightarrow \qquad V_1^2 = 2V_2^2 \Rightarrow V_1 = \sqrt{2}V_2$$

Where the total flow rate is

$$Q = Q_1 + Q_2 = \frac{\pi}{4}d^2V_1 + \frac{\pi}{4}\left(\frac{d}{2}\right)^2 V_2$$

Substituting the expression for V_2, we obtain the relationship between the total flow rate Q and velocity V_1 as:

$$Q = \pi d^2 V_1 + \pi\left(\frac{d^2}{4}\right)\left(\frac{V_1}{\sqrt{2}}\right) = \frac{\pi}{4}d^2V_1\left(1+\frac{1}{4\sqrt{2}}\right) = 1.177\frac{\pi}{4}d^2V_1$$

Therefore, we obtain the following relationship for head loss h in case of parallel flow:

$$h = f\left(\frac{L}{d}\right)\frac{Q^2}{2g\left(\frac{\pi}{4}\right)^2 d^4}\left(\frac{1}{1.177^2}\right) = 0.722f\left(\frac{L}{d}\right)\frac{Q^2}{2g\left(\frac{\pi}{4}\right)^2 d^4}$$

The ratio $\frac{H}{h}$ is then

$$H/h = \frac{\left\{33f\left(\frac{L}{d}\right)\frac{Q^2}{2g\left(\frac{\pi}{4}\right)^2 d^4}\right\}}{\left\{0.722f\left(\frac{L}{d}\right)\frac{Q^2}{2g\left(\frac{\pi}{4}\right)^2 d^4}\right\}} = 45.7$$

Example 5.7

Consider the pipeline assembly shown in the figure below. Water at 50 *lpm* flows into a tee joint where it gets "branched" into loop 1. Flow through loop 2 is "in-line" with the main incoming flow at point A. The loop-1 has three straight pipelines, each of diameter 12.5 mm and length 10 m, whereas loop-2 has only one straight pipeline of diameter 12.5 mm and length 10 m. Furthermore, the loop-1 has two identical 90 elbows of 12.5 mm each having a head loss coefficient of $K = 2$. The head loss coefficient for each tee joint in the "branched" line is $K = 2.4$ whereas for the "in-line" flow it is $K = 0.9$. Assume that the friction factor for all four straight pipes is $f = 0.005$. What is the water flow rate (lpm) through loop-1 and loop-2?

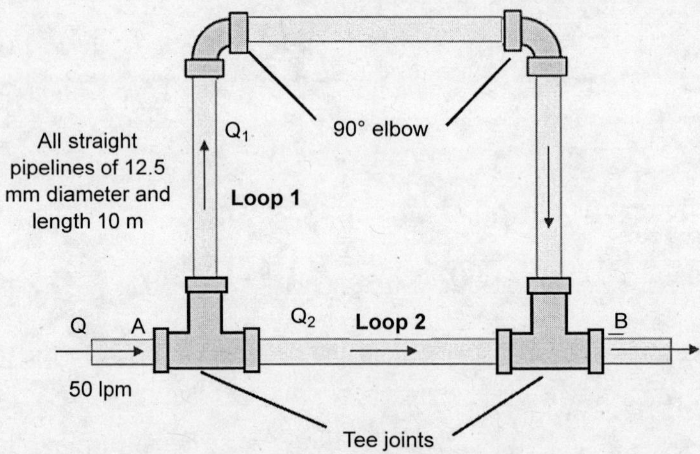

Solution 2:

Loop-1 calculation:

The major loss is given by the frictional losses through the 3 straight pipes:

$$H_{major,loop-1} = f\left(\frac{3l}{d}\right)\left(\frac{V_1^2}{2g}\right) = 12\left(\frac{V_1^2}{2g}\right)$$

Minor loss is given by the summation of losses through the 2 tee "branched" junctions and 2 elbows:

$$H_{minor,loop-1} = 2K_{tee,branched}\left(\frac{V_1^2}{2g}\right) + 2K_{elbow}\left(\frac{V_1^2}{2g}\right) = 2\times 2.4\left(\frac{V_1^2}{2g}\right) + 2\times 2\left(\frac{V_1^2}{2g}\right) = 8.8\left(\frac{V_1^2}{2g}\right)$$

Total head loss in loop-1 is:

$$H_{total,loop-1} = 12\left(\frac{V_1^2}{2g}\right) + 8.8\left(\frac{V_1^2}{2g}\right) = 20.8\left(\frac{V_1^2}{2g}\right)$$

Loop-2 calculation:

The major loss is given by the frictional losses through the 1 straight pipe:

$$H_{major,loop-2} = f\left(\frac{l}{d}\right)\left(\frac{V_2^2}{2g}\right) = 4\left(\frac{V_2^2}{2g}\right)$$

Minor loss is given by the summation of losses through the 2 tee junctions having in-line flow:

$$H_{minor,loop-2} = 2K_{tee,in-line}\left(\frac{V_2^2}{2g}\right) = 2\times 0.9\left(\frac{V_2^2}{2g}\right) = 1.8\left(\frac{V_2^2}{2g}\right)$$

Total head loss in loop-2 is:

$$H_{total,loop-1} = 4\left(\frac{V_2^2}{2g}\right) + 1.8\left(\frac{V_2^2}{2g}\right) = 5.8\left(\frac{V_2^2}{2g}\right)$$

As loop-1 and loop-2 are parallel, the head loss between A to B is same across the loops. Therefore, we have:

$$H_{total,loop-1} = H_{total,loop-2} \Rightarrow 20.8\left(\frac{V_1^2}{2g}\right) = 5.8\left(\frac{V_2^2}{2g}\right)$$

$\Rightarrow$ $V_2 = 1.894\ V_1$

$$Q = \frac{50\times 10^{-3}}{60} = \frac{\pi d^2}{4}(V_1 + V_2) = 0.7235\pi d^2 V_1$$

$\Rightarrow$ $0.833 \times 10^{-3} = 0.000355 V_1$

$\Rightarrow$ $V_1 = 2.36$ m/s

$\Rightarrow$ $V_2 = 4.44$ m/s

Hence, flow rate through the individual loops are:

$\Rightarrow$ $$Q_1 = \frac{\pi d_1^2}{4} V_1 = 17.3\ lpm$$

$\Rightarrow$ $$Q_2 = \frac{\pi d_1^2}{4} V_2 = 32.7\ lpm$$

Example 5.8

Consider the previous problem with an electronic water flowmeter placed in the straight pipeline of loop-2 as shown in the figure below. This flowmeter offers minor head loss coefficient $K = 2$. Assume that the flow meter is extremely small and the straight pipeline length on which it is installed is still 10 m in length. Determine the flow rate measured by this flowmeter and compare it with the flow rate (lpm) predicted in the previous problem for loop-2. How much is the difference (in lpm)?

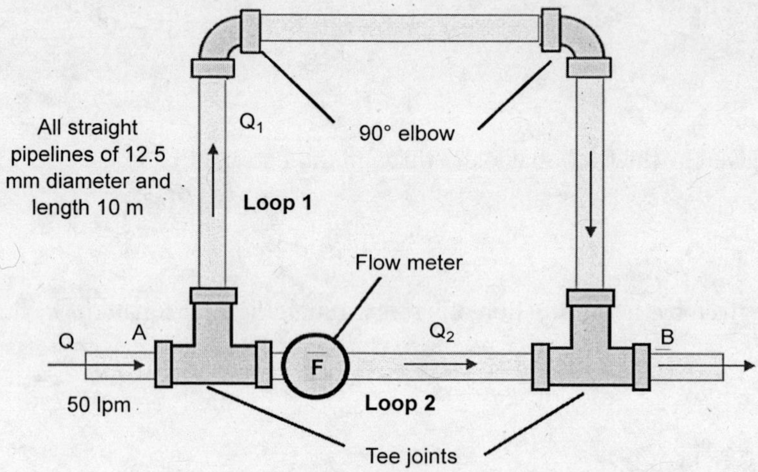

Solution:

Loop-1 calculation:

The entire calculation remains identical to the previous problem as there is no change in this loop arrangement.

Loop-2 calculation:

The major loss through loop-2 remains identical to the previous problem, only the minor head loss calculation changes. Minor loss in this problem is given by the summation of losses through the 2 tee junctions having in-line flow and the flow meter installed in this loop:

$$H_{minor,loop-2} = 2K_{tee,in-line}\left(\frac{V_2^2}{2g}\right) + K_{flowmeter}\left(\frac{V_2^2}{2g}\right) = 2\times 0.9\left(\frac{V_2^2}{2g}\right) + 2\left(\frac{V_2^2}{2g}\right) = 3.8\left(\frac{V_2^2}{2g}\right)$$

Total head loss in loop-2 is:

$$H_{total,loop-1} = 4\left(\frac{V_2^2}{2g}\right) + 3.8\left(\frac{V_2^2}{2g}\right) = 7.8\left(\frac{V_2^2}{2g}\right)$$

As loop-1 and loop-2 are parallel, the head loss between A to B is same across the loops. Therefore, we have:

$$H_{total,loop-1} = H_{total,loop-2} \Rightarrow 20.8\left(\frac{V_1^2}{2g}\right) = 7.8\left(\frac{V_2^2}{2g}\right)$$

$\Rightarrow \qquad V_2 = 1.633V_1$

$$Q = \frac{50\times10^{-3}}{60} = \frac{\pi d^2}{4}(V_1 + V_2) = 0.658\pi d^2 V_1$$

$\Rightarrow \qquad 0.833\times10^{-3} = 0.000323V_1$

$\Rightarrow \qquad V_1 = 2.6 \text{ m/s}$

$\Rightarrow \qquad V_2 = 4.2 \text{ m/s}$

Hence, flow rate through the individual loops are:

$$Q_1 = \frac{\pi d_1^{\,2}}{4}V_1 = 19.1\ lpm$$

$$Q_2 = \frac{\pi d_1^{\,2}}{4}V_2 = 30.9\ lpm$$

The difference between flow rate measured by the electronic flowmeter and the actual flowrate without the flowmeter is 32.7 – 30.8 = 1.9 lpm. This reduction in flowrate is due to the additional resistance to fluid flow offered by the electronic flowmeter. Unfortunately, this is unavoidable. (This phenomenon is very similar to drop in electric current observed when ammeters are used.)

5.9 WATER HAMMER: CAUSES AND REMEDIES

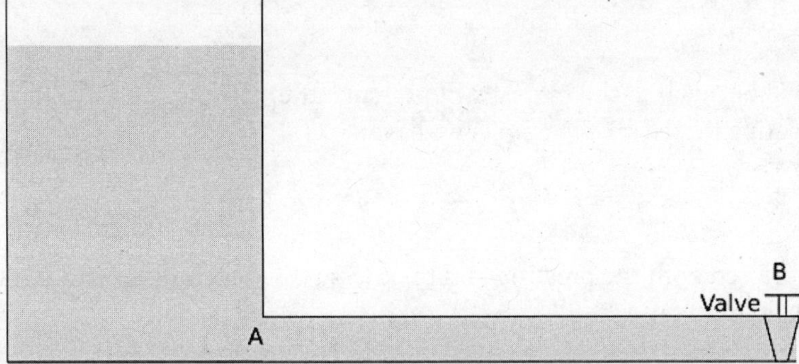

Figure 5.22: A schematic of a pipe discharging to the ambient and fed by a large reservoir.

Let us consider a long pipe as shown in Fig. 5.22. One end of this pipe, *A*, is connected to a large reservoir while the other end *B* is open to discharge to the atmosphere via a valve. The length of the pipe is *L*. Initially the valve is completely open and water flows through it with a velocity *V*. If the valve is now suddenly closed, the inertia of the flowing water is suddenly lost and therefore the pipe has to make a suddent transition from being at the discharge pressure to the pressure which is suddenly built up. The sudden change in pressure leads to a pressure wave which is transmitted along the pipe with the speed of sound in the medium, in this case, the fluid carried by the pipe. This may additionally create a knocking sound in the piping. The travelling high pressure wave leads to a hammering on the walls of the pipe due to the sudden change in pressure and is known as water hammer.

The pressure rise during the phenomenon of water hammer depends on the

- velocity of the flow before the valve is shut
- length of the pipe
- speed at which the valve is closed
- properties of the pipe such as the Young's modulus

The last property is particularly important from an engineering viewpoint and from a bio-engineering viewpoint. Having thin walled flexible pipes can often lead to a dampened pressure wave. Moreover, the flexibility of the arteries and veins in the body also lead to a dampening of the pressure wave when the valves of the heart open and close suddenly.

Let us now consider the different cases of water hammer.

5.9.1 Valve is Gradually Closed

Consider the same pipe shown above. Let the time required to close the valve be T. Let the intensity of the pressure wave be p. The mass of liquid held in the pipe at any time instant is

$$m = \rho AL \tag{5.63}$$

If the valve is gradually closed, the deceleration faced by the mass of liquid is

$$|a| = \frac{V}{T} \tag{5.64}$$

The force experienced is therefore

$$F = m|a| = \rho AL\frac{V}{T} \tag{5.65}$$

This force would be equal to the intensity of pressure after closing the valve times the area of cross section, i.e. pA. Equating these two expessions we have

$$p = \frac{\rho LV}{T} \Rightarrow h = \frac{p}{\rho g} = \frac{LV}{gT} \tag{5.66}$$

i.e. we have expressed both the pressure and the pressure in terms of the head. If C is the velocity of sound in the liquid, then we say that the valve closure is sudden when

$$T < \frac{2L}{C} \tag{5.67}$$

i.e. the time required to close the valve is faster the time required by the pressure wave to move up and down the pipe, i.e. across a length, $2L$. Alternately, the valve closure is said to be gradual when

$$T > \frac{2L}{C} \tag{5.68}$$

5.9.2 Pipe is Rigid and Valve is Suddenly Closed

Given that most fluids are compressible, it is only natural that the bulk modulus of water plays an important role in the transmission of the wave. Let us analyze the situation for a rigid pipe. The loss of kinetic energy when the fluid is brought to rest is

$$\Delta KE = \frac{1}{2} \times \rho AL \times V^2 \tag{5.69}$$

which is equal to the gain in the strain energy of the fluid, i.e. the strain energy per unit volume, $\frac{1}{2}\frac{p^2}{K}$ times the volume of the fluid, i.e. AL.

$$\Delta SE = \frac{1}{2}\left(\frac{p^2}{K}\right) \times AL \tag{5.70}$$

When the two energies are equated, we obtain an expression for the pressure as

$$p = \rho V\sqrt{\frac{K}{\rho}} = \rho VC, \quad \because C = \sqrt{\frac{K}{\rho}} \left(By\ definition\ of\ speed\ of\ sound\right) \tag{5.71}$$

The case where the pipe is elastic and the valve is suddenly closed is beyond the scope of this book.

UNIT SUMMARY

- Actual flow rate through a venturimeter

$$Q_{actual} = C_d Q = C_d \frac{A_1 A_2}{\sqrt{A_1{}^2 - A_2{}^2}} \left(\sqrt{2g(\frac{\rho_m}{\rho} - 1)\Delta h}\right)$$

- Coefficient of contraction

$$C_c = \frac{A_c}{A_0} = \frac{Area\ of\ vena\ contracta}{Area\ of\ the\ orifice}$$

- Volumetric flow rate through a orificemeter

$$Q = A_c V_c = C_c A_0 C_v \sqrt{\frac{2g\left(\frac{\rho_m}{\rho} - 1\right)\Delta h}{\left(1 - \frac{A_c^2}{A_1^2}\right)}} = C_v C_c A_0 \sqrt{\frac{2g}{1 - \frac{C_c^2 A_0^2}{A_1^2}}} \sqrt{\left(\frac{\rho_m}{\rho} - 1\right)\Delta h}$$

- Reynolds number

$$\mathrm{Re} = \frac{\rho UL}{\mu}$$

- Hagen-Poiseuille flow, velocity distribution

$$u = -\frac{R^2}{4\mu}\frac{dp}{dx}\left(1 - \frac{r^2}{R^2}\right)$$

- Pressure drop in a constant cross section circular pipe

$$\Delta p = \frac{8\mu LQ}{\pi R^4}$$

- Frictional head loss

$$h_f = f\frac{L}{D}\frac{V^2}{2g} = \frac{8\mu LQ}{\pi R^4}$$

- Friction factor for laminar flow

$$f = \frac{64}{\text{Re}_d}$$

- head loss due to sudden expansion

$$h_e = \frac{(V_1 - V_2)^2}{2g} = \frac{V_1^2}{2g}\left(1 - \frac{A_1}{A_2}\right)^2$$

- Head loss due to sudden contraction

$$h_c = \frac{V_2^2}{2g}\left(\frac{A_2}{A_c} - 1\right) \Rightarrow h_c = K\left(\frac{V_2^2}{2g}\right)$$

- Head loss due to sudden exit

$$h_i = 0.5\frac{V^2}{2g}$$

EXERCISES

Multiple Choice Questions

1. For a laminar flow of water through a straight circular pipe, keeping all other parameters constant, the major head loss (h_f)) is:
   (a) Independent of velocity of water inside the pipe
   (b) Varies linearly with velocity of water inside the pipe
   (c) Varies as V_2 where V is the velocity of water inside the pipe
   (d) Cannot say
2. Consider any liquid with density ρ flowing with velocity V inside a straight circular pipe. Suppose we change the liquid and measure the major head loss (h) for each case. Assuming laminar flow for each case and assuming all other parameters being constant, the major head loss (h_f) is:
   (a) Independent of ρ
   (b) Varies linearly with ρ
   (c) Varies as ρ^2
   (d) Varies as $\frac{1}{\rho}$

3. For a siphon, cavitation of liquid may happen:
(a) At the inlet of the siphon
(b) At the outlet of the siphon
(c) At the topmost point in siphon
(d) Cavitation is not possible inside a siphon

4. A venturimeter works on the principle of
(a) Conservation of static pressure head
(b) Conservation of dynamic pressure head
(c) Conservation of piezometric pressure head
(d) Conservation of total pressure head

5. Which of the following is/are TRUE?
(a) In an orificemeter, the velocity is maximum at the orifice
(b) In a venturimeter, the converging section cone angle is higher than the diverging section
(c) Static pressure at any point in flow field is greater than or equal to stagnation pressure
(d) Stagnation pressure at any point in flow field is greater than or equal to static pressure

6. Two pipelines of diameter D and d are attached in series where $D > d$. Water first enters the pipeline of diameter D and then passes through a sudden contraction to enter into pipeline of diameter d. The pressure at the inlet and outlet of this arrangement are P_1, P_2 respectively. Similarly, the velocity of water at inlet and outlet are V_1, V_2 respectively. Taking all the pressure head losses into account, which of the following is/are TRUE?
(a) $P_1 > P_2$
(b) $P_1 < P_2$
(c) Cannot comment on relationship between static pressures
(d) $V_1 > V_2$
(e) $V_1 < V_2$

7. Two pipelines of diameter D and d are attached in series where $D > d$. Water first enters the pipeline of diameter D and then passes through a sudden contraction to enter into pipeline of diameter d. The pressure at the inlet and outlet of this arrangement are P_1, P_2 respectively. Similarly, the velocity of water at inlet and outlet are V_1, V_2 respectively. Taking all the pressure head losses into account, which of the following is/are TRUE?
(a) $P_1 > P_2$
(b) $P_1 < P_2$
(c) Cannot comment on relationship between static pressures
(d) $V_1 > V_2$
(e) $V_1 < V_2$

8. Water hammer effect is severe in which of the following case(s)?
(a) Flexible pipelines
(b) Rigid pipelines
(c) Liquid flow
(d) Gaseous flow

ANSWERS

1.	(a)	2.	(d)	3.	(c)	4.	(d)	5.	(b), (d)	6.	(a), (e)	7.	(c), (d)	8.	(b), (c)

Unsolved Questions

Level - I

1. Water pass through a pipe of 12 mm internal diameter at a velocity of 1 m/s and then it enters a pipe of 25 mm diameter. What is the velocity of water in this pipe?
2. Water flows through a tube of diameter 10 mm with velocity of 0.1 m/s. What is fluid flow regime? For this flow regime determine the pressure drop (Pa) for a pipe length of 10 m? (Assume water density as 1000 kg/m^3 and dynamic viscosity as 9×10^{-4} Pa.s.)
3. For a hydraulic siphon shown in Fig. 5.1, determine the water discharge rate (lpm) from the siphon if the point B is $\Delta z = 1$ m below the free surface A. The siphon tube diameter is 5 mm. (Assume acceleration due to gravity as 9.81 m/s^2).
4. Water flows through a horizontal venturimeter of inlet pipe diameter 400 mm and throat diameter of 200 mm with a discharge coefficient of 0.96. Gauge liquid in the differential manometer shows a height difference of 0.1 m. Specific gravity of manometric liquid is 13.6. What is the flow rate of water (m^3/s)?
5. A pitot static tube is placed on the nose of a racing car to estimate its speed. The probe measures the difference in stagnation and static pressure as 1500 Pa. Estimate the speed of the car (in m/s). (Assume air density as 1.2 kg/m^3).
6. Air at 0.3 kg/s passes through a straight circular duct of diameter 250 mm and length 25 m. Estimate the pressure drop (Pa) due to friction for a smooth pipe. The friction factor can be estimated by: $f = \dfrac{0.079}{Re^{0.25}}$. (Assume air density as 1.2 kg/m^3 and dynamic viscosity as 2×10^{-5} Pa.s.)
7. Water flowing through a short pipe at 1 m/s suddenly expands into another short pipe having twice the diameter. The pressure at the exit of the larger diameter pipe is 100 kPa. Assume the minor head loss coefficient for sudden expansion is $K = 0.5$. Neglecting major losses, what is the pressure drop (Pa)?
8. Water flowing through a pipe of 25 mm diameter, 5 m length and having friction factor $f = 0.003$ encounters a sudden contraction to another tube of 12.5 mm diameter tube and 5 m in length. Assume identical friction factor for both pipes. The velocity of water in the larger pipe is 0.5 m/s and the minor head loss coefficient due to sudden contraction is $K = 0.4$. What is the total pressure drop (Pa) through the entire pipe assembly of 10 m?

Level - II

1. Redo the problem no. 3 in above section (Level I) also taking into account the frictional pressure drop inside siphon tube. The siphon tube length is 1.5 m and the friction factor is $f = 0.004$. Neglect any minor losses.
2. Water flows through a pipeline of 25 mm at a velocity of 1 m/s. An orifice plate with an orifice diameter of 12.5 mm (Fig. 5.4) and a coefficient of contraction $C_c = 0.7$ is placed in this pipeline. Estimate the water velocity at the vena contracta? If coefficient of viscosity is 0.9 what is the height difference (mm) between the manometer limbs if the manometric fluid is mercury with specific gravity = 13.6?

3. A large overhead reservoir with water at height H (from ground level) is used to supply water to a far off reservoir on the ground level. The diameter of supply pipe is D and length is L with a friction factor f. After calculation it is observed that the estimated flow rate delivered by this arrangement is Q, whereas the desired flow rate is $2Q$. In order to attain the flow rate $2Q$, there are two proposals. Proposal 1: increase the height of the overhead tank keeping the pipeline diameter unchanged; Proposal 2: change the pipeline diameter without changing the overhead tank height (assume the friction factor f of pipe remains unchanged and neglect minor losses). Determine the new height of overhead tank for proposal 1 and new pipeline diameter for proposal 2.
4. Air flows through a circular duct of diameter D_1 and length L. In another scenario, water flows through a circular pipe of diameter D_2 and length L . Assume that the mass flow rate of air and water is identical, and the pipe friction factor f for both flows are also same. Determine the ratio of diameter $\frac{D_1}{D_2}$, for same head loss across the circular duct and pipe. (Neglect minor losses and density of air is 1.2 kg/m³)
5. Water flows through two loops (loop-1 and loop-2) as shown in the figure below. Determine the head loss coefficient offered by the "Globe valve" placed in loop-2 such that flow rate ratio $\frac{Q_1}{Q_2} = 0.6$.

   Assume that all straight pipeline lengths are 10 m long and diameter 25 mm. The friction factor offered by each of these straight pipes is $K = 2.5$. The head loss coefficient for each elbow is . The loss coefficient for tee joints in loop-1 (branched flow) is 2.4 while in loop-2 (in-line flow) is 0.9.

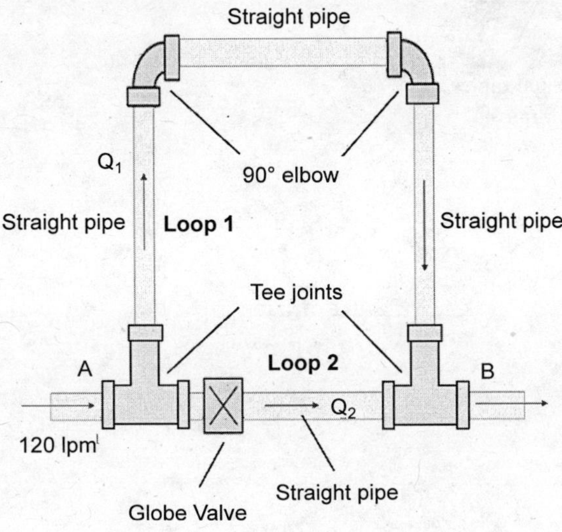

6. Water flows from tank 1 to tank 2 through an entire piping layout as shown in figure below. Tank 1 has water level at a height of 36 m from ground level. The minor loss coefficients associated with various bends and connectors are provided in the table. The flow rate of water through the pipeline is 300 lpm. Pipe roughness ratio $\frac{\in}{D} = 0.001$ and the friction factor for straight pipe has

to be obtained from the relation $f = \dfrac{0.25}{\left[\ln\left(\dfrac{\varepsilon/D}{3.7} + \dfrac{5.4}{\mathrm{Re}^{0.9}}\right)\right]^2}$. What is the height of water in tank 2?

Determine total power lost (W) in this process? (Assume dynamic viscosity of water as 9.0 10-4 Pa.s)

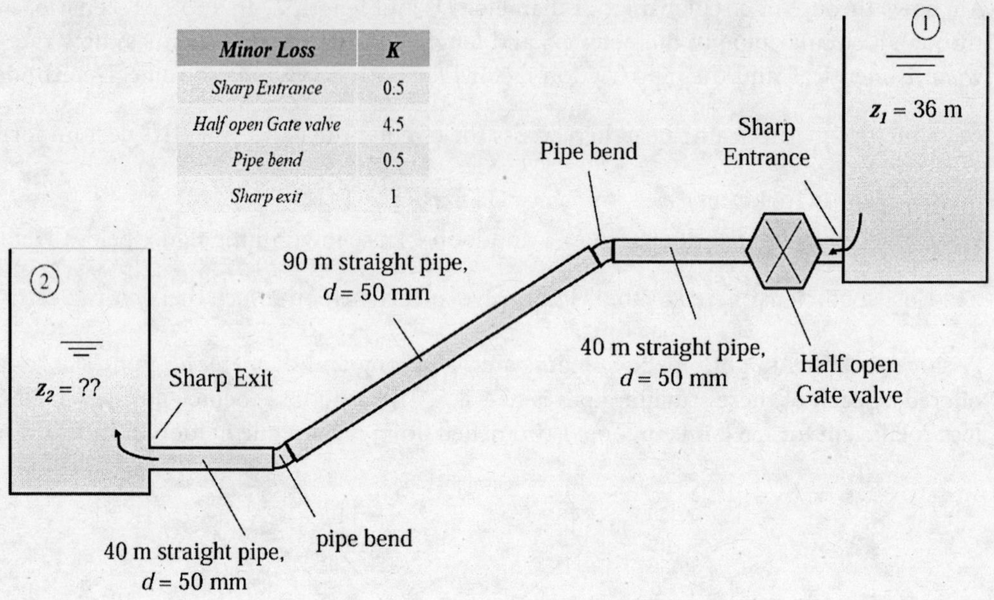

Answer Key

Level - I

1. 0.23 m/s
2. Laminar
3. 5.2 lpm
4. 0.161 m³/s
5. 50 m/s
6. 7.4 Pa
7. 250 Pa
8. 4075 Pa

LEVEL - II

1. 3.5 lpm
2. 146.2 mm
3. Proposal 1: 4 *H*, Proposal 2: 1.32 *D*
4. 3.84
5. 1.824
6. z_2=29.1 m, Power = 338 W

PRACTICAL

Aim

Understanding pressure loss due to a sudden obstruction in a pipe carrying water.

Apparatus

Tap supply from overhead tank, PVC pipe 1" diameter, gate valve, bucket, and stopwatch.

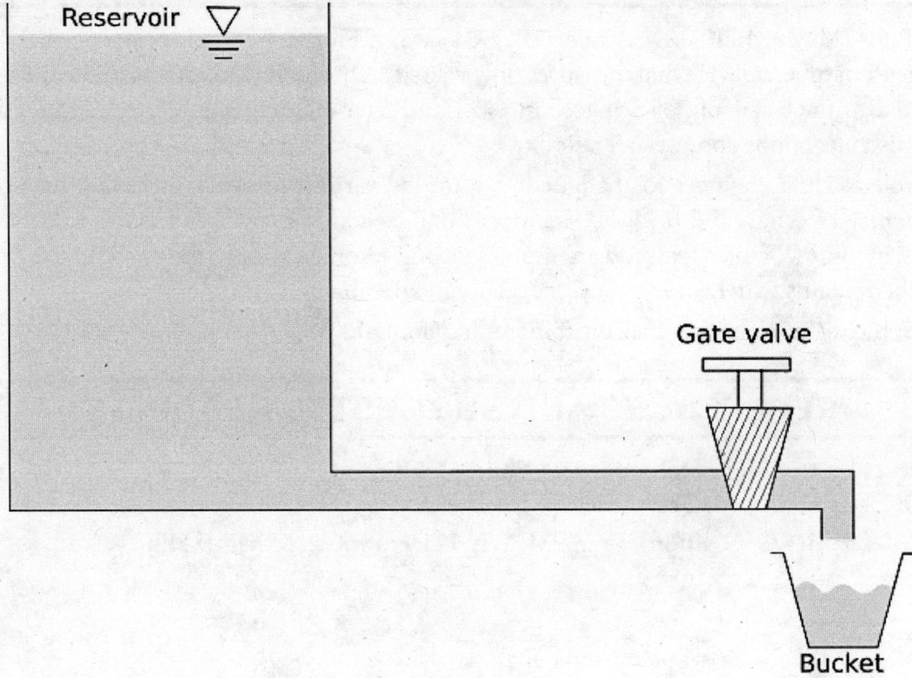

Figure 5.23: A simplified schematic of the practical.

Theory

In the particular experiment, we will measure the different flow rates through the pipe as the pipe obstruction is increased across successive experiments. The fundamental idea is that there is a source of water that we have which is initially at a large height (say the top of the house). This is the total head at that location. Now, water enters into the pipe through the tap and moves through the pipe. If now the gate valve is fully opened, it will experience no loss in head. If the gate valve is kept nearly closed, then there will be a significant loss in head and thus, the flow rate obtained will be poor. The loss due to the sudden obstruction in the flow has been discussed in the Unit and is tantamount to considering a sudden contraction and a sudden expansion; both of these losses depend on V^2. A simplified schematic is shown in Fig. 5.23. The student performing the experiment can, however, simply use the supply line from a tap.

Procedure

1. Open the gate valve fully and the incoming tap fully
2. Measure the time required for the bucket to fill completely.

3. Turn the gate valve so that the area reduces slightly.
4. Measure the time required for the bucket to fill completely.
5. Repeat this until the gate valve is now completely closed.

The plot of the flow rate vs gate closure is equivalent to the plot of the velocity vs blockage area and can be used to infer the amount of head loss as found out from the theoretical predictions.

KNOW MORE

Henri Pitot (May 3, 1695 – December 27, 1771) was a French hydraulic engineer best known for inventing the Pitot tube, a device that measures flow velocity. Originally a mathematician and astronomer, he was elected to the Academy of Sciences in 1724 and later studied water flow in rivers and canals, correcting misconceptions about water velocity.

In his role as chief engineer for Languedoc, Pitot led various projects, including the construction and maintenance of canals and bridges. His most notable achievement was the Montpellier aqueduct, built from 1753 to 1786, which featured an impressive one-kilometer-long stone arch in the Roman style. The Pitot tube remains widely used today, especially in anemometers for measuring wind speed.

[Source: https://www.britannica.com/biography/Henri-Pitot]

REFERENCES AND SUGGESTED READINGS

1. **Introduction to Fluid Mechanics and Fluid Machines** (3rd Edition), S K Som, Gautam Biswas, Suman Chakraborty, McGraw Hill
2. **Fluid Mechanics** (9th Edition), Frank M. White, Henry Xue, McGraw Hill

❑❑

6 Engineering Applications 2: Open Channel Flows

UNIT SPECIFICS

In this unit we will discuss about the following topics:

1. Applications of Bernoulli's theorem
2. Concept of laminar and turbulent flow
3. Losses in flow through pipes
4. Concept of friction factor

RATIONALE

Often in several natural systems and engineering systems, we encounter flows where it is driven by the slope of the surface and moreover, the flow bears a free surface. This could be a flow of a river or transport of a chemical in a partially full pipe. In such flows, the pressure gradient at the interface between the fluid and air is negligible and the flow is entirely driven by gravity. Therefore, such flows, termed as open-channel flows, are immensely useful for civil, mechanical, and environmental engineers. In general such flows can be quite complex owing to the large dimensions and three dimensional nature. However, several theories have been devised in order to make these problems tractable and helps us to obtain engineering estimates and correlations for flows through such open channels. Basic ideas of friction loss, hydraulic radius, and head loss are vital towards making sense of the theory discussed in this particular unit.

PRE-REQUISITES

Basic calculus.

UNIT OUTCOMES

U6-O1: Flow in open channel

U6-O2: Chézy formula for uniform flow

U6-O3: Most economical open flow channel

Unit-6 Outcomes	EXPECTED MAPPING WITH PROGRAMME OUTCOMES (1 - Weak Correlation; 2 - Medium Correlation; 3 - Strong Correlation)						
	CO-1	CO-2	CO-3	CO-4	CO-5	CO-6	CO-7
U6-O1	-	-	-	2	-	3	2
U6-O2	-	-	-	2	-	3	2
U6-O3	-	-	-	2	-	3	2

6.1 FLOW CLASSIFICATION

While in the previous Unit we have analyzed flows through closed conduits, we shift our attention to flows in open channels, i.e. flows with a free surface. The free surface, by virtue of being exposed to the atmosphere, is at the atmospheric pressure. From an engineering perspective, flow of water in open channels is by far the most important applications of such kinds of flows. Such flows can be seen in canals, rivers etc. The flow in such cases takes place primarily due to the slope of the bed of the channel and due to the action of gravity. The observant reader will note that the hydraulic grade line will coincide with the free surface of the water for open channel flows.

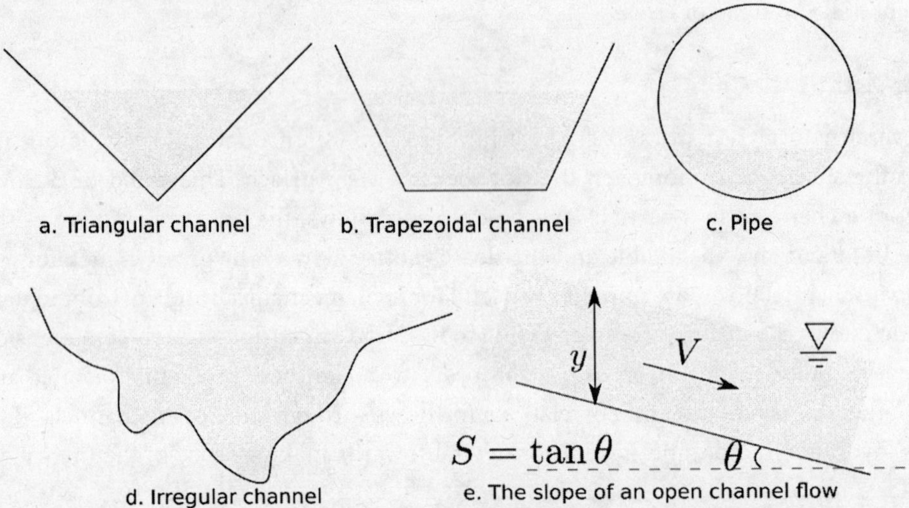

Figure 6.1: (a-d) Various cross sections of commonly encountered open channel flows. (e) The slope of the open channel flow and depth is shown in the figure.

An open channel flow has at least two sides and one bottom bed, barring a circular ditch or a partially filled pipe. At all these surfaces, the flow satisfies the no-slip boundary condition owing to the presence of viscosity. Some typical cross sections are shown in Fig. 6.1.

The flows in open channels can be classified into the following

5. Steady and unsteady flow
6. Uniform and non-uniform flow
7. Critical, sub-critical, and supercritical flow

In the study of open channel flows, the important flow variables that we will deal with are the velocity, V, the flow rate, Q, and the depth of the flow, which we will denote with the coordinate, y. The expression for the flow rate along the channel may be evaluated at any cross section at a bed location, x, through

$$Q = V(x)A(x) = \text{constant} \tag{6.1}$$

The reason the flow rate remains constant at all axial locations is because of the continuity equation, which is essentially the conservation of mass.

We can naturally apply Bernoulli's theorem between two points in the flow, say points 1 and 2, which are upstream and downstream respectively. They are related through

$$\frac{V_1^2}{2g} + z_1 = \frac{V_2^2}{2g} + z_2 + h_F \tag{6.2}$$

where, h_F represents the head loss due to friction. Note that we have not written down the pressure at these two axial locations because the pressures at those sections are equal to the atmospheric pressure, i.e. $p_1 = p_2 = p_a$. We may apply a friction factor based expression for the head loss and write down

$$h_F = f\left(\frac{x_2 - x_1}{D_H}\right)\frac{V_a^2}{2g} \tag{6.3}$$

where V_a represents the average velocity of the flow between the two cross sections. The parameter D_H appearing in the expression above represents the hydraulic diameter and is given by

$$D_H = \frac{4A}{P} \qquad (6.4))$$

where A represents the average cross section between 1 and 2 and P represents the *wetted perimeter*.

6.1.1 Steady and unsteady flow

A flow is termed as a steady flow when the velocity, flow rate, and the death of the flow do not change in time, i.e.

$$\frac{\partial V}{\partial t} = 0 \quad \frac{\partial Q}{\partial t} = 0 \quad \frac{\partial y}{\partial t} = 0 \tag{6.5}$$

Flows where the above conditions are not true are known as unsteady flows.

6.1.2 Uniform Flow and Non-uniform Flow

Uniform flows are those where the velocity and depth of the flow at a given cross section remain constant over the entire cross section. Contrary to this, non-uniform flows are characterized by velocity and depth of the flow to be spatially non-uniform at a certain cross section.

Non-uniform flows and depth variation

In non-uniform flows, which are also called as *varied flows,* classification can be done as

1. Gradually varied flow
2. Rapidly varied flow

It has been observed that uniform flows are often separated from rapidly varied flows by gradually varied flows. Rapidly varied flows are those where the depth of the flow changes suddenly over a small length of the flow. This can be caused by the presence of an obstacle in an open-channel flow. A gradually varied flow does not have such rapid variations.

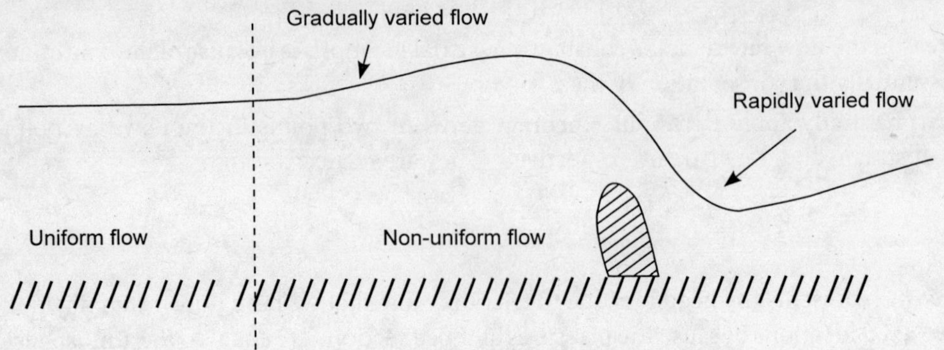

Figure 6.2: Schematic of a uniform and non-uniform flow. The non-uniform flows are also classified into gradually and rapidly varied flows.

6.1.3 Critical Flows - Froude Number

Froude number is defined as $Fr = \frac{V}{\sqrt{gy}}$. Physically, the number represents the ratio of the velocity of the flow to the velocity of the shallow water surface wave. This number helps to classify the flow further as subcritical flow, critical flow, and supercritical flow, i.e.

$$\text{Fr} < 1 \ldots\text{subcritical flow} \tag{6.6}$$

$$\text{Fr} = 1\ldots\text{critical flow} \tag{6.7}$$

$$\text{Fr} > 1.0\ldots\text{supercritical flow} \tag{6.8}$$

6.2 CHÉZY FORMULA FOR UNIFORM FLOW

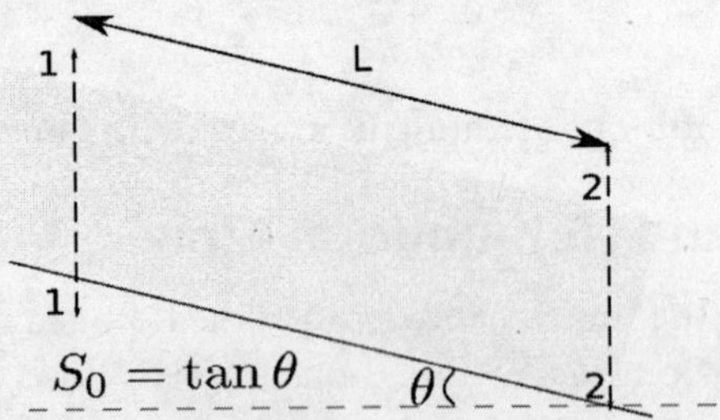

Figure 6.3: An open-channel flow down a slope.

For long channels of constant slope and constant cross section areas, a uniform flow will occur. For such a situation, we can obtain the discharge through the channel. For this, we consider an open channel as shown in Fig. 6.3 and cross sections 1-1 and 2-2. Let L be the length of channel, D_H be the hydraulic diameter, and S_0 be the slope (where S_0 = tanθ). In this case, the velocity is equal at both the cross sections. Let us denote this velocity as V_0. The head difference between the two cross sections are

$$h_f = z_1 - z_2 = S_0\, L) \quad (6.9$$

If we assume that the flow is fully developed, then we can write down the head loss between cross sections 1-1 and 2-2 as

$$h_f = f\frac{L}{D_H}\frac{V_0^2}{2g} \quad (6.10)$$

noting that $D_H = 4\frac{A}{P}$, where A and P represent the area of cross section and wetted perimeter respectively. When the two aforementioned expressions are combined, we obtain an expression for the uniform velocity as

$$V_0 = \left(\frac{8g}{f}\right)^{\frac{1}{2}}\left(\frac{A}{P}\right)^{\frac{1}{2}} S_0^{\frac{1}{2}} = \left(\frac{8g}{f}\right)^{\frac{1}{2}} (R_H)^{\frac{1}{2}} S_0^{\frac{1}{2}} \quad (6.11)$$

where we have written down the expression in terms of the hydraulic radius $R_H = \frac{A}{P}$. For a given shape of the channel, the expression $\left(\frac{8g}{f}\right)^{\frac{1}{2}}$ is constant and thus, we can write down the expression for the uniform velocity as

$$V_0 = C\left(R_H S_0\right)^{\frac{1}{2}} \Rightarrow Q = AV_0 = CA\left(R_H S_0\right)^{\frac{1}{2}}\,(6.12) \quad (6.12)$$

These formulae are known as Chézy formulae and are named after the French engineer Antoine Chézy. The constant C is known as the Chézy constant.

The above formulae do not take into consideration the roughness of the bed, the bed shape, the cross section area etc. Thus, there have been several efforts towards developing correlations which can render the formula compatible with the aforementioned variables. A widely used correlation was given by Manning where it is proposed that the coefficient C is given by

$$C = \alpha\frac{R_H^{\frac{1}{6}}}{n} \quad (6.13)$$

where n represents the roughness parameter and α represents a dimensional constant to render C dimensionless.

Example 6.1

Water flows through a rectangular channel of 4 m wide and 2 m deep. The channel is having a bed slope of 1 in 1500. Determine the velocity of flow and rate of flow of water, when the channel is running full, considering the Chezy's constant to be $C = 60$.

Solution

The above question specifies that the width of the rectangular channel (b) is 4 m and the depth of the channel (d) is 2 m.

Hence the area of the channel becomes A = 4 × 2 = 8 m².

The bed slope is given as $S_0 = 1 \text{ in } 1500 = \frac{1}{1500}$ and the Chezy's constant is given by $C = 60$.

Perimeter of the rectangular channel is given as $P = b + 2d = 4 + 2 \times 2 = 8$ m.

Hence the hydraulic mean depth is calculated as, $R_H = \frac{A}{P} = \frac{8}{8} = 1$ m

Velocity of the flow is given by equation $V_0 = C\sqrt{R_H S_0} = 60\sqrt{1 \times \frac{1}{1500}} = 1.5491$ m/s.

Example 6.2

Determine the discharge through a trapezoidal channel of width 6 m and side slope of 1 horizontal to 4 vertical. The depth of flow of water is 2.8 m and value of Chezy's constant, C = 65. The slope of the bed of the channel is given 1 in 2000.

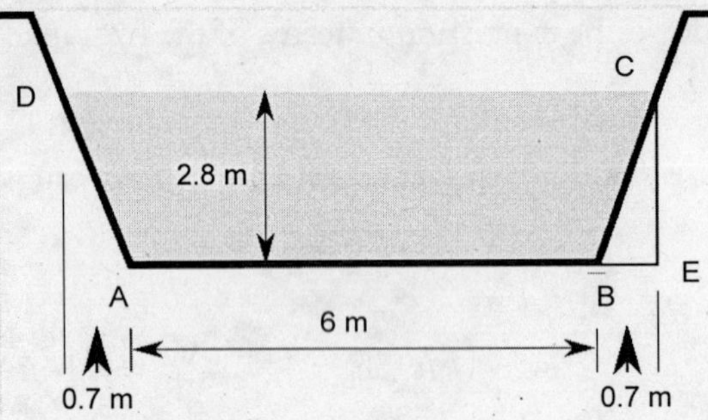

Solution:

It is given in the question that the width of the trapezoidal channel (b) is 6 m, side slope is 1 horizontal to 4 verticals, depth (d) is 2.8 m and the Chezy's constant $C = 65$. Bed slope S_0 is specified to be $S_0 = \frac{1}{2000}$.

It is obvious that for a trapezoidal channel when the depth is 2.8 m the horizontal distance BE becomes $BE = 2.8 \times \frac{1}{4} = 0.7$ m.

The top width of the channel becomes:

CD = AB + 2 × BE = 6 + 2 × 0.7 = 7.4 m

The area of the trapezoidal channel, ABCD is given as:

$$A = (AB + CD) \times \frac{CE}{2} = (6 + 7.4) \times \frac{2.8}{2} = 13.4 \times 1.4 = 18.76\, m^2$$

Wetted perimeter, (P = AB + BC + AD = AB + 2BC) (∵ BC = AD)

But $BC = \sqrt{BE^2 + CE^2} = \sqrt{0.7^2 + 2.8^2} = 2.886$ m

∴ P = 6 + 2 × 2.886 = 11.772 m

The hydraulic mean depth is defined as, m

The discharge, is given by equation $R_H = \frac{A}{P} = \frac{18.76}{11.772} = 1.5936$ m/s.

6.3 MOST ECONOMICAL CHANNEL SECTIONS

Often for many engineering applications, we must minimize the amount of construction required for canals etc. Such a situation implies that given a discharge rate, the wetted perimeter must be minimum. Such a task is tantamount to finding out the most economical channel section, also known as the best section. Let us consider a cross section area A, and the slope of the bed, S_0. We can write down the flow rate as

$$Q = CA\sqrt{R_H S_0} \tag{6.14}$$

where, when the area is a constant, we can write down the volume flow rate as

$$Q = \frac{K}{\sqrt{P}} \quad \text{where } K = CA\sqrt{AS_0} \tag{6.15}$$

It is clear from the above equation that the given a fixed area, the maximum flow rate is obtained for minimum wetted perimeter. We can now use this condition to find out the most economical channel. We consider certain kinds of channels:

1. Rectangular channel
2. Trapezoidal channel
3. Circular channel

6.3.1 Rectangular Channel

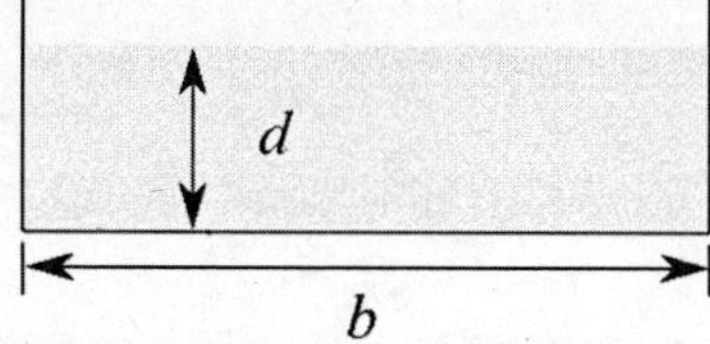

Figure 6.4: An open channel with rectangular cross section.

Let us consider a channel of cross section as shown in Fig. 6.4, which has a breadth of b and a depth of d. Using the figure as a reference, we can write down the following geometric quantities

$$A = bd, \quad P = b + 2d \Rightarrow P = \frac{A}{d} + 2d \tag{6.16}$$

where we have simply expressed the breadth in terms of the depth from the expression for the area in order to find the third expression. In order to minimize the perimeter, we can find out the derivative of the perimeter with respect to the diameter first and set it to zero. We see that

$$P' = -\frac{A}{d^2} + 2 = 0 \Rightarrow d = \sqrt{\frac{A}{2}} \tag{6.17}$$

where we can now ascertain whether this value of d yields the minimum perimeter through the sign of the second derivative

$$P'' = \frac{2A}{d^3} \Rightarrow P'' > 0 \Rightarrow d = \sqrt{\frac{A}{2}} \tag{6.18}$$

hence, we see that indeed we have a minimum wetted perimeter.

We can now find out the value of the breadth of the channel as

$$A = bd = 2d^2 \Rightarrow b = 2d \tag{6.19}$$

We can write down the expression of the hydraulic radius, R_H as

$$R_H = \frac{A}{P} = \frac{bd}{b+2d} = \frac{d}{2} \tag{6.20}$$

This means that the hydraulic radius is half the depth of the flow for minimum wetter perimeter.

Example 6.3

A rectangular channel of width, 10 m is having a bed slope of 1 in 2400. Find the maximum discharge through the channel. Take value of .

Solution:

Given the width of the channel (b) is 10 m, bed slope is $R_H = \frac{1}{2400}$ and the Chezy's constant is 55. Let us recall that the discharge will be maximum when the channel is most economical. The conditions for most economical rectangular channel are:

$$b = 2d \text{ or } d = \frac{b}{2} = \frac{10}{2} = 5m; \quad \text{and} \quad R_H = \frac{d}{2} = \frac{5}{2} = 2.5m$$

Area of most economical rectangular channel becomes, $A = b \times d = 10 \times 5 = 50$ m^2

Hence, the discharge through the channel is given by:

$$Q = AC\sqrt{R_H S_0} = 50 \times 55 \times \sqrt{2.5 \times \frac{1}{2400}} = 88.75\ m^3/s$$

6.3.2 Trapezoidal Cross Section

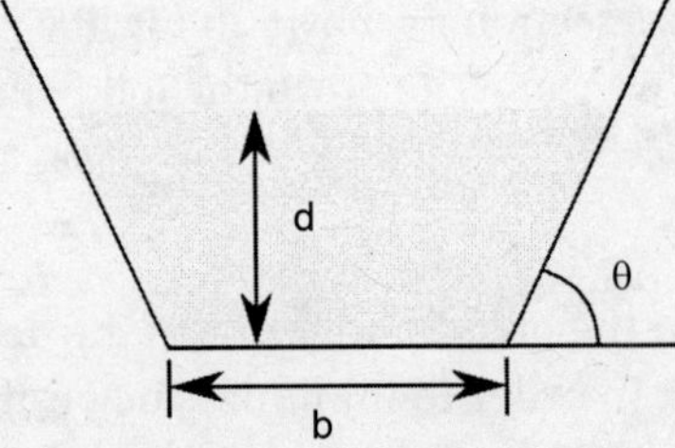

Figure 6.5: Open channel with trapezoidal cross section.

In much the similar fashion, we can find out the most economical trapezoidal cross section, i.e. finding out a condition for the minimum perimeter. Let us consider the cross-section area as Fig. 6.5. Once again, the depth is d, the breadth of the base is b while the angle made by the side walls with the horizontal is θ. We denote $n = \tan\theta$. Using elementary geometry, one can show that

$$A=(b+nd)d \quad P=b+2d\sqrt{n^2+1} \tag{6.21}$$

The corresponding expression for the perimeter only in terms of depth can be thus written by noting that $b=\frac{A}{d}-nd$ and thus we obtain

$$P=\frac{A}{d}-nd+2d\sqrt{n^2+1} \tag{6.22}$$

As done in the previous section, we can find out the condition for minimum area as

$$P'=0=-\frac{A}{d^2}-n+2\sqrt{n^2+1}\Rightarrow\frac{A}{d^2}+n=2\sqrt{n^2+1} \tag{6.23}$$

Using this, we can find out the solution for the depth of the channel through the solution of the quadratic equation for d above. However, we can simplify the expressions by noting that $A = (b + nd)d$ and thus, we have from the previous equation

$$\frac{(b+nd)d}{d^2}+n=2\sqrt{n^2+1}\Rightarrow\frac{b+2nd}{2}=d\sqrt{n^2+1} \tag{6.24}$$

From the geometry of the problem, we note that $\frac{b+2nd}{2}$ represents the half of the width at the top surface. Similarly, $d\sqrt{n^2+1}$ represents the length of the side surface. Thus, the condition for the most economical trapezoidal cross section implies that the half of the width of the top surface is equal to the length of the side wall, i.e.

$$\frac{b+2nd}{2}=d\sqrt{n^2+1} \tag{6.25}$$

From this, we can now write down the expression of the hydraulic radius for the most economic section as

$$R_H=\frac{A}{P}=\frac{(b+nd)d}{2(b+nd)}=\frac{d}{2}$$

Example 6.4

A rectangular channel carries water at the rate of 200 liters/s when bed slope is 1 in 4000. Find the most economical dimensions of the channel if $C = 50$.

Solution:

Given the discharge Q = 200 liters/s = 0.2 m³/s, bed slope is 1 in 4000 and Chezy's constant $C = 50$.

For the rectangular channel to be most economical

$$b=2d \;;\; and \quad R_H=\frac{d}{2}$$

∴ Area of the flow, $A = b \times d = 2d \times d = 2d^2$

Using the equation for discharge $Q = AC\sqrt{R_H S_0}$

$$\Rightarrow \quad 0.2 = 2d^2 \times 50 \times \sqrt{\frac{d}{2} \times \frac{1}{4000}} = 2 \times 50 \times \sqrt{\frac{1}{2 \times 4000}}\, d^{5/2} = 1.118\, d^{5/2}$$

$$\Rightarrow \quad d^{5/2} = \frac{0.2}{1.118} = 0.17889$$

$$\Rightarrow \quad d = (0.17889)^{2/5} = 0.5023 \text{ m}$$

$$\Rightarrow \quad b = 2 \times d = 2 \times 0.5023 = 1.004 \text{ m}$$

Example 6.5

A trapezoidal channel has side slopes of 2 horizontal to 5 vertical and slope of its bed is 1 in 2500. Determine the optimum dimensions of the channel, if it is to carry water at 0.4 m³/s. Take Chezy's constant as 75.

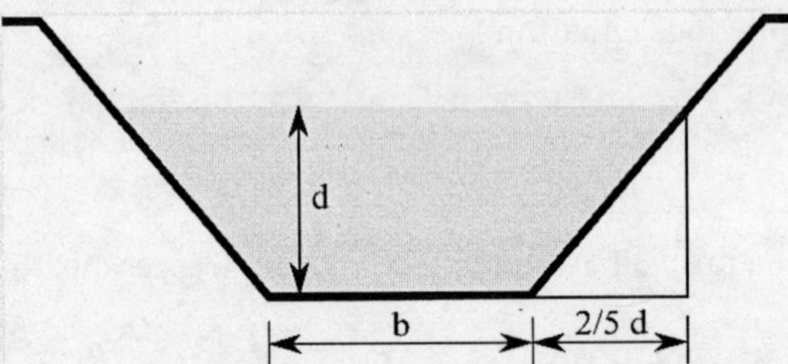

Solution:

Given the side slope $n = 2/5$, bed slope $S_0 = 1/2500$, discharge Q=0.4 m³/s, and Chezy's constant $C = 75$.

For the most economical section, the condition is given by the equation as $\frac{b + 2nd}{2} = d\sqrt{n^2 + 1}$

$$\Rightarrow \quad \frac{b + 0.8d}{2} = 1.077d$$

$$\Rightarrow \quad b = 2 \times 1.077d - 0.8d = 1.354d$$

For the most economical section the hydraulic mean depth $R_H = \frac{d}{2}$.

Area of the trapezoidal section is given as: $A = (b + nd) \times d = \left(d + \frac{2}{5}d\right) \times d = 1.4d^2$

For discharge Q is calculated using $Q = AC\sqrt{R_H S_0}$

$$\Rightarrow \quad 0.4 = 1.4d^2 \times 75 \times \sqrt{\frac{d}{2} \times \frac{1}{2500}} = 1.4849\, d^{5/2}$$

$$\Rightarrow \quad d = \left(\frac{0.4}{1.4849}\right)^{2/5} = 0.591 m$$

Hence the optimum dimensions of the channel are depth (d) = 0.591 m and width $b = 1.354\, d = 0.8$ m.

Example 6.6

An open channel of most economical section, having the form of a half hexagon with horizontal bottom is required to give a maximum discharge of 25 m³/s of water. The slope of the channel bottom is 1 in 1000. Taking Chezy's constant, $C = 65$ determine the dimensions of the cross section.

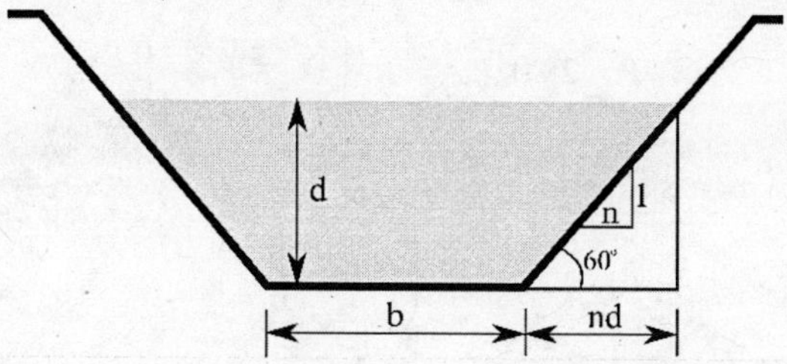

Solution:

Given maximum discharge $Q = 25$ m³/s, bed slope is 1 in 1000 and Chezy's constant is $C = 65$.

The channel is the form of a half hexagon as shown in figure. This means that the angle made by the sloping side with horizontal will be 60°.

$$\Rightarrow \qquad \tan\theta = \tan 60^\circ = \sqrt{3} = \frac{1}{n}$$

$$\Rightarrow \qquad n = \frac{1}{\sqrt{3}}$$

As the channel given is the most economical section, hence the conditions (i) half of the top width = one of the sloping side, and (ii) hydraulic mean depth = half of depth of flow $\left(R_H = \frac{d}{2}\right)$, should be satisfied.

$$\Rightarrow \qquad \frac{b + 2nd}{2} = d\sqrt{n^2 + 1}$$

Substituting $n = \frac{1}{\sqrt{3}}$ in the above equation, we obtain $b = \frac{2d}{\sqrt{3}}$

Area of the flow becomes: $A = (b + nd)d = \left(\frac{2d}{\sqrt{3}} + \frac{d}{\sqrt{3}}\right)d = \sqrt{3}d^2$

Using equation for discharge $Q = AC\sqrt{R_H S_0}$

$$\Rightarrow \qquad 25 = \sqrt{3}d^2 \times 65 \times \sqrt{\frac{d}{2} \times \frac{1}{1000}} = 2.5174\, d^{5/2}$$

$$\Rightarrow \qquad d = (9.93)^{2/5} = 2.50 \text{ m}$$

Hence the value of becomes $b = \frac{2d}{\sqrt{3}} = b = \frac{2}{\sqrt{3}} \times 2.50 = 2.8924$ m

6.3.3 Circular Channel

Let us consider the flow of a liquid through a circular channel. Let the radius of the channel be b and the angle subtended by the free surface with the center of the circle be 2θ. In this case we can note that the geometric parameters as

$$P = 2R\theta \quad A = R^2\left(\theta - \frac{\sin 2\theta}{2}\right)(6.27) \tag{6.27}$$

The hydraulic diameter for this flow is

$$R_H = \frac{A}{P} = \frac{R}{2\theta}\left(\theta - \frac{\sin 2\theta}{2}\right) \tag{6.28}$$

We can now find out the condition for the most economic cross section for maximum flow rate through

$$Q = CA\sqrt{R_H S_0} = C\sqrt{\frac{A^3}{P}S_0} \tag{6.29}$$

We can now find out the value of θ for which $\sqrt{\frac{A^3}{P}}$ is maximum. This can be easily done by finding out the condition for which

$$\frac{dQ}{d\theta} = 0 \tag{6.30}$$

Substituting the expression for A and P in terms of θ, we obtain

$$4\theta - 6\theta\cos 2\theta + \sin 2\theta = 0 \tag{6.31}$$

The solution for this equation is given by

$$\theta = 154° \tag{6.32}$$

Example 6.7

Find the discharge through a circular pipe of diameter 2.5 m, if the depth of water in the pipe is 0.8 m and the pipe is laid at a slope of 1 in 2000. Take the value of Chezy's constant as 55.

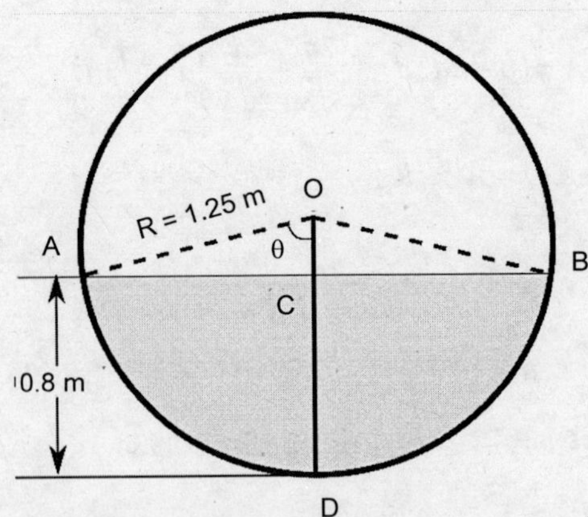

Solution:

Given the dia of pipe $D = 2.5$ m, depth of water in pipe $d = 0.8$ m, bed slope is 1 in 2000, Chezy's constant is 55.

From the figure, we have OC = OD – CD = R – 1 = 1.25 – 0.8 = 0.45 m and AO = R = 1.25 m

Also, $\cos\theta = \frac{OC}{AO} = \frac{0.45}{1.25} = \frac{9}{25}$, so $\theta = 68.89° = 1.2025$ radians

Wetted perimeter is given as : $P = 2R\theta = 2 \times 1.25 \times 1.2025 = 3.00$ m

Wetted area is given as: $A = R^2\left(\theta - \frac{\sin 2\theta}{2}\right) = 1.25^2\left(1.2025 - \frac{\sin\left(2\times 68.89^\circ\right)}{2}\right) = 1.3539\ \text{m}$

Thus, the hydraulic mean depth, $R_H = \frac{A}{P} = \frac{1.3539}{3.00} = 0.4513$

The discharge is given by, $Q = AC\sqrt{R_H S_0}$

$$\Rightarrow \quad Q = 1.3539 \times 55 \times \sqrt{0.4513 \times \frac{1}{2000}} = 1.1185\, m^3/s$$

Example 6.8

Calculate the quantity of water that will be discharged at a uniform depth of 0.7 m in a 1.1 m diameter pipe which is laid at a slope 1 in 2500. Assume Chezy's $C = 64$.

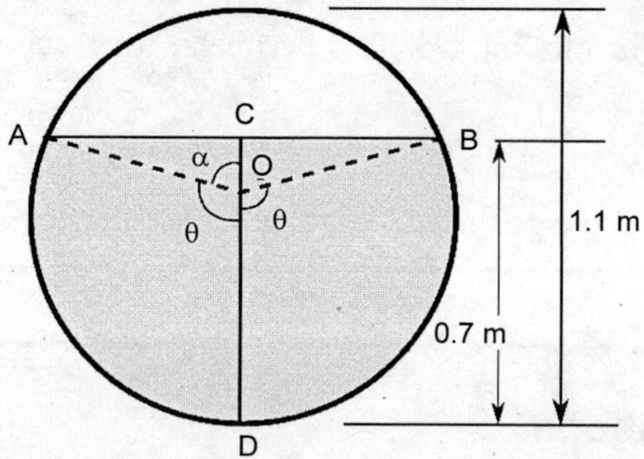

Solution:

Given the dia of pipe $D = 1.1$ m, depth of water in pipe $d = 0.7$ m, bed slope is 1 in 2500, Chezy's constant is 64.

From the figure, we have OC = CD – OD = 0.7 – R = 0.7 – 0.55 = 0.15 m and OA = R = 0.55 m

Now, in triangle AOC, $\cos\alpha = \frac{OC}{OA} = \frac{0.15}{0.55} = \frac{3}{11} \Rightarrow \alpha = \cos^{-1}\left(\frac{3}{11}\right) = 74.17^\circ$

θ = Angle DOA = 180° – α = 180° – 74.17° = 105.82° = 0.588 π radians

Now, the wetted perimeter is given as $P = 2R\theta = 2 \times 0.55 \times 0.588\pi = 2.032$ m.

Area of the flow is given as: $A = R^2\left(\theta - \frac{\sin 2\theta}{2}\right) = 0.55^2\left(0.588\pi - \frac{\sin\left(2\times 105.82^\circ\right)}{2}\right) = 0.638\ \text{m}^2$

Now, the discharge is given by, $Q = AC\sqrt{R_H S_0}$

$$\Rightarrow \quad Q = 0.638\times 64\times\sqrt{\frac{A}{P}\times\frac{1}{2500}} = 0.638\times 64\times\sqrt{\frac{0.638}{2.032}\times\frac{1}{2500}} = 0.4575\, m^3/s$$

UNIT SUMMARY

- Hydraulic diameter

$$D_H = \frac{4A}{P}$$

- Froude number

$$Fr = \frac{V}{\sqrt{gy}}$$

- Chézy constant

$$C = \alpha\frac{R_H^{\frac{1}{6}}}{n}$$

- The hydraulic diameter for minimum wetted perimeter for rectangular, trapezoidal and circular cross section open channels

$$R_H = \frac{d}{2}$$

EXERCISES

Multiple Choice Questions

1. When Froude number is 1, the flow is
   (a) Supercritical flow (b) Subcritical flow
   (c) Critical flow (d) None of these
2. In the formula for discharge, $Q = C\sqrt{R_H S_0}$, S_0 is known as
   (a) Slope of the bed (b) Wetted perimeter of cross section
   (c) Hydraulic mean depth (d) None of these
3. Froude number of a flow in an open channel is given by
   (a) Inertia force / Viscous force (b) Inertia force / Surface tension force
   (c) Inertia force / Gravity force (d) Inertia force / Drag force

4. For most economical rectangular section the hydraulic depth is given by
   (a) Half of the depth of the flow
   (b) Two times the depth of flow
   (c) Equal to depth of flow
   (d) None of these
5. Dimension of Chezy's constant is
   (a) $L^2 T^{-1}$
   (b) $L^{\frac{1}{2}} T^{-1}$
   (c) $L^2 T^{-2}$
   (d) None of these
6. Which one of the following is a most common formula while calculating the discharge of flow in open channel
   (a) $Q = C\sqrt{R_H}$
   (b) $Q = CR_H S_0$
   (c) $Q = CS_0\sqrt{R_H}$
   (d) $Q = C\sqrt{R_H S_0}$

ANSWERS															
1.	(c)	2.	(a)	3.	(c)	4.	(a)	5.	(b)	6.	(d)				

Unsolved Questions

Level - I

1. Find the discharge through a trapezoidal channel of width 6 m and side slope 1 horizontal and 2 vertical. The depth of flow of water is 2.2 m and the value of Chezy's constant, (C) = 45. The slope of the bed of the channel is given 1 in 3000.

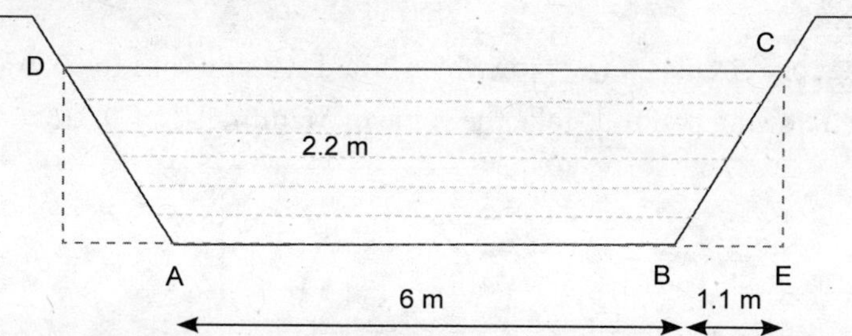

2. Find the discharge of water through a channel shown in the figure. Take the value of Chezy's constant is 50 and the slope of the bed is 1 in 2500.

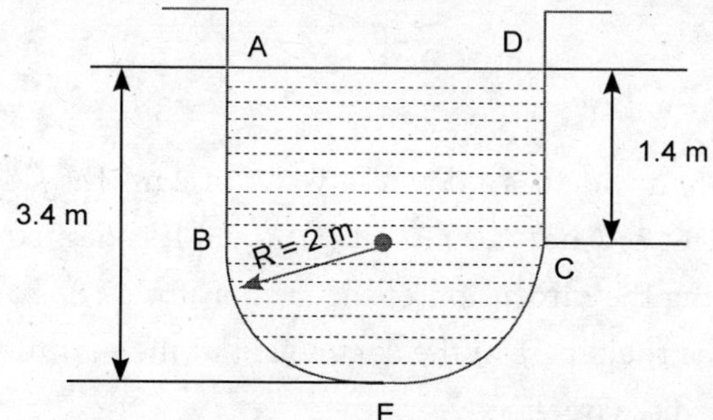

3. Find the discharge through a V-shaped channel as shown in the figure. Take the value of Chezy's constant (C) is 40 and the slope of the bed is 1 in 1500.

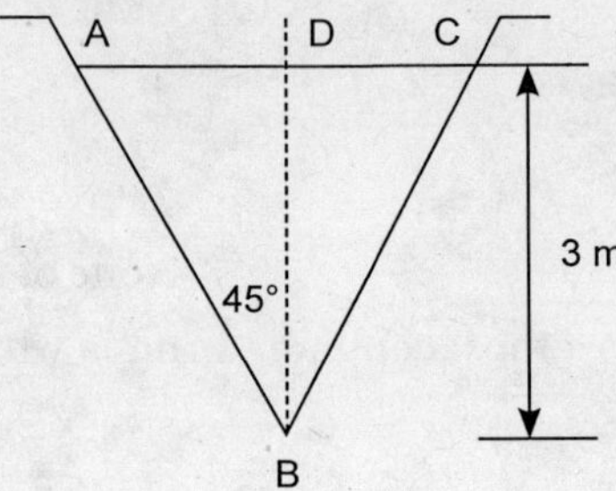

4. Find the bed slope of a trapezoidal channel of bed width 3 m, depth of water 2 m and the side slope of 1 horizontal to 2 vertical, when the discharge through the channel is 15 m³/s. Take Manning's N = 0.02 in Manning's formula $C = \frac{1}{N} R_H^{1/6}$.

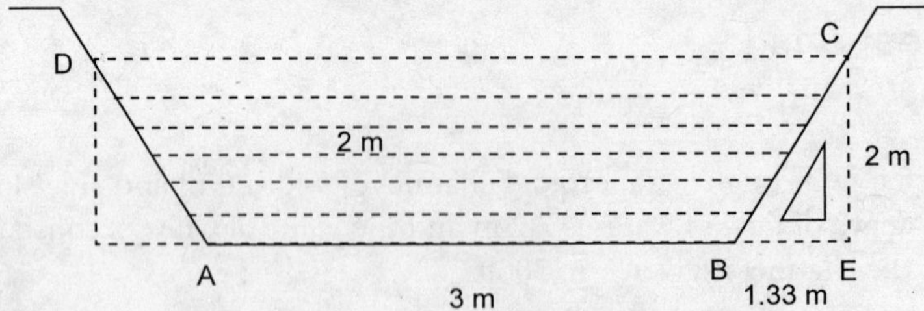

5. Find the diameter of a circular sewer pipe which is laid at a slope of 1 in 6000 and carries a discharge of 0.7 m³/s when flowing half full. Take the value of Manning's N = 0.015

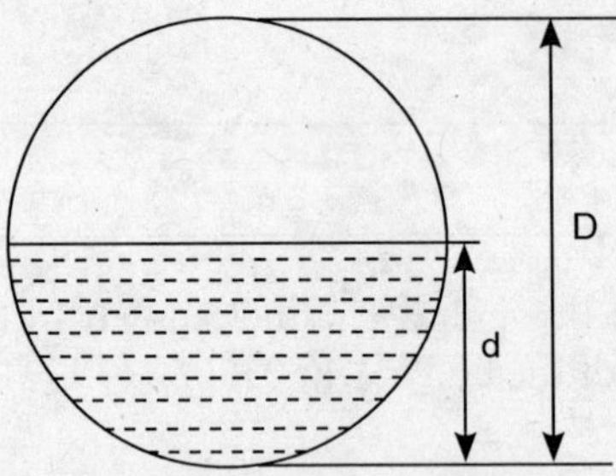

Level - II

1. A rectangular channel 5 m wide has a depth of water of 2 m. The slope of the bed of the channel is 1 in 1200 and the value of Chezy's constant (C) is 50. It is desired to increase the discharge to a maximum by changing the dimension of the section for a constant area of cross-section, the slope of the bed and the roughness of the channel. Find the new dimensions of the channel and percentage increases in discharge.

2. A trapezoidal channel has side slopes of 1 horizontal and 2 vertical and the slope of the bed is 1 in 2000. The area of the section is 50 . Find the dimensions of the section if it is most economical. Determine the discharge of the most economical section if Chezy's constant (C) is 45.

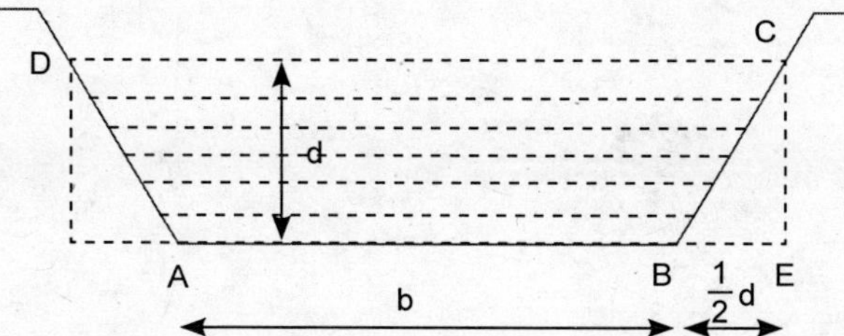

3. Find the discharge through a circular pipe of diameter 4 m. If the depth of water in the pipe is 1.5 m and the pipe is laid at a slope of 1 in 1500. Take the value of Chezy's constant (C) as 60.

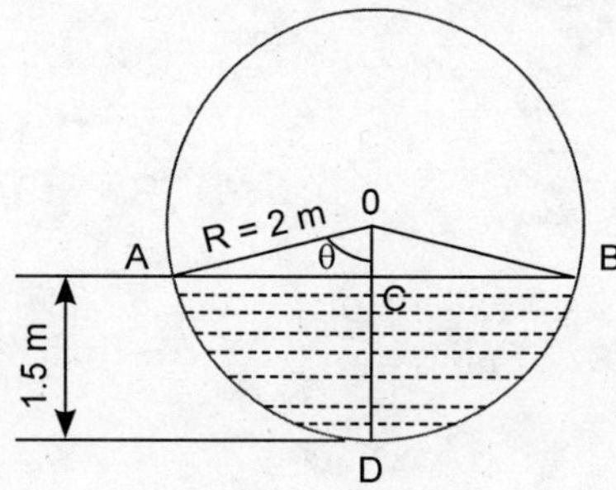

4. A flow of water of 0.2 m³/s flows down in a rectangular flame of width 500 mm and having adjustable bottom slope. If Chezy's constant C is 50, find the bottom slope necessary for uniform flow with a depth of flow of 300 mm. Also find the conveyance $K = \frac{AC}{R_H}$. of the flame.

5. A trapezoidal channel has side slopes of 1 horizontal and 2 vertical has to be designed to convey 9 m³/s at a velocity of 3 m/s so that the amount of concrete lining for the bed and sides is the minimum. Determine the dimensions of the channel.

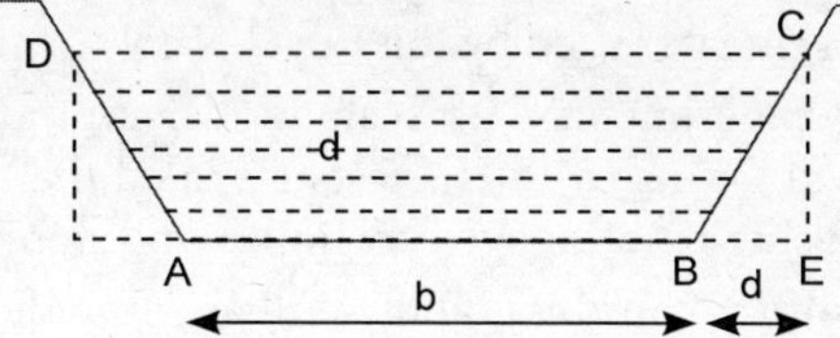

6. Find the discharge through a V-shaped channel as shown in the figure When the value of Chezy's constant (C) is 50 and the slope of the bed is 1 in 1500. Also, determine the percentage change in discharge value when the value of Chezy's constant (C) is increased by 20% and all other parameters will remain the same.

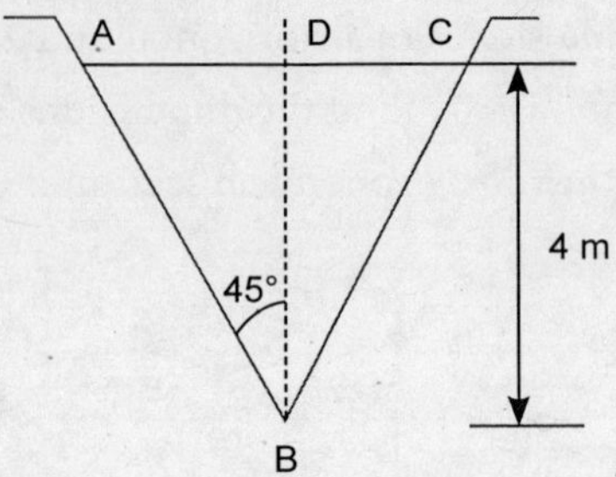

Answer Key

Level - I

1. Q = 15.34 m³/s
2. Q = 13.58 m³/s
3. Q = 9.57 m³/s
4. $S_0 = \frac{1}{840}$
5. D = 1.85 m

Level - II

1. $b_1 = 4.47$ m, $d_1 = 2.23$ m, $Q_{increase} = 0.3096\%$
2. b = 6.63 m, d = 5.36 m, Q = 82.41 m³/s
3. Q = 6.02 m³/s
4. $S_0 = \frac{1}{191}$, Conveyance (K) = 2.76 m³/s
5. $b = 1.62$ m, $d = 1.31$ m
6. Q = 24.56 m³/s, Percentage change in discharge (Q) = 20

To know more about open channel flows

KNOW MORE

William Froude (1810–1879), an English engineer and naval architect, revolutionized ship design by developing a method to test scale models and apply the findings to full-sized vessels. After beginning his career in railway engineering, he shifted to ship hydrodynamics in 1846, discovering that a deep bilge keel reduced ship rolling—a design later used by the Royal Navy.

In 1868, Froude proposed experiments with scale models to the British Admiralty, leading to the construction of a model-testing tank in 1870. His tests revealed that skin friction and wave formation were the main forces resisting a ship's motion, greatly improving ship efficiency.

[Source: https://www.britannica.com/biography/William-Froude]

REFERENCES AND SUGGESTED READINGS

1. **Introduction to Fluid Mechanics and Fluid Machines** (3rd Edition), S K Som, Gautam Biswas, Suman Chakraborty, McGraw Hill
2. **Fluid Mechanics** (9th Edition), Frank M. White, Henry Xue, McGraw Hill

❑❑

7 Fluid Machinery: Turbines

UNIT SPECIFICS

In this unit we will discuss about the following topics:

1. Force due to jets on stationary and moving vanes
2. Concept of velocity triangle to analyze flow through turbine blades
3. Classification of hydraulic turbines
4. Components of hydro-electric power plant
5. Use of non-dimensional numbers in turbines
6. Performance characteristics of hydraulic turbines

RATIONALE

The previous units discuss the features of water flowing internally through a pipe or through external channels. This unit will discuss how the energy carried by water flow can be harnessed to derive mechanical shaft work by means of a device called "Turbine". Hydraulic turbines are frequently used for power production in hydro-electric power plants. Potential energy of hydrostatic water head can be easily converted to the kinetic energy of water jet. This jet can strike a series of vanes to impart motion which can be used to turn the shaft of a generator. In order to analyze hydraulic turbines, one needs to first analyze the momentum transfer by jet striking vanes in different orientations. This understanding will lend to the introduction of a technique called velocity triangle, which allows us to analyze all the velocity components of water flow inside a moving vane and relate it to work production. This unit will also discuss the various types of hydraulic turbines, their performance characteristics and applications.

PRE-REQUISITE

1. Elementary vector analysis
2. Trigonometric relationships

UNIT OUTCOMES

U7-O1: Force balance for jets striking stationary and moving vanes at different angles

U7-O2: Different types of turbines based on the water flow pattern through the runner

U7-O3: Understanding the various components of velocity as water flows through the turbine runner

U7-O4: Relationship of velocity components to turbine runner geometry

U7-O5: Relationship of work output of turbine to the runner geometry

U7-O6: Specific speed and other scaling laws of a turbine

U7-O7: Performance curves of a turbine

Unit-7 Outcomes	EXPECTED MAPPING WITH PROGRAMME OUTCOMES (1 - Weak Correlation; 2 - Medium Correlation; 3 - Strong Correlation)						
	CO-1	CO-2	CO-3	CO-4	CO-5	CO-6	CO-7
U7-O1	-	-	-	3	-	-	1
U7-O2	-	-	-	3	-	-	1
U7-O3	-	-	-	3	-	-	1
U7-O4	-	-	-	3	-	-	2
U7-O5	-	-	-	3	-	-	2
U7-O6	-	-	-	3	-	-	2
U7-O7	-	-	-	3	-	-	2

7.1 INTRODUCTION

Hydraulic machinery form a backbone of modern day power generation and process engineering. These are machines which enable us to convert hydraulic energy into mechanical energy. Eventually this mechanical energy is converted into electrical energy with the help of generators. Hydraulic machinery which convert hydraulic energy into mechanical energy are classified as turbines whereas hydraulic machinery which convert mechanical energy into hydraulic energy are defined as pumps. This unit and the unit after this deal with the two aforementioned types of hydraulic machines.

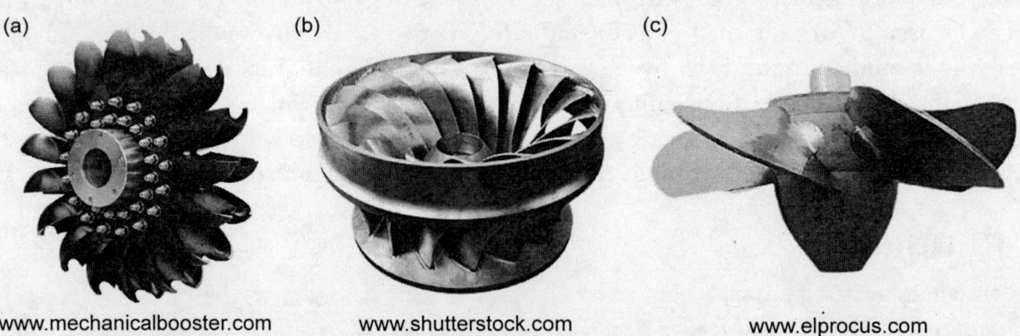

Figure 7.1: Different types of turbines.

Turbines are typically found in any kind of power plant. Typically they handle water or steam, depending on the source of energy. These turbine shafts are then typically coupled to the electric generator, which help in the generation of high voltage AC, which can then be transmitted over large distances.

Before delving into the specifics of the various kinds of turbines, it is instructive to first discuss the fundamental ideas which can help us to obtain some formulae pertaining to the performances of turbines etc. The following sections make use of the ideas of momentum conservation which were established in Unit 4.

7.2 JET IMPINGING ON A STATIC CURVED SURFACE

7.2.1 Jet Impinging on a Curved Surface and Leaving Tangentially

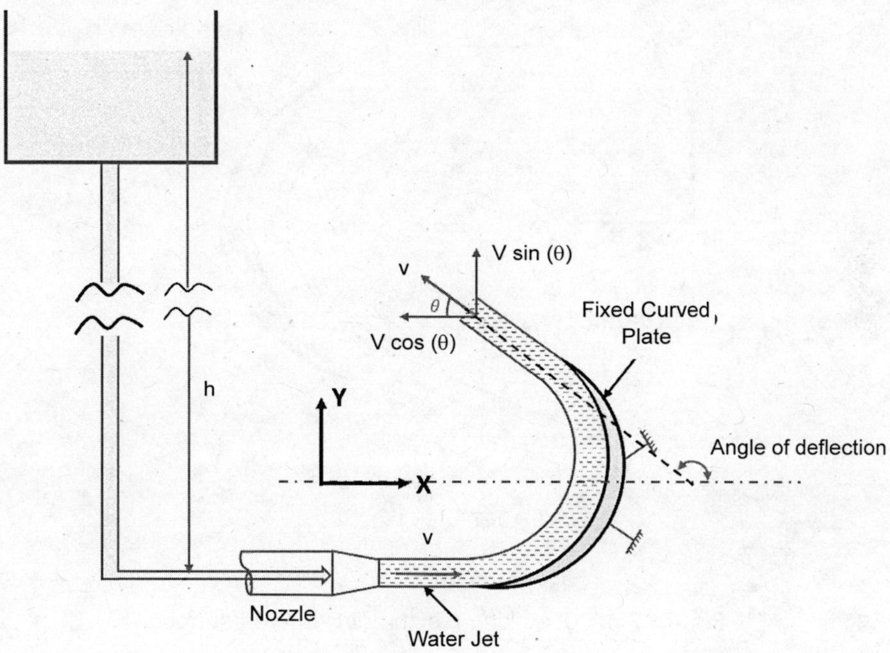

Figure 7.2: Flow impinging a curved surface. The water jet is obtained using a head *h*.

Let us begin our analysis with that of a jet impinging on a static curved surface. We consider a jet shown in the figure. There is a fixed curved plate at which the jet strikes and gets deflected at an angle of θ with respected to the horizontal. We assume that the jet comes out with the same velocity owing to the smoothness of the plate and no impact loss of the jet with the static curved plate. The velocity at the outlet can therefore be decomposed into two components. One in the horizontal direction, i.e. in the direction of the jet, and the other in the orthogonal direction of the jet, i.e. normal to the direction of the jet. We can therefore write

$$V_x = -V\cos\theta, \quad V_y = V\sin\theta \tag{7.1}$$

In the above expression we have made use of the sign convention that the jet is initially ejecting fluid in the positive x direction. The net force acting on the fixed curved plate can therefore be found out through the change in momentum of the jet and is given by

$$Fx = \rho AV(V - (-V\cos\theta)) = \rho AV^2(1 + \cos\theta) \tag{7.2}$$

The corresponding force acting on the curved surface in the y direction can be correspondingly found out by noting that the inlet momentum in the y direction is zero, and the exiting momentum is what contributes to the force. The force is given by

$$Fy = \rho AV(0 - V\sin\theta) = -\rho AV^2 \sin\theta) \tag{7.3}$$

7.2.2 Jet Strikes a Curved Symmetric Surface and Leaves Tangentially

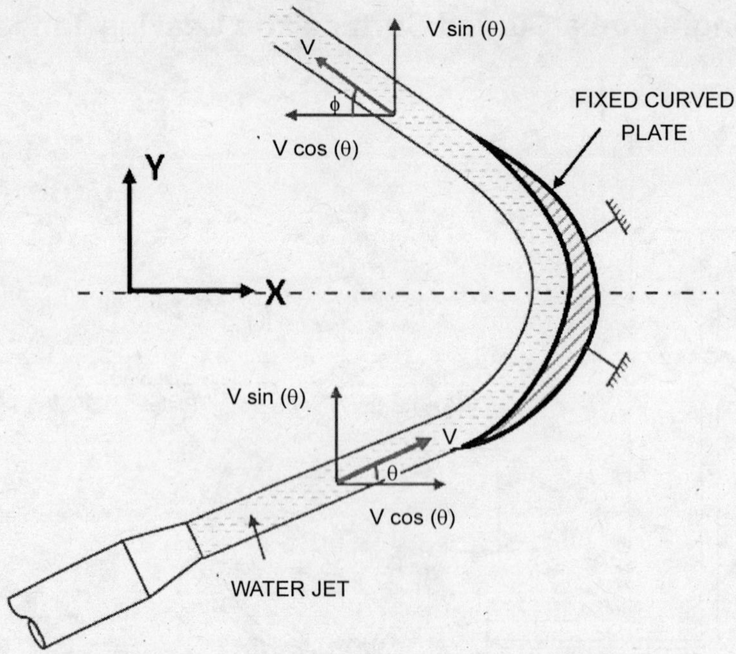

Figure 7.3: Jet striking a symmetric curved surface.

We now consider the situation as shown in the figure. The jet is now tangentially impinging onto the static curved plate. If we can consider the coordinate system shown in the figure, we can decompose the incoming velocity into the respective components and also the outgoing velocity into its respective components along the x and y direction.

We can write down the force exerted by the jet on the static curved plate as

$$F_x = \rho AV(V_{ix} - V_{ex}) = \rho AV(V\cos\theta - (-V\cos\theta)) = 2\rho AV^2 \cos\theta) \tag{7.4}$$

where in the expression above, V_{ix} and V_{ex} represent the inlet and exit velocity in the x direction respectively. Similarly, we can write down the force exerted by the jet in the direction in terms of the velocity components at the inlet and exit in the direction as

$$F_y = \rho AV(V_{iy} - V_{ey}) = \rho AV(V\sin\theta - V\sin\theta) = 0 \tag{7.5}$$

which is an obvious result given that the incoming and outgoing momentum of the fluid does not change as it is turned by the static blade. Therefore, we can conclude that the blade only encounters a transfer of momentum in the x direction.

7.2.3 Jet strikes an Asymmetric Curved Surface Tangentially

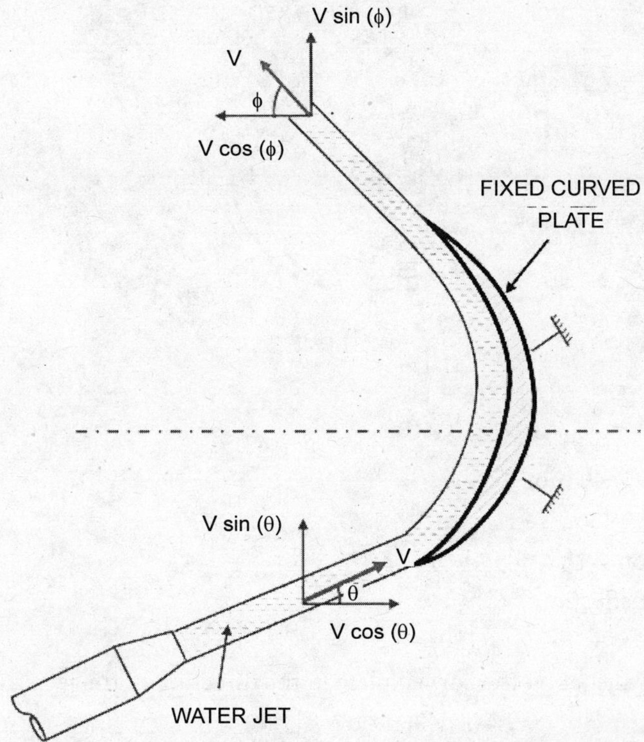

Figure 7.4: Jet striking an asymmetric curved surface.

The last case we consider is that of an asymmetric curved surface. The jet strikes the curved surface symmetrically but the incipient angle and the exit angle are no the same. At the inlet we assume that the jet (and hence the tangent of the curved surface) makes an angle θ with the horizontal x axis. At the exit we assume that the jet (and hence the tangent of the curved surface at that point) makes an angle ϕ with respect to the horizontal x axis. We can therefore write down the expressions by generalizing the results obtained in the previous subsection and obtain the force acting on the surface in x direction as

$$F_x = \rho AV(V_{ix} - V_{ex}) = \rho AV(V\cos\theta - (-V\cos\phi)) = \rho AV^2 (\cos\theta + \cos\phi) \qquad (7.6)$$

while the force acting in the direction can be written as

$$F_y = \rho AV(V_{iy} - V_{ey}) = \rho AV(V\sin\theta - V\sin\phi) = \rho AV^2 (\sin\theta - \sin\phi) \ (7.7)$$

Example 7.1

A horizontal jet of water of diameter 40 mm with a jet velocity of 40 m/s strikes a stationary curved symmetrical plate at the bottom edge in a manner as shown in the figure. Find the force exerted by the jet of water in the direction of the jet and perpendicular to it. Assume that the jet gets deflected by 135° between the inlet and outlet.

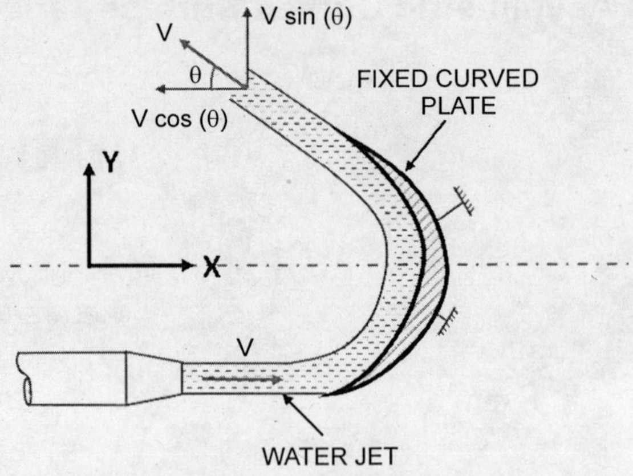

Solution:

Given data:

Diameter of the jet $d = 40$ mm

Velocity of the jet $V = 40$ m/s

Angle of jet deflection = 135°

The actual angle of deflection:

$$= 180° - 135° = 45°$$

The force exerted by the jet on the curved plate in the direction of the jet is given by:

$$F_x = \rho A V^2 \, [1 + \cos \theta]$$

$$F_x = 1000 \times \frac{\pi}{4} \times (0.040)^2 \times (40)^2 \times \left[1 + \cos(45^{\circ})\right] = 3.43\,kN$$

Example 7.2

A water jet diameter of 75 mm and velocity 30 m/s strikes a stationary curved plate tangentially at an angle of 30° to horizontal. What is the force on the plate in the horizontal and vertical direction if (*a*) the curved plate is symmetric and (*b*) the curved plate is such that the water exits at 40° to horizontal?

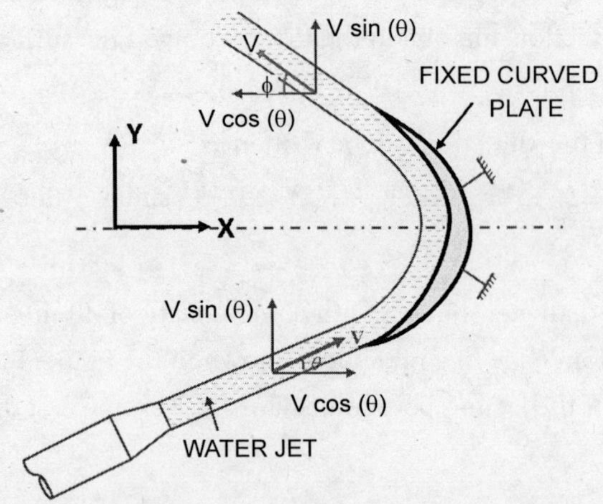

Solution:

Given data:

Diameter of the jet $d = 75$ mm

Velocity of the jet $V = 30$ m/s

Angle made by the jet at inlet tip with horizontal, $\theta = 30°$

Angle made by the jet at outlet tip with horizontal, $\phi = 40°$

(a) Force exerted by the jet when the curved plate is symmetric:

$$F_x = \rho A V^2 [1 + \cos\theta]$$

$$F_x = 1000 \times \frac{\pi}{4} \times (0.075)^2 \times (30)^2 \times \left[1 + \cos(30^\circ)\right] = 6.89\,kN$$

$$F_y = 0 \text{ kN}$$

(b) Force exerted by the jet when the curved plate is such that the water exits at 40 to horizontal:

$$F_x = \rho A V^2 \left[\cos\theta + \cos\phi\right]$$

$$F_x = 1000 \times \frac{\pi}{4} \times (0.075)^2 \times (30)^2 \times \left[\cos(30^\circ) + \cos(40^\circ)\right] = 6.49\,kN$$

$$F_y = \rho A V^2 [\sin\theta - \sin\phi]$$

$$F_y = 1000 \times \frac{\pi}{4} \times (0.075)^2 \times (30)^2 \times \left[\sin(30^\circ) - \sin(40^\circ)\right] = -0.57\,kN$$

Example 7.3

A water jet diameter of 75 mm and velocity 30 m/s strikes a curved plate tangentially at an angle of 30° to horizontal. What are the resultant force on the curved plate and its direction if the jet gets deflected by 60° between the inlet and outlet?

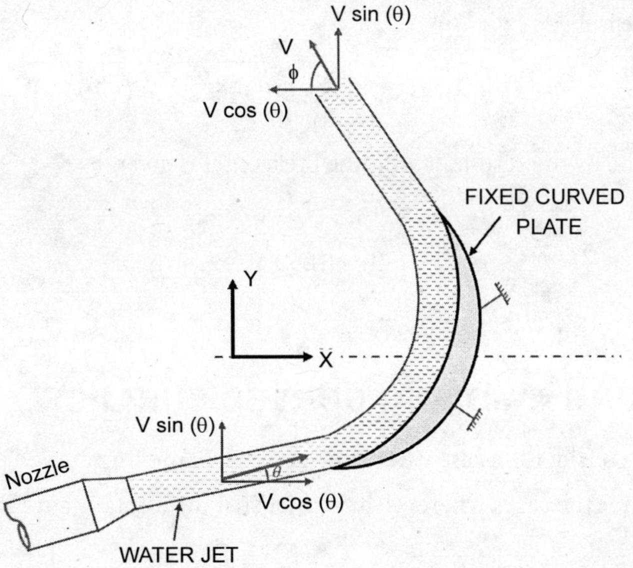

Solution:

Given data:

Diameter of the jet d =75 mm

Velocity of the jet $V = 30$ m/s

Angle made by the jet at inlet with horizontal, $\theta = 30°$

Angle of deflection = 60°

The angle made by the jet at an outlet with horizontal is given by,

$$\phi = \theta + \text{angle of deflection} = 30° + 60° = 90°$$

The force exerted by the jet of water in the direction of x is given by,

$$Fx = \rho AV[V_{1x} - V_{2x}]$$

$$V_{1x} = V\cos 30° = 30 \times \cos 30° = 25.98 \text{ m/s}$$

$$V_{2x} = V\cos 90° = 30 \times \cos 90° = 0 \text{ m/s}$$

$$F_x = 1000 \times \frac{\pi}{4} \times (0.075)^2 \times 30 \times [25.98 - 0] = 1062.6\,N$$

The force exerted by the jet of water in the direction of y is given by,

$$F_y = \rho AV[V_{1y} - V_{2y}]$$

$$F_y = 1000 \times \frac{\pi}{4} \times (0.075)^2 \times 30 \times \left[30 \sin 30^\circ - 30 \sin 90^\circ\right] = -594.8\,N$$

The negative sign shows that force F_y is acting in the downward direction.

The resultant force is given by,

$$F_R = \sqrt{F_x^2 + F_y^2} = \sqrt{(1062.6)^2 + (-594.8)^2} = 1217.75\,N$$

And the angle made by the resultant with the horizontal is given by,

$$tan\alpha = \frac{F_y}{F_x} = \frac{-594.8}{1062.6} = -0.5597$$

$$\alpha = -29.2°$$

7.3 JET STRIKING A MOVING CURVED SURFACE

In this section we shall consider the case where the jet strikes a moving curved plate. As one can imagine this is quite important for the cases where we have a jet striking a rotating turbine blade.

7.3.1 Jet striking a Moving Curved Surface Along the Direction of Motion

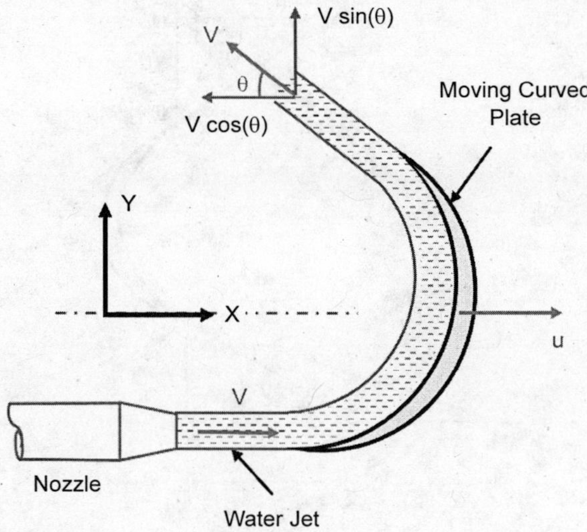

Figure 7.5: Jet strikes a moving curved plate along the direction of motion of the blade.

Let us consider a curved surface which is moving with a velocity u in the same direction as the jet. Let V be the jet velocity, A be the jet area. In this case we must account for the relative momentum of the jet which is interacting with the curved surface. In this case, like the previous section, we assume that the plate is smooth and that there is no loss of energy due to the impact of the jet with the curved surface.

First, the mass flux striking the curve surface is given by

$$\dot{m} = \rho A\left(V - u\right) \tag{7.8}$$

which is essentially the relative velocity with which the jet is approaching the curved surface in the frame of the curved surface. Extending the same logic, the velocity with which the jet is exiting the surface is given by $V - u$. Therefore, we can then decompose this velocity into the x component as $(V - u)\cos\theta$. We can therefore write down the expression for the force acting on the curved surface due to the jet as

$$F_x = \rho A(V - u)[(V - u) - (-(V - u)\cos\theta)] \tag{7.9}$$

Note that the flux exiting from the surface points in the negative x direction and hence has a negative sign associated with it. The above expression can be simplified to obtain

$$F_x = \rho A(V - u)^2 (1 + \cos\theta) \tag{7.10}$$

We can therefore find out the power of the jet, i.e. the work done in moving the plate along the direction of the jet per second as

$$P = F_x u = \rho A(V - u)^2 u(1 + \cos\theta) \tag{7.11}$$

7.3.2 Asymmetric Curved Plate is Moving and Jet Strikes Tangentially

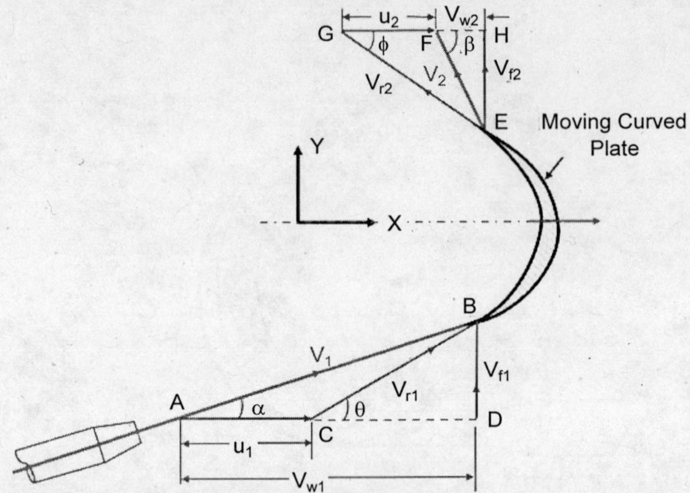

Figure 7.6: Jet strikes a moving curved plate tangentially.

We move on to the case where we have a moving curved surface which is moving as shown in the figure. At the same time, there is a jet which strikes the curved surface tangentially to the surface. The jet is tangential in the case where the moving surface is static. When the surface moves, then the relative velocity, obviously, will change both at the inlet and at the exit of the surface.

Like the cases considered earlier, we will consider that the loss of energy due to the jet striking the surface is zero. At the point of impact, the jet strikes the surface with the relative velocity with respect to the plate. We must therefore first analyze the velocity fields at the inlet and exit of the surface. To analyze this, we must first describe some of the notations that will appear in the analysis.

We will denote the inlet by the subscript 1 and the exit by the subscript 2. At the inlet we have

V_1 = Inlet velocity of the jet

u = Velocity of the moving surface

V_{r1} = Relative velocity of the jet with respect to moving surface

α = Guide blade angle – angle between jet and motion of surface

θ = Vane angle at inlet – angle made by relative velocity and motion of surface

V_{w1} = Whirl velocity at the inlet; component of V_1 in the direction of motion

V_{f1} = Flow velocity at inlet

By extending the same logic to the outlet section, i.e. 2, we can define the whirl velocity, the flow velocity etc. The only changes would be in the guide blade angle and the vane angle at inlet. Instead of these we have

β = exit angle made by jet with moving surface

ϕ = exit angle made by relative velocity and motion of surface

To our analysis, we must have the relative velocities at the inlet and exit apart from the mass flux. Let us consider them one by one.

Inlet velocity triangle

The pertinent velocities at the inlet are shown in the figure above. The inlet jet is shown by the line A_1B_1. The velocity of the moving frame is denoted by C_1A_1. The corresponding relative velocity vector is therefore found out by making use of the vector addition rules and is given by C_1B_1. Essentially, we have $A_1C_1 = -\vec{u}$, $A_1B_1 = \vec{V_1}$ $A_1B_1 = \vec{V_1}$ and $C_1B_1 = \vec{V_{r_1}}$ which conveys the vector addition of $\vec{V_{r_1}} = \vec{V_1} - \vec{u}$.

The angle made by the inlet jet velocity with the direction of motion, i.e. the horizontal direction, is given by α and the dangle made by the relative velocity vector with the horizontal is given by θ. The projection of the jet velocity with the x axis is V_{w1} while that on the axis is given by V_{f1}.

For the flow to occur in a loss-less manner, we must have the condition that the relative velocity is perfectly tangential to the moving surface, i.e. in terms of the turbine we must have that *the relative velocity should be tangential to the vane at the inlet.*

Exit velocity triangle

In fact, given that there is no loss occurring at the surface, we must also have the condition that the magnitude of the relative velocity at the exit must be equal to the magnitude of the relative velocity at the inlet, i.e. $V_{r1} = V_{r2}$.

The pertinent velocity vectors are shown in the figure. The relative exit velocity is shown as B_2C_2. The velocity of the reference frame is shown as A_2C_2. The absolute velocity is denoted by B_2A_2. One can verify from the vector addition rules that $\vec{V_{r_2}} = \vec{V_2} - \vec{u}$. Note that the vector $\vec{V_{r_2}}$ denoted by the segment B_2C_2 must be tangential to the surface owing to the no-loss condition. The angle made by the relative velocity vector with the horizontal is given as ϕ. The velocity made by the absolute jet velocity with respect to the horizontal is β. The projections of the absolute jet velocity on the y and x axis is given by V_{f2} and V_{w2} respectively.

The mass flux arriving at the moving surface is

$$\dot{m} = \rho A V_{r_1} \tag{7.12}$$

which is basically the mass flux due to the relative velocity at the inlet (which, by virtue of $V_{r2} = V_{r1}$, is also the mass flux at the exit). We can now determine the force acting on the moving surface due to the jet impinging on it. It is fundamentally equal to the mass flux arriving at the moving surface times the difference in the velocity between the inlet and exit. We must then first find out the expressions desired. At the inlet and exit, the absolute velocity in the direction of motion is given by

$$\text{Inlet x-velocity} = V_{w1} \tag{7.13}$$

$$\text{Exit x-velocity} = -V_{w2} \quad \text{Direction accounted by the – sign} \tag{7.14}$$

Using this we can write down the force as

$$F_x = \rho A V_{r1} \times \mathrm{Vw1} - (-V_{w2})] = \rho A V_{r1} \times (V_{w1} + V_{w2}) \tag{7.15}$$

In the above expression, we have accounted for the fact that the swirl velocity at the exit faces the opposite direction to that of the motion of the surface. We can have two more cases:

7. When $\beta = 0$, there is no swirl component and we have $F_x = \rho A V_{r1} V_{w1}$.
8. When B > 90°, then the swirl component at the exit faces the same direction as that of the motion (the keen reader can easily verify this), and therefore the expression for the force is modified to $F_x = \rho A V_{r1} (V_{w1} - V_{w2})$.

We can therefore find out the power output of this process as

$$P = \rho A V_{r1} \, (V_{w1} + V_{w2} \times u) \tag{7.16}$$

We can also find out the work done per second per weight of the fluid impinging on the surface as

$$= \frac{P}{\rho A V_{r1} g} = \frac{1}{g}\left(V_{w1} + V_{w2}\right) \tag{7.17}$$

We note that in both the expressions above we have used the expression for the force to be $F_x = \rho A V_{r2} \times (V_{w1} + V_{w2})$. For other specific cases depending on the angle β one must use the appropriate expression.

We can define the efficiency of the jet, η as the ratio of the power output from the moving surface to the power input of the jet (which is the incoming kinetic energy of the jet, i.e. $\dot{m} \times V_1^2 / 2$):

$$\eta = \frac{\rho A V_{r1}\left(V_{w1} + V_{w2}\right) \times u}{\frac{1}{2}\left(\rho A V_1\right) V_1^2} \tag{7.18}$$

Example 7.4

Consider a water jet of velocity 50 m/s striking a moving vane horizontally as shown in figure. The vane moves with a velocity of 10 m/s in the horizontal direction. The jet leaves the vane at an angle of 30° with the horizontal (with respect to the moving vane). What is the resultant force and direction on the vane if the jet mass flow rate is 1 kg/s?

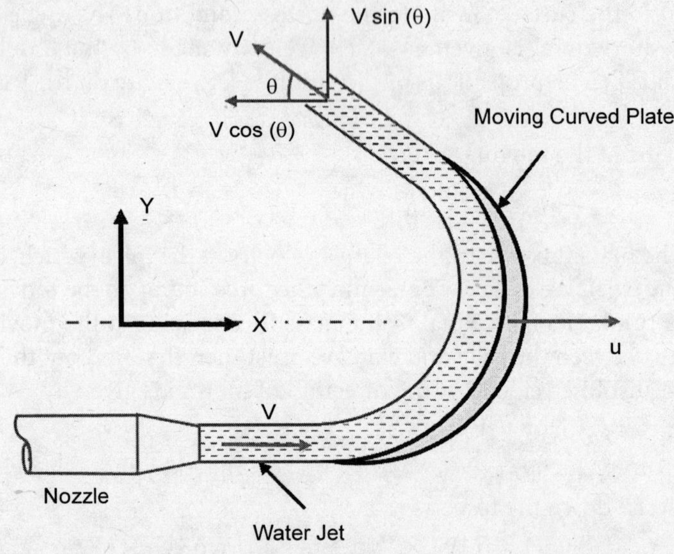

Solution:

Given data:

Velocity of the jet $V = 50$ m/s

Velocity of the vane $u = 10$ m/s

Initial velocity in the direction of x,

$$V_{1x} = V - u = 40 \text{ m/s}$$

Final velocity in the direction of x,

$$V_{2x} = -(V - u) \times \cos 30° = -(50 - 10) \times \cos 30° = -34.64 \text{ m/s}$$

Initial velocity in the direction of y,

$$V_{1y} = 0 \text{ m/s}$$

Final velocity in the direction of y,

$$V_{2y} = (V - u) \times \sin 30° = (50 - 10) \times \sin 30° = 20 \text{ m/s}$$

Force in the direction of jet per unit weight,

$$F_x = \dot{m}\left[V_{1x} - V_{2x}\right] = 1 \times \left[40 + 34.64\right] = 74.64\,N$$

The force exerted by the jet in the direction perpendicular to the direction of the jet, per unit weight,

$$F_y = \dot{m}\left[V_{1y} - V_{2y}\right] = 1 \times \left[0 - 20\right] = -20\,N$$

The resultant force is given by,

$$F_R = \sqrt{F_x^2 + F_y^2} = \sqrt{(74.64)^2 + (-20)^2} = 77.27\,N$$

And the angle made by the resultant with the horizontal is given by,

$$tan\alpha = \frac{F_y}{F_x} = \frac{-20}{74.64} = -0.2679$$

$$\alpha = -15° \text{ with horizontal}$$

Example 7.5

Consider a water jet of diameter 40 mm and velocity 25 m/s striking a curved vane moving at 10 m/s in the same direction as the jet. The jet at the inlet makes 0° with the direction of motion of the vane whereas it makes 60° with the direction of motion of the vane at the outlet. Determine the: (a) vane angle at the outlet, (b) work done by the jet and (c) jet efficiency.

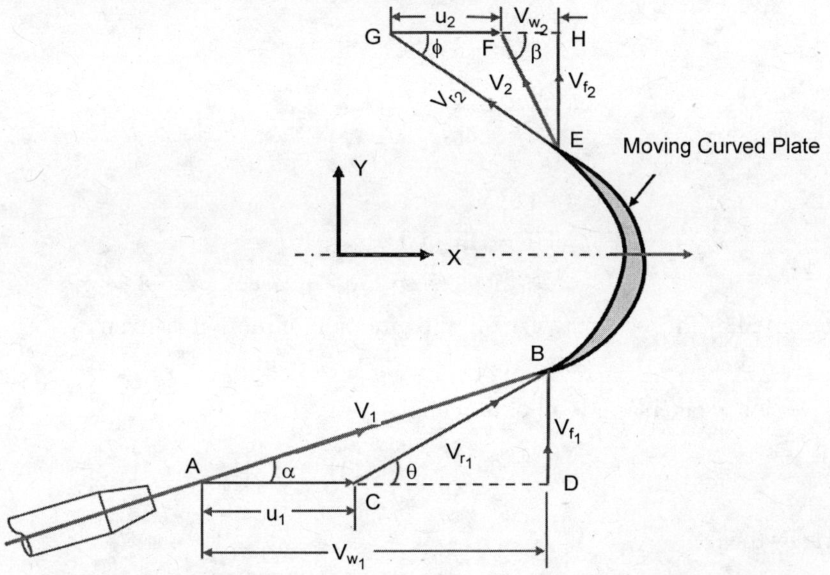

Solution:

Given data:

Diameter of the jet $d = 40$ mm

Velocity of the jet $V_1 = 25$ m/s

Velocity of the vane $u_1 = u_2 = u = 10$ m/s

As jet and vane are moving in the same direction,

$$\alpha = 0°$$

Angle made by the leaving jet, with the direction of motion = 60°

$$\beta = 80° - 60° = 120°$$

For the given problem,

$$V_{r1} = V_{r2}$$

$$V_{r1} = AB - AC = V_1 - u_1 = 25 - 10 = 15 \text{ m/s}$$

$$V_{w1} = V_1 = 25 \text{ m/s}$$

$$V_{r2} = V_{r1} = 15 \text{ m/s}$$

Now in ΔEFG,

$$EG = V_{r2} = 15 \text{ m/s}$$

$$GF = u_2 = 10 \text{ m/s}$$

(a) Vane angle at the outlet,

$$\angle GEF = 180° - (60° + \phi) = (120° - \phi)$$

From sine rule, we have

$$\frac{EG}{\sin 60^\circ} = \frac{GF}{\sin\left(120^\circ - \phi\right)}$$

or

$$\frac{15}{\sin 60^\circ} = \frac{10}{\sin\left(120^\circ - \phi\right)}$$

$$\phi = 84.8°$$

$$V_{w2} = HF = GF - GH$$

$$V_{w2} = u_2 - Vr_2 \cos\phi = 10 - 15 \times \cos 60° = 8.64 \text{ m/s}$$

(b) The force exerted by the jet on the vane in the direction of motion is given by,

$$F_x = \rho A Vr_1 \left[V_{w1} - V_{w2}\right]$$

The negative sign is taken as β is an obtuse angle.

$$F_x = 1000 \times \frac{\pi}{4} \times (0.040)^2 \times 15 \times \left[25 - 8.64\right] = 308.37 N$$

Work done by the jet,

$$W = F_x \times u = 308.37 \times 10 = 3083.7 \text{ W}$$

(c) The jet efficiency,

$$\eta = \frac{\text{Work done by the jet}}{\text{Initial K.E per second of the jet}}$$

$$\eta = \frac{3083.7}{\frac{1}{2} \times 1000 \times \frac{\pi}{4} \times (0.040)^2 \times 25^3} = \frac{3083.7}{9817} = 31.4\%$$

Example 7.6

A jet at 24 m/s strikes a curved vane moving at 10 m/s. The jet deflects by 120° between the inlet to the outlet of the vane. The vane angle at the outlet is 30°. What should be the jet angle at the inlet with respect to the vane's direction of motion so that there is no shock as it strikes the moving vane? What is the absolute velocity of the jet at the exit of the moving vane?

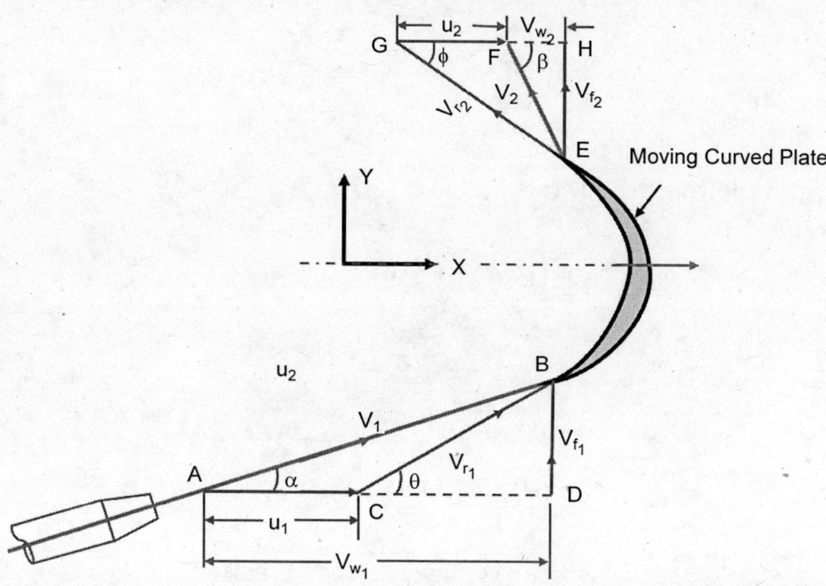

Solution:

Given data:

Velocity of the jet $V_1 = 24$ m/s

Velocity of the vane $u_1 = u_2 = u = 10$ m/s

The angle of deflection of jet = 120°

$$\theta + \phi = 180° - 120° = 60°$$

It is not given that the vane is symmetrical and without this condition, the problem cannot be solved. Assuming the vane to be symmetrical, we have,

$$\theta = \phi = 30°$$

The angle of a jet at the inlet with the direction of motion of the vane = α

In Δ ABC, applying the sine rule, we have

$$\frac{AB}{\sin\left(180^\circ - \theta\right)} = \frac{AC}{\sin\left(30^\circ - \alpha\right)}$$

or

$$\frac{V_1}{\sin\theta} = \frac{u_1}{\sin\left(30^\circ - \alpha\right)}$$

or

$$\frac{24}{\sin 30^\circ} = \frac{10}{\sin\left(30^\circ - \alpha\right)}$$

or

$$\sin\left(30^\circ - \alpha\right) = \frac{10}{48}$$

$$\alpha = 18^\circ$$

In Δ ABC, again applying the sine rule, we have

$$\frac{V_1}{\sin\left(180^\circ - \theta\right)} = \frac{V_{r1}}{\sin\alpha}$$

or

$$\frac{24}{\sin\theta} = \frac{V_{r1}}{\sin 18^\circ}$$

$$V_{r1} = V_{r2} = 14.83 \text{ m/s}$$

In ΔABD,

$$V_{w1} = V_1 \times \cos\alpha = 24 \times \cos 18^\circ = 22.82 \text{ m/s}$$

At outlet, from ΔEGH, we have,

$$V_{r2} \cos\phi = u_2 + V_{w2}$$

or

$$14.83 \times \cos 30^\circ = 5 + V_{w2}$$

$$V_{w2} = 2.8 \text{ m/s}$$

$$V_{f2} = V_{r2} \sin 30^\circ = 14.83 \times \sin 30^\circ = 7.415 \text{ m/s}$$

$$V_2 = \sqrt{V_{f2}^2 + V_{w2}^2}$$

$$V_2 = \sqrt{(7.415)^2 + (2.8)^2} = 7.926 m/s$$

7.4 JET STRIKING A SERIES OF VANES/PADDLES

7.4.1 Jet Striking a Rotating Wheel Consisting of Vanes

We can extend the analysis to that of a wheel on which a series of paddles are connected. Let there be only one jet striking the paddles as the wheel begins to rotate. We can assume that at some point the wheel reaches equilibrium, i.e. the rotational speed reaches constant. As usual, let us follow the standard nomenclature of V being the velocity of the jet, A being the area of cross section of the jet, and u be the tangential velocity of the wheel at the point where the jet impinges on the paddles.

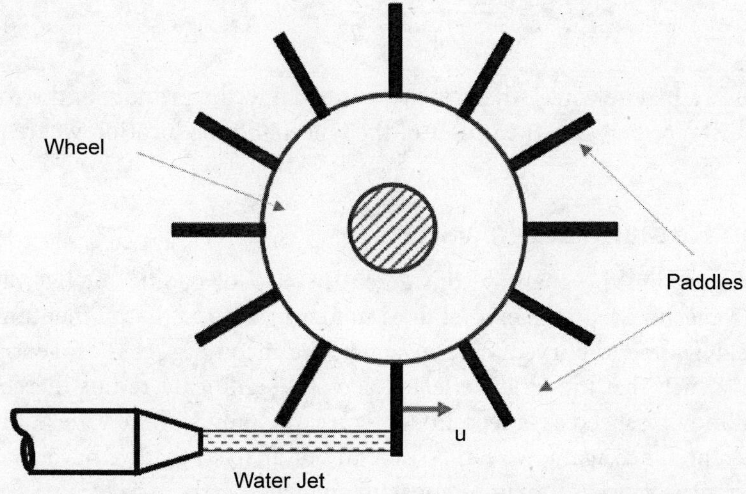

Figure 7.7: Jet striking a series of paddles or vanes.

We further consider that after the jet strikes the paddle, the jet is deflected in such a way that it is tangential to the paddle. Therefore, the exit jet causes no change in momentum of the paddle and the wheel in the tangential direction of the jet. The force exerted by the jet in the direction of the jet is given by

$$Fx = \rho AV(V - u) \tag{7.19}$$

where we have already taken the velocity at the exit as 0, while the relative velocity at the inlet is $(V - u)$

The power transferred by the jet is given by

$$P = F_x u = \rho AV(V - u)u \tag{7.20}$$

The incoming kinetic energy of the jet per unit time is

$$P_i = \frac{1}{2}\dot{m}V^2 = \frac{1}{2}\rho VAV^2 = \frac{1}{2}\rho AV^3 \tag{7.21}$$

We thus obtain the efficiency of such a paddle wheel as

$$\eta = \frac{\text{Power output}}{\text{Power input}} = \frac{2u(V-u)}{V^2} \tag{7.22}$$

Maximum efficiency condition

With the expression for the efficiency, we can find out a relationship between the jet velocity V and the paddle velocity u. This can be done by finding out the derivative of the efficiency with respect to u as shown below

$$\frac{d\eta}{du} = 0 \Rightarrow \frac{2V - 4u}{V^2} = 0 \Rightarrow u = \frac{V}{2} \tag{7.23}$$

For this condition, the value of the efficiency is

$$\eta = \frac{2u(2u - u)}{(2u)^2} = \frac{1}{2} = 50\% \tag{7.24}$$

In the case of such a paddle wheel, the velocity u is equal to the product of the rotational speed of the wheel times the distance between the center of the wheel and the location where the jet strikes the paddle, i.e. $u = \omega R$.

7.4.2 Force on Radial Curved Vanes

While in the previous section the analysis considered the case of paddles or flat vanes, we saw that the jet impinges the vanes in a way the exit jet does not contribute to the momentum transfer. In this section we will consider a radial curved vane in which the incoming jet is tangential to the surface. The moving surface is such that the jet is deflected across a significant radius thereby making the jet encounter a different moving speed as is seen in the figure; the radius of the vane at the inlet and outlet are significantly different. Once again, we can assume in our analysis that we have reached steady state, i.e. the wheel which has the vanes is moving a constant angular speed.

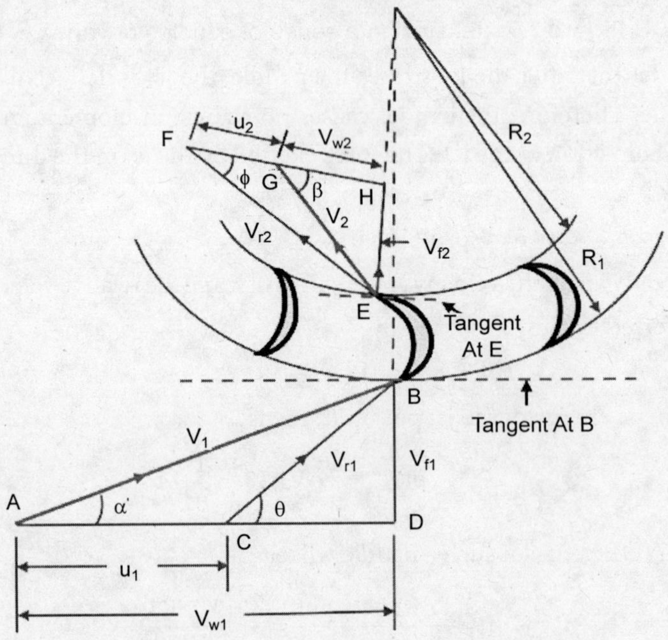

Figure 7.8: Velocity triangle for radial curved vane.

Let R_1 and R_2 be the radius of the wheel at the inlet vane and outlet vane respectively. If the angular speed of the wheel is ω. In that case, the vane speed seen by the jet at the inlet will be $u_1 = \omega R_1$ while the vane speed seen by the jet at the outlet will be $u_2 = \omega R_2$. The corresponding velocity triangles are shown in the figure.

The mass flux striking the vanes is given by

$$\dot{m} = \rho A V_1 \tag{7.25}$$

The momentum of the incoming jet can therefore be written in terms of the swirl component as

$$\rho A V_1 \, V_{w1} \,, \; V_{w1} = V_1 \cos\alpha \tag{7.26}$$

whereas the momentum of the exiting jet can be written as

$$\rho A V_1 \, (-V_{w2}) \quad , \quad V_{w2} = V_2 \cos\beta \tag{7.27}$$

where we have appropriately accounted for the sign. We have also expressed the swirl components at the inlet and outlet.

We can now express the angular momentum at the inlet and outlet as

$$\text{At inlet:} = \rho A V_1 \, V_{w1} \, R_1 \tag{7.28}$$

$$\text{At outlet:} = -\rho A V_1 \, V_{w2} \, R_2 \tag{7.29}$$

The difference in the angular momentum between the inlet and exit is therefore equal to the torque exerted by the jet on the wheel and can be therefore expressed as

$$\mathrm{T} = \rho A V_1 (V_{w1} \, R_1 - (-V_{w2} \, R_2)) = \rho A V_1 \, (V_{w1} R_1 + V_{w2} \, R_2) \tag{7.30}$$

The power transferred to the wheel containing the vanes can be expressed as

$$P = T\omega = \rho A V_1 (V_{w1} R_1 + V_{w2} \, R_2)\omega = \rho A V_1 (V_{w1} \, u_1 + V_{w2} u_2) \tag{7.31}$$

where we have made use of the linear velocities at the two radii of the wheel.

The above expression for power is true for the case of an acute angle β. For obtuse angle, the sign inside the parenthesis will change. In fact, we can express the expression for power in terms of a single expression as

$$P = \rho A V_1 \, (Vw_1 \, u_1 \pm V_{w2} \, u_2) \tag{7.32}$$

We can now proceed to determine the efficiency of such a radial curved vane.

Efficiency of a radial curved vane

The efficiency can be defined, as done earlier, as the ratio of the power transferred the vane carrying wheel to the power contained in the incoming jet. Therefore, we have

$$\eta = \frac{\text{Power transferred to wheel}}{\text{Power incoming with jet}} = \frac{\rho A V_1 \left(V_{w_1} u_1 \pm V_{w_2} u_2 \right)}{\frac{1}{2} \dot{m} V_1^2} = \frac{2\left(V_{w_1} u_1 \pm V_{w_2} u_2 \right)}{V_1^2} \tag{7.33}$$

In case we consider the situation where there is no loss in energy of the jet when it flows over the radial vanes, then we can state, from an overall conservation of energy, that the difference in the kinetic energy held by the jet per second would be equal to the power transferred to the wheel carrying the radial vanes. Essentially then we have

$$P = \frac{1}{2}\dot{m}\left(V_1^2 - V_2^2\right) = \frac{1}{2}\rho A V_1^2\left(V_1^2 - V_2^2\right) \tag{7.34}$$

In such a case, the efficiency can be found by as

$$\eta = \frac{\frac{1}{2}\rho A V_1^2\left(V_1^2 - V_2^2\right)}{\frac{1}{2}\rho A V_1^2 V_1^2} = \frac{V_1^2 - V_2^2}{V_1^2} = \left(1 - \frac{V_2^2}{V_1^2}\right) \tag{7.34}$$

Example 7.7

A jet of water at 45 m/s strikes a series of radial curved vanes mounted on a wheel rotating at 300 r.p.m. The jet makes an angle of 20° with the tangent to the wheel at the inlet. At the outlet, the jet has an absolute velocity of 7.5 m/s and it makes 130° with respect to the tangent to the wheel. The wheel diameter at the inlet and outlet are 1.0 m and 0.5 m respectively. Determine the: (a) Vane angles at the inlet and outlet and (b) Efficiency.

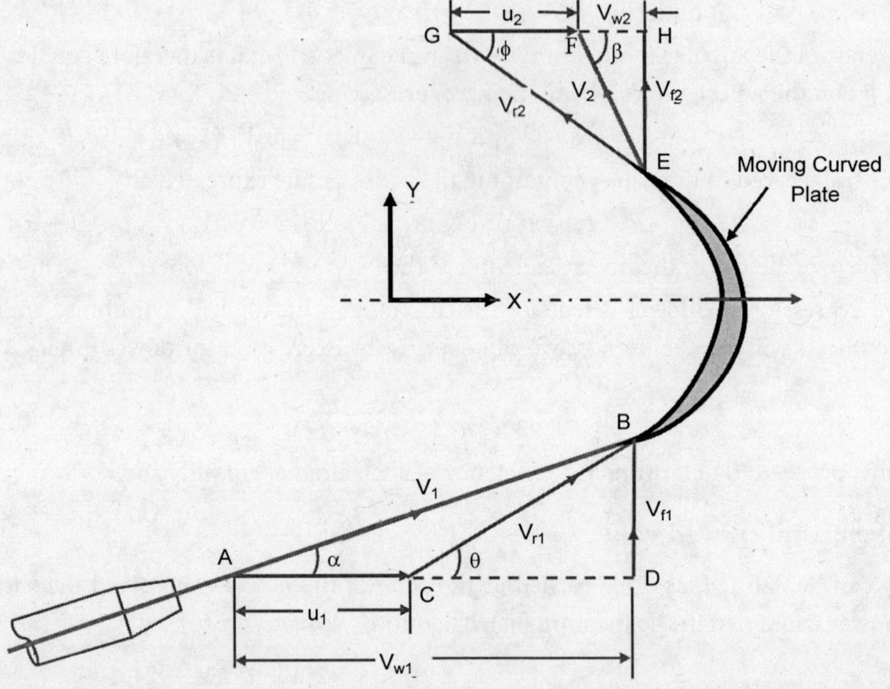

Solution:

Given data:

Velocity of jet at inlet $V_1 = 45$ m/s

Velocity of jet at outlet $V_2 = 7.5$ m/s

Rotating speed of the wheel N = 300 r.p.m

Jet angle at inlet $\alpha = 20$

Jet angle at outlet $\beta = 180 - 130 = 50$

Wheel diameter at inlet $D_1 = 1.0$ m

Wheel diameter at outlet $D_2 = 0.5$ m

Wheel velocity at inlet,

$$u_1 = \frac{\pi D_1 N}{60} = \frac{\pi \times 1.0 \times 300}{60} = 15.7\, m/s$$

Wheel velocity at outlet,

$$u_2 = \frac{\pi D_2 N}{60} = \frac{\pi \times 0.5 \times 300}{60} = 7.85\, m/s$$

At inlet, from ΔABD, we have,

$$V_{f1} = V_1 \sin\alpha = 45 \times \sin(20) = 15.23 \text{ m/s}$$

$$V_{w1} = V_1 \cos\alpha = 45 \times \cos(20) = 42.28 \text{ m/s}$$

(a) Vane angle at inlet is given by,

$$\tan\theta = \frac{V_{f1}}{V_{w1} - u_1} = \frac{15.23}{42.28 - 15.7} = 0.5729$$

$$\theta = \tan^{-1}(0.5729) = 30°$$

At outlet, from ΔEFH, we have,

$$V_{f2} = V_2 \sin\beta = 7.5 \times \sin(50) = 5.75 \text{ m/s}$$

Again, at outlet, from EFH, we have,

$$V_{w2} = V_2 \cos\beta = 7.5 \times \cos(50) = 4.28 \text{ m/s}$$

Vane angle at outlet is given by,

$$\tan\phi = \frac{V_{f2}}{V_{w2} + u_2} = \frac{5.75}{4.28 + 7.85} = 0.4740$$

$$\phi = \tan^{-1}(0.4740) = 25.36°$$

(b) Efficiency is given by,

$$\eta = \frac{2(u_1 V_{w1} + u_2 V_{w2})}{V_1^2}$$

$$\eta = \frac{2(15.7 \times 42.28 + 7.85 \times 4.82)}{45^2} = 69.3\%$$

7.5 LAYOUT OF A HYDROELECTRIC POWER PLANT

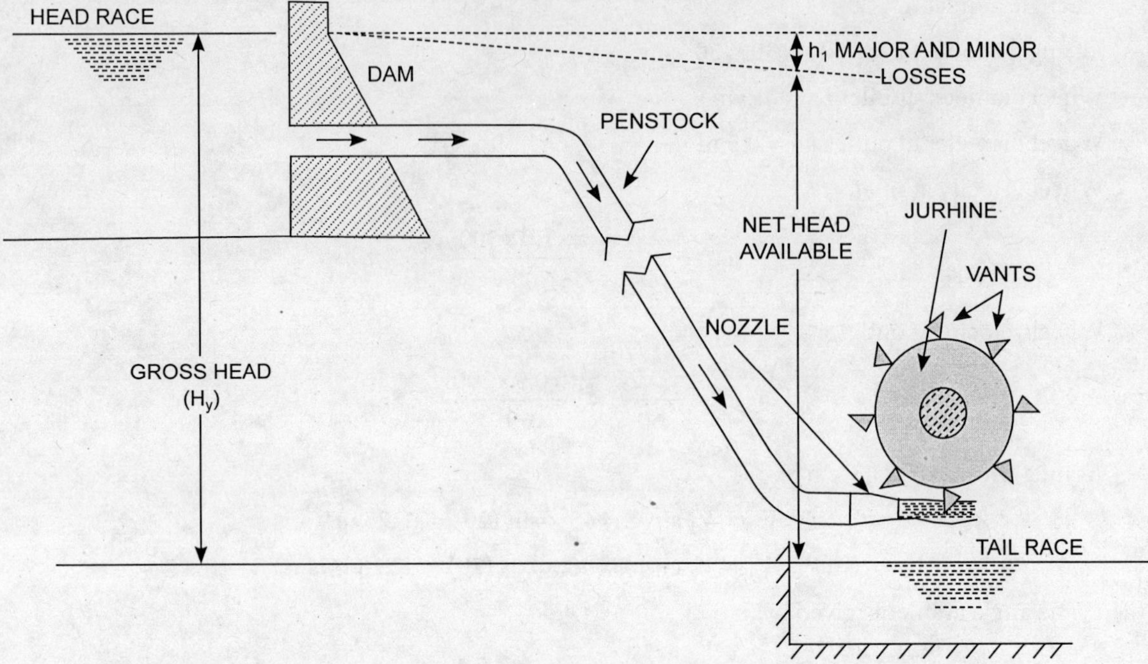

Figure 7.9: Schematic of a power plant showing various components.

Typically, a hydroelectric power plant consists of

9. Dam and reservoir - to hold a large amount of water
10. Penstock - Large diameter pipes which carry water from the reservoir to the turbines
11. Turbine - to generate power via coupling to a generator
12. Tail race - Discharges water from the turbine

7.6 TYPES OF HYDRAULIC TURBINES

Depending on the inlet conditions such as the energy available, direction of flow, head at inlet, and specific turbine speed we can classify turbines. The most common classification is done based on the energy available at the inlet. They are classified as (a) Impulse turbine, (b) Reaction turbine.

Impulse turbine

When the energy of water at the inlet of the turbine is available as kinetic energy, then an impulse turbine is used. As the water flows over the vanes of the turbine, the pressure is atmospheric.

Reaction turbine

When, at the inlet of the turbine, the incoming water has significant amount of kinetic energy, and pressure head, we employ a reaction turbine. As the water flows over the vanes of the turbine, there is a transformation of the pressure head into kinetic energy as well. In this case, the runner is not open to the atmosphere and is sealed.

Apart from these two types, there are other classification of turbines as well. Vanes are attached to the shaft of the turbine on a part called as a runner. When water flows tangent to the runner, the turbine is called as a tangential flow turbine. When the water flows radially through the runner, the turbine is called as a radial flow turbine. For radial turbines, there are further two classifications. When the water flows radially inwards into a turbine, it is called as an inward radial flow turbine. Conversely, when the water flows radially outwards into the turbine, the turbine is called as an outward radial flow turbine. An axial turbine is one in which the flow is coaxial to the runner direction.

7.7 PELTON TURBINE

A Pelton wheel is the most commonly employed impulse turbine. In this kind of turbine there is a jet of water which strikes the runner which carries a series of buckets. As the incoming water has only kinetic energy with it (the inlet and outlet of the turbine is at atmospheric pressure), this kind of turbine works by converting the incoming energy of the jet into rotation of the wheel through the momentum transfer at the buckets. Hence this kind of turbine is known as an impulse turbine. A Pelton wheel typically consists of

1. Nozzle with flow regulation provision
2. Runner with buckets
3. casing
4. breaking jet

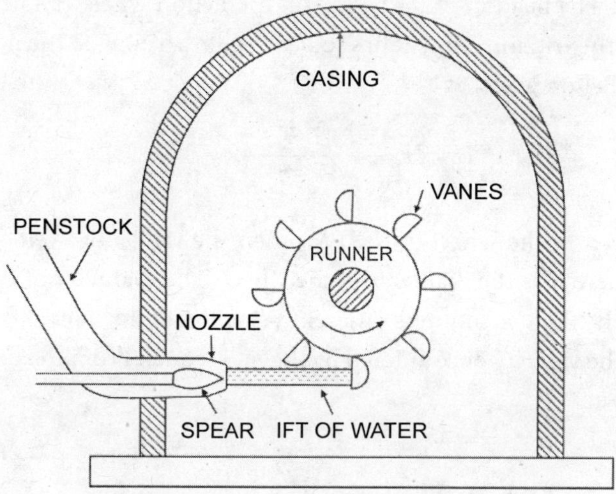

Figure 7.10: Schematic of a Pelton turbine.

The nozzle is responsible for the incoming jet of water. This is fed by the penstock and converts the head at the reservoir of the dam into the high kinetic energy jet. A conical needle, known as a spear, is arranged in such a way at the exit that rotating the wheel causes the spear to move in and out, thereby changing the area of the exit of the nozzle, thereby allowing a degree of control over the flow rate. The Pelton wheel contains a circular disc, i.e. the runner, with a number of buckets. The shape of the buckets is kept in such a way that the jet striking the bucket is split and turned in the opposite direction of the incoming jet. The single jet is split into two symmetric parts by a partition on the bucket wall, which is called as a splitter. This causes the large transfer of momentum from the jet to the rotating wheel.

The casing of the turbine is primarily present to prevent water splashing and any untoward accident. As such, the casing does not have any other function. The water eventually discharges out from the casing into the tail race.

When the turbine is to be stopped, there is a nozzle which faces the opposite direction to that of the main jet. The main jet is turned off and then the breaking jet is activated. This decelerates the wheel, thereby allowing the jet to stop within a short span of time.

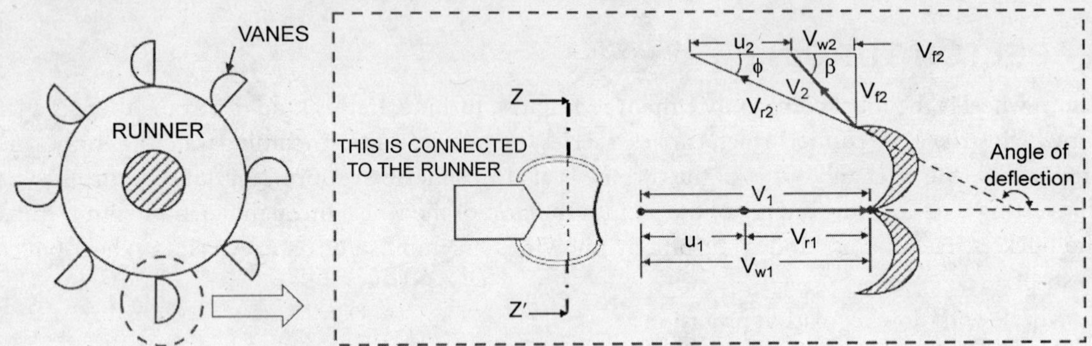

Figure 7.11: Velocity triangle at the bucket.

The Pelton wheel is characterized by the bucket shape which has an angle of deflection, ϕ, with respect to the horizontal. The head of water acting on the Pelton wheel can be obtained by the head at the penstock reduced by the frictional loss it has to bear while arriving at the nozzle. Thus, we can write the available head at the Pelton wheel as

$$H = H_p - f\frac{4L}{D_p}\frac{V^2}{2g} \tag{7.36}$$

If the rotational speed of the wheel is N RPM, then the tangential velocity of the bucket will be given as $u = \pi DN/60$, where D is the diameter of the wheel. The diameter of the jet is d. As opposed to reaction turbines where the fluid encounters different velocities at the inlet and exit section owing to the difference in diameters, the velocity encountered by the jet is equal. From the derivations in the previous section, we can write down

$$V_{r1} = V_1 - u \tag{7.37}$$

$$V_{w1} = V_1 \tag{7.38}$$

where, the inlet angles α and θ are both 0°.

At the outlet velocity triangle, we have

$$V_{r2} = V_{r1} \tag{7.39}$$

$$V_{w2} = V_{r2}\cos\phi - u \tag{7.40}$$

The transfer in momentum between the incoming and outgoing jet to the wheel is given by

$$F = \rho A V_1 (V_{w1} + V_{w2}) \tag{7.41}$$

where we have already accounted for the sign of V_{w2}.

The power generated at the wheel, i.e. the power transferred to the runner by the jet, is therefore

$$P = Fu = \rho A V_1 (V_{w1} + V_{w2}) \tag{7.42}$$

The work done per unit weight of water can therefore be written as

$$= \frac{P}{\dot{m}g} = \frac{\rho A V_1 \left(V_{w_1} + V_{w_2}\right)}{\rho A V_1 g} = \frac{\left(V_{w_1} + V_{w_2}\right)u}{g} \tag{7.43}$$

Similarly, the hydraulic efficiency of the turbine is given by

$$\eta = \frac{P}{\dot{m}\frac{V_1^2}{2}} = \frac{2\left(V_{w_1} + V_{w_2}\right)u}{V_1^2} \tag{7.44}$$

We can express the swirl velocities at the inlet and exit as

$$V_{w1} = V_1 - u; \; V_{w2} = V_{r2} \cos\phi - u = V_{r1} \cos\phi - u(V_1 - u)\cos\phi - u) \tag{7.45}$$

where we have made use of the fact that $V_{r1} = V_{r2}$, i.e. the flow velocities are equal. By substituting these values in the expression for the efficiency, we obtain

$$\eta = \frac{2(V_1 - u)(1 + \cos\phi)u}{V_1^2} \tag{7.46}$$

We leave it as an exercise to the reader to show that the maximum efficiency for a fixed V_1 is obtained when

$$u = \frac{V_1}{2} \tag{7.47}$$

The corresponding efficiency for this tangential velocity is given by

$$\eta = \frac{1 + \cos\phi}{2} \tag{7.48}$$

Typically for Pelton wheel, we have a parameter called as the jet ratio which is the ratio of the diameter of the wheel, more generally, it is the pitch diameter of the buckets, to the diameter of the jet. i.e. $m = \frac{D}{d}$. Typically, this value is kept around 12. The number of buckets on a runner is also given by a rule of thumb as $Z = 15 + \frac{D}{2d}$.

Example 7.8

A Pelton wheel has a bucket speed of 10 m/s with a jet of water flowing at 700 liter/sec under a head of 60 m. The bucket deflects the jet by 150°. Assume coefficient of velocity as 0.96. Calculate the (a) power generated and (b) hydraulic efficiency.

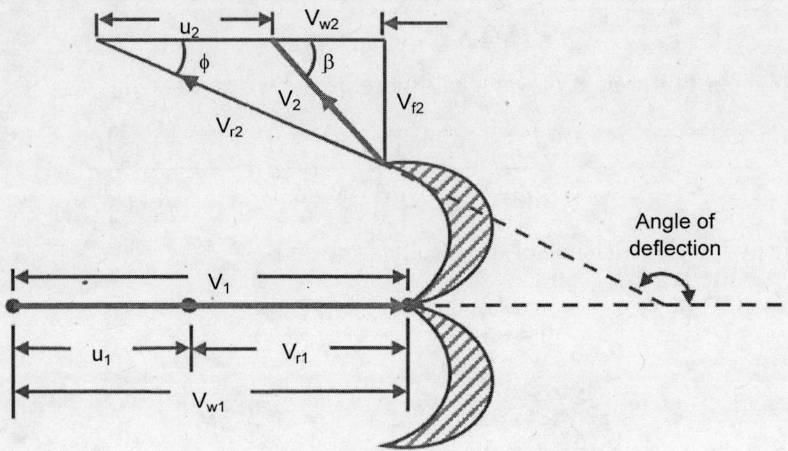

Solution:

Given data:

Velocity of bucket $u_1 = u_2 = 10$ m/s

Discharge Q = 700 liter/sec

Head of water H = 60 m

Jet angle of deflection = 150°

= 180 – 150 = 30

Coefficient of velocity $C_v = 0.96$

Velocity of jet at inlet is given by,

$$V_1 = C_v\sqrt{2gH} = 0.96\times\sqrt{2\times 9.81\times 60} = 32.94\, m/s$$

At inlet,

$$Vr_1 = V_1 - u_1 = 32.94 - 10 = 22.94 \text{ m/s}$$

From the Pelton wheel properties,

$$V_{w1} = V_1 = 32.94 \text{ m/s}$$

Assume that there is no friction loss from inlet to outlet, we have,

$$V_{r1} = V_{r2} = 22.94 \text{ m/s}$$

From outlet velocity triangle,

$$V_{w2} = V_{r2}\cos\phi - u_2 = 22.94 \times \cos 30° - 10 = 9.87 \text{ m/s}$$

(a) Power generated,

$$P = \frac{\rho Q\left[V_{w1} + V_{w2}\right]u}{1000} = \frac{1000\times 0.7\times\left[32.94 + 9.87\right]\times 10}{1000} = 299.7\, kW$$

(b) Hydraulic efficiency of Pelton wheel,

$$\eta_h = \frac{2\left[V_{w1} + V_{w2}\right]u}{V_1^2} = \frac{2\times\left[32.94 + 9.87\right]\times 10}{32.94^2} = 79\%$$

7.8 FRANCIS TURBINE

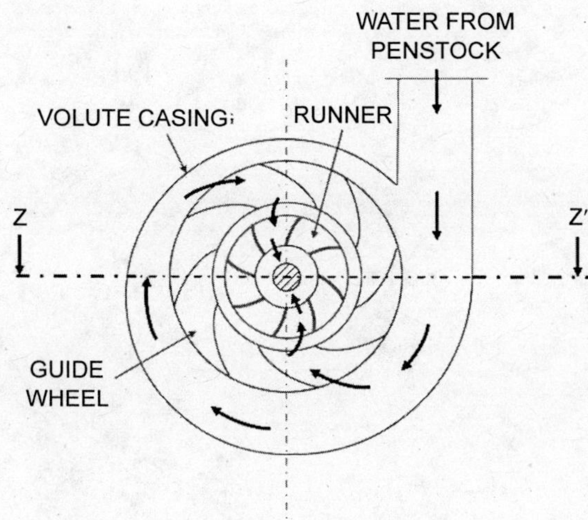

Figure 7.12: Schematic of a Francis turbine.

A Francis turbine is an inward flow reaction turbine. In this turbine water enters the runner of the turbine at the outer periphery after which it flows over the vanes towards the center of the turbine. Once there it then bends by 90° at which point the flow then leaves the runner along the same direction as the axis of the turbine. There are some important parameters which are useful for describing the Francis turbines. They are the aspect ratio of the turbine, i.e. $n = \frac{B}{D}$, where represents the breadth of the turbine, and represents the diameter of the turbine. This ratio is typically around 0.1 – 0.4. The flow ratio is given by the flow velocity at the inlet to the velocity obtained if the entire head at the inlet were to be converted into kinetic energy, i.e. Flow ratio $= \frac{V_{f_1}}{\sqrt{2gH}}$ which is typically between 0.15 to 0.3. The speed ratio is the ratio of the rotational speed of the rpm to the velocity obtained if the entire head at the inlet were to be converted into kinetic energy i.e. Speed ratio $= \frac{u_1}{\sqrt{2gH}}$ which varies between 0.6 to 0.9.

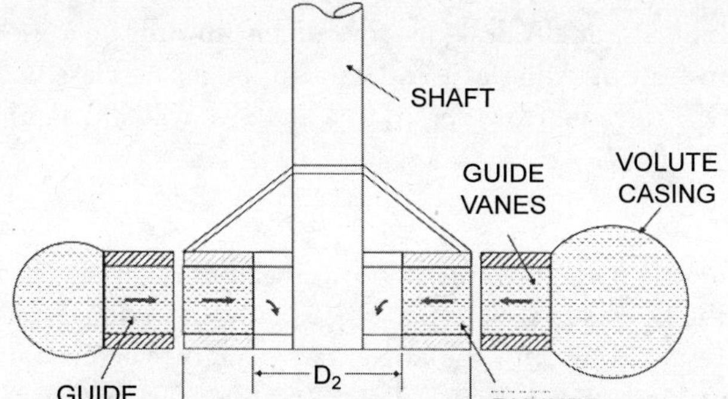

Figure 7.13: Flow of the jet through a Francis turbine.

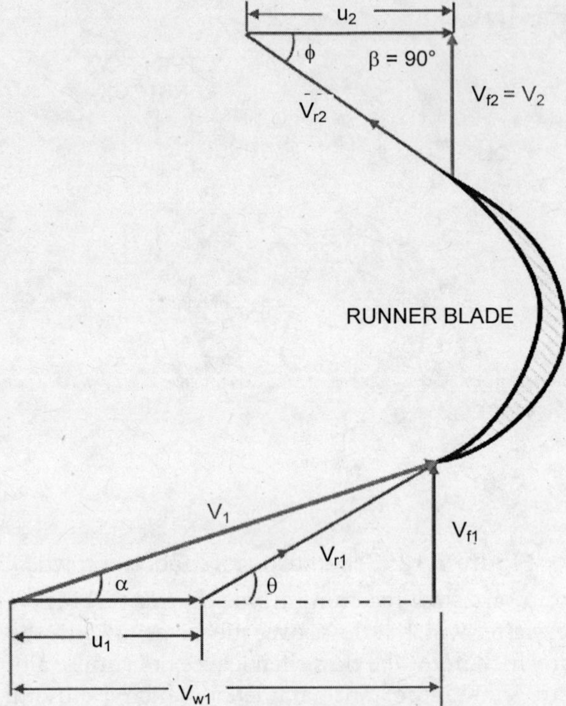

Figure 7.14: Velocity triangle at the vanes of the Francis turbine.

Figure 7.12 and 7.13 show radial exit of water from the runner which ensures that the absolute velocity (V_2) at exit of blade makes 90° with U_2. It must be noted that the blade angle, which is the angle between V_{r2} and U_2 is not at exit. At the inlet there are guide vanes to ensure that the angle V_{r1} makes with the U_1 matches the blade angle. This minimises losses (like shock loss) during the entry of flow into the turbine blades. The velocity triangle is similar to what has been discussed earlier.

Volume flow rate through the blade at inlet of runner blade is given by $B_1\ \pi D_1\ V_{f1}$ (assuming negligible blade thickness). Where B_1 is the blade width at inlet and D_1 is the runner diameter at inlet. Similarly, the volume flow rate at the exit is $B_2\ \pi D_2 V_{f2}$. V_f is the flow velocity which is sometimes called the meridional velocity.

Though Francis turbine was initially designed for radially inwards flow, modern day Francis turbine runners allow axial exit instead of radial exit and can no longer be classified as radial type turbine. They are actually mixed type turbine. However, in this book we will still assume radial exit for solving numerical problems.

Example 7.9

A Francis turbine works at 450 r.p.m. under an unknown head of water H. The diameter of the impeller at the inlet is 2.4 m and the flow area is 0.8 m^2. The absolute and relative velocities at the inlet make an angle of 20° and 60° respectively with the tangential velocity. The absolute velocity at the exit of the turbine is at 90° to the tangential velocity. Determine the (a) power generated and (b) hydraulic efficiency if the coefficient of velocity for the nozzle is 0.78.

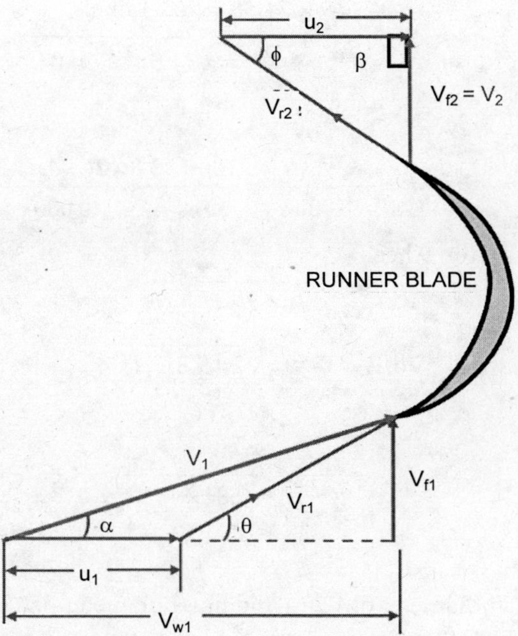

Solution:

Given data:

Speed of turbine N = 450 r.p.m

Diameter of impeller at inlet D_1 = 2.4 m

Flow area A_f = 0.8 m^2

Angle made by absolute velocity at inlet, $\alpha = 20°$

Angle made by relative velocity at inlet, $\theta = 60°$

Whirl at outlet, $V_{w2} = 0$

Tangential velocity of the turbine at inlet,

$$u_1 = \frac{\pi D_1 N}{60} = \frac{\pi \times 2.4 \times 450}{60} = 56.54\, m/s$$

From inlet velocity triangle,

$$Vf_1 = Vw_1 \tan\alpha = V_{w1} \times \tan 20°$$

$$Vf_1 = 0.3639\ V_{w1} \tag{7.49}$$

$$\tan\theta = \frac{V_{f1}}{V_{w1} - u_1}$$

$$\tan 60^{\circ} = \frac{0.3636 \times V_{w1}}{V_{w1} - 56.54}$$

$$V_{w1} = 71.58 \text{ m/s}$$

$$Vf_1 = 0.3639 \times 71.58 = 26.04 \text{ m/s}$$

$$V_1 = \sqrt{V_{w1}^2 + V_{f1}^2} = \sqrt{71.58^2 + 26.04^2} = 76.17\, m/s$$

(a) Power generated,

$$P = \frac{\rho Q V_{w1} u_1}{1000} = \frac{1000 \times 0.8 \times 26.04 \times 71.58 \times 56.54}{1000} = 85475.45\, kW$$

(b) Hydraulic efficiency of Pelton wheel,

$$V_1 = C_v \times \sqrt{2 \times g \times H}$$

$$76.17 = 0.78 \times \sqrt{2 \times 9.81 \times H}$$

$$H = 486 \text{ m}$$

$$\eta_h = \frac{V_{w1} u_1}{gH} = \frac{71.58 \times 56.54}{9.81 \times 486} = 84.88\%$$

Example 7.10

A Francis turbine with an outer diameter of 1.2 m and inner diameter of 0.6 m operates with a head of 44 m. The blade is designed such that the velocity of flow is constant = 3.5 m/s throughout the runner. The guide blade angle at the inlet is 10 and the runner vane angle is 90. At the outlet, the absolute velocity makes 90 to the tangential velocity of the runner. Determine the (a) angular speed of the turbine runner, (b) vane angle at the outlet and (c) hydraulic efficiency.

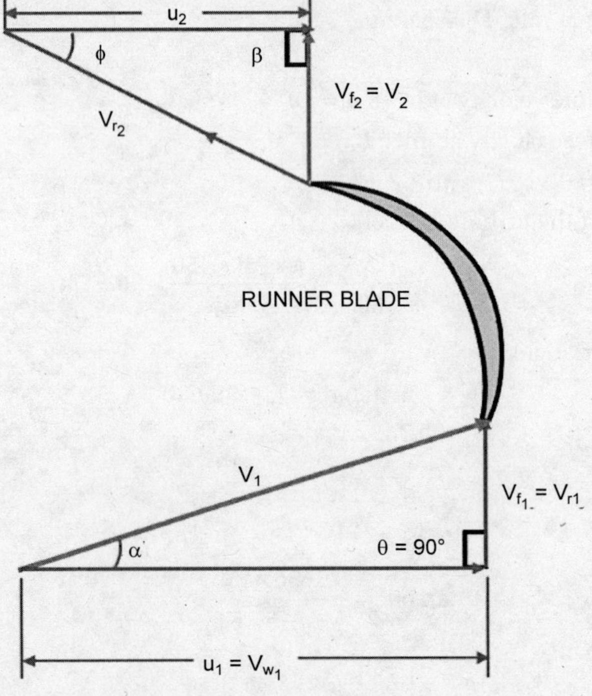

Solution:

Given data:

Diameter of impeller at inlet $D_1 = 1.2$ m

Diameter of impeller at outlet $D_2 = 0.6$ m

Head = 44 m

The velocity of flow from the inlet to the outlet is constant,

$$V_{f1} = V_{f2} = 3.5 \text{ m/s}$$

Guide blade angle at inlet $\alpha = 10$

From the inlet velocity triangle,

$$\tan\alpha = \frac{V_{f1}}{V_{w1}}$$

or

$$V_{w1} = \frac{V_{f1}}{\tan 10^\circ} = 19.84\, m/s$$

Also from inlet velocity triangle,

$$V_{w1} = u_1 = 19.84 \text{ m/s}$$

$$V_1 = \sqrt{V_{w1}^2 + V_{f1}^2} = \sqrt{19.84^2 + 3.5^2} = 20.14\, m/s$$

(a) Turbine speed,

$$u_1 = \frac{\pi D_1 N}{60}$$

$$19.84 = \frac{\pi \times 1.2 \times N}{60}$$

$$N = 316 \text{ r.p.m}$$

Now we determine the magnitude of velocity u_2

$$u_2 = \frac{\pi D_2 N}{60} = \frac{\pi \times 0.6 \times 316}{60} = 9.92\ m/s$$

(b) Vane angle at outlet,

$$\tan\phi = \frac{V_{f2}}{u_2} = \frac{3.5}{9.92} = 0.3528$$

$$\phi = \tan^{-1}(0.3528) = 19.43^\circ$$

(c) Hydraulic efficiency,

$$\eta_h = \frac{V_{w1} u_1}{gH} = \frac{19.84 \times 19.84}{9.81 \times 44} = 91.19\%$$

7.9 KAPLAN TURBINE

Axial flow turbines are a class of reaction turbines where the direction of the flow is parallel to the axis of rotation of the shaft. In reaction turbines, the pressure head is converted into kinetic energy. A Kaplan turbine is one such turbine whose axis of rotation is kept vertical. At the end of the turbine there is a hub onto which the vanes are attached. Such turbines are more generally classified as a propeller turbine. However, when the vanes are adjustable, then the turbine is called as a Kaplan turbine. The main parts of a Kaplan turbine are

1. Scroll casing
2. Guide vanes
3. Turbine runner vanes
4. Draft tube

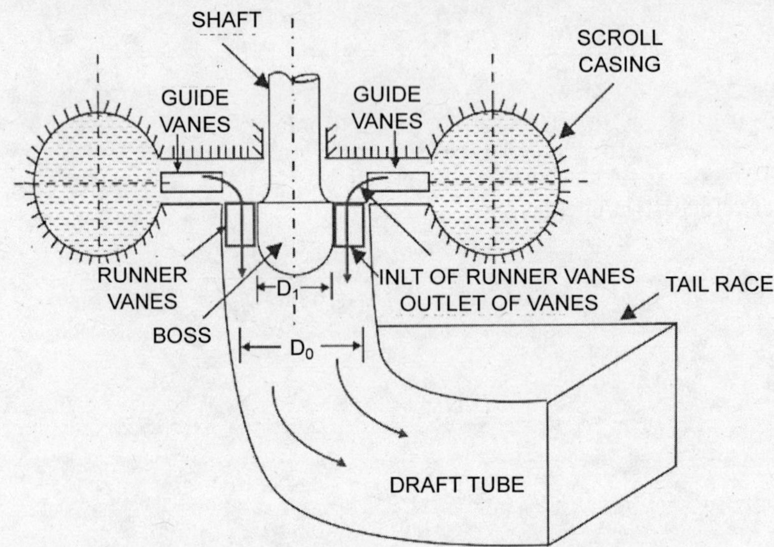

Figure 7.15: Schematic of a Kaplan turbine.

Water from the penstock enters the scroll casing and is guided through the guide vanes into the runner. After flowing over the vanes, the water is discharged through the draft tube. If the D_o represents the diameter of the runner, and D_b represents the diameter of the boss or hub, then the area of flow available is $\frac{\pi}{4}\left(D_o^2 - D_b^2\right)$. For such an axial turbine, the flow velocity at the inlet and outlet are also equal.

Example 7.11

A Kaplan turbine working under a head of 20 m develops a shaft power of 11772 kW. The outer diameter of the runner is 3.5 m and the hub diameter is 1.75 m. The inlet guide blade angle at the runner's outer tip is such that there is no shock and the inlet vane angle at the outer tip is 79°. At the outlet, the vane angle is 39° and the whirl velocity is negligible at the outer tip of the runner. The hydraulic efficiency of the turbine is 88%. Determine (a) the inlet guide vane angle at the outer tip, (b) the overall efficiency, and (c) the angular speed (r.p.m) of the turbine runner.

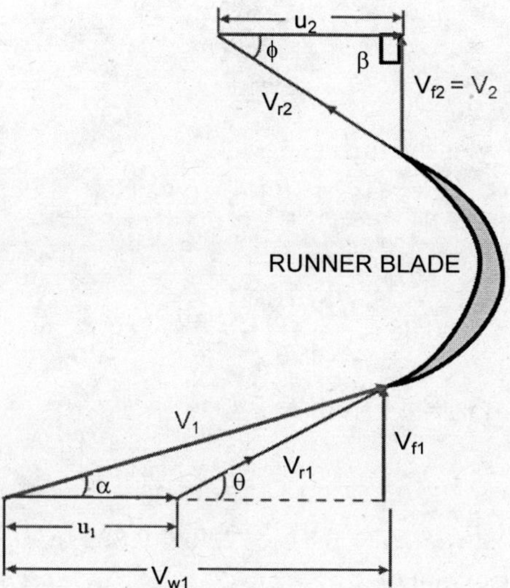

Solution:

Given data:

Head of turbine H = 20 m

shaft power = 11772 kW

Outer diameter of the runner D_0 = 3.5 m

Hub diameter D_h = 1.75 m

At inlet,

$\theta = 79$

At outlet,

$\phi = 39$

Hydraulic efficiency of the turbine is $\eta_h = 88\%$

From the properties of the Kaplan turbine we know,

$u_1 = u_2$, and $V_{f1} = V_{f2}$

Based on the outlet velocity triangle:

$$\tan\phi = \frac{V_{f2}}{u_2}$$

$$\tan 39^\circ = \frac{V_{f2}}{u_2}$$

$$V_{f2} = 0.81\, u_2 \tag{7.50}$$

Based on the inlet velocity triangle:

$$\tan\theta = \frac{V_{f1}}{V_{w1} - u_1} = \frac{V_{f2}}{V_{w1} - u_1} = \frac{0.81 \times u_2}{V_{w1} - u_1}$$

$$\tan 79^\circ = \frac{0.81 \times u_2}{V_{w1} - u_1}$$

$$u_1 = 0.864 \times V_{w1} \tag{7.51}$$

From the hydraulic efficiency of the turbine,

$$\eta_h = \frac{V_{w1} u_1}{gH} = \frac{V_{w1} \times 0.864 \times V_{w1}}{9.81 \times 20}$$

$$0.88 = \frac{0.864 \times V_{w1}^2}{196.2}$$

$$Vw_1 = 14.13 \text{ m/s}$$

Thus,

$$u_1 = 0.864 \times 14.13 = 12.20 \text{ m/s}$$

$$Vf_1 = V_{f2} = 0.81 \times 12.20 = 9.882 \text{ m/s}$$

(a) Based on the inlet velocity triangle:

$$\tan\alpha = \frac{V_{f1}}{V_{w1}} = \frac{9.882}{14.13} = 0.6993$$

$$\alpha = 34.96^\circ$$

Now we determine the value of discharge Q,

$$Q = \frac{\pi}{4} \times \left(D_o^2 - D_h^2\right) \times V_{f1}$$

$$Q = \frac{\pi}{4} \times \left(3.5^2 - 1.75^2\right) \times 9.882 = 71.29\, m^3/s$$

(b) Overall efficiency,

$$\eta_0 = \frac{\text{shaft power}}{\rho g Q H} = \frac{11772 \times 1000}{1000 \times 9.81 \times 71.29 \times 20} = 84.16\%$$

(c) Turbine speed,

$$u_1 = \frac{\pi D_o N}{60}$$

$$12.20 = \frac{\pi \times 3.5 \times N}{60}$$

$$N = 66.57 \text{ r.p.m}$$

Example 7.12

The hub diameter of a Kaplan turbine is 0.35 times the runner's outer diameter. The turbine is running at 200 r.p.m under a head of 48 m. The flow ratio is 0.6 and the hydraulic efficiency is 85%. The vane angle at the runner outlet is 15°. The velocity of the whirl is zero at the outlet. Determine (a) the outer diameter of the runner, (b) the water flow rate (m^3/s) through the runner, (c) the vane angle at the inlet, and (d) the specific speed of the turbine.

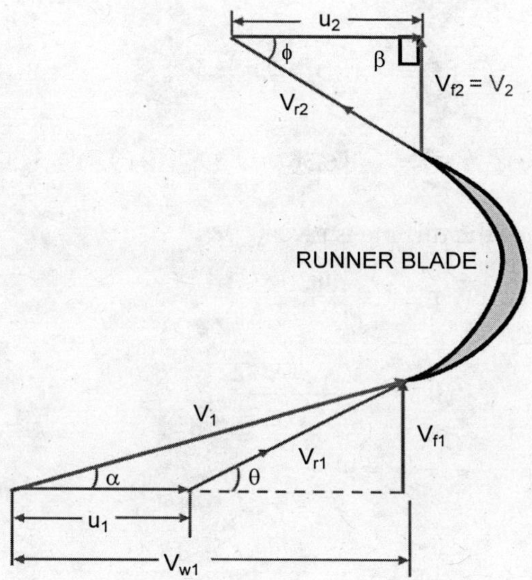

Solution:

Given data:

Head of turbine H = 48 m

turbine speed N = 200 r.p.m

$D_h = 0.35\ D_0$

flow ratio = 0.6

Hydraulic efficiency $\eta_h = 0.85$

$\phi = 15$

$$\text{Flow ratio} = \frac{V_{f1}}{V_{Jet}} = \frac{V_{f1}}{\sqrt{2\times 9.81\times 48}}$$

$$V_{f1} = V_{f2} = 18.41 \text{ m/s}$$

$$\tan\phi = \frac{V_{f2}}{u_2}$$

$$\tan 15 = \frac{18.41}{u_2}$$

$$u_1 = u_2 = 68.72 \text{ m/s}$$

(a) Outer diameter of the runner,

$$u_2 = \frac{\pi D_o N}{60}$$

$$68.72 = \frac{\pi \times D_o \times 200}{60}$$

$$D_o = 6.56 \text{ m}$$

$$D_h = 0.35 \times D_o = 0.35 \times 6.56 = 2.296 \text{ m}$$

(b) Water flow rate Q is given by,

$$Q = \frac{\pi}{4} \times \left(D_o^2 - D_h^2\right) \times V_{f1}$$

$$Q = \frac{\pi}{4} \times \left(6.56^2 - 2.296^2\right) \times 18.41 = 545.74\, m^3/s$$

The hydraulic efficiency of the turbine is given,

$$\eta_h = \frac{V_{w1} u_1}{gH}$$

$$0.85 = \frac{V_{w1} \times 68.72}{9.81 \times 48}$$

$$V_{w1} = 5.82 \text{ m/s}$$

(c) Vane angle at the inlet,

$$\tan\theta = \frac{V_{f1}}{V_{w1} - u_1} = \frac{18.41}{5.82 - 68.72}$$

$$\theta = 163.68°$$

Now we determine the shaft power,

$$P = \rho Q u_1 V_{W1} = 1000 \times 545.74 \times 68.72 \times 5.82 = 218136 \text{ kW}$$

(d) Specific speed of the turbine is given by,

$$N_s = \frac{N\sqrt{P}}{H^{5/4}} = \frac{200 \times \sqrt{218136}}{(48)^{5/4}} = 739.34\, r.p.m$$

7.10 DRAFT TUBE

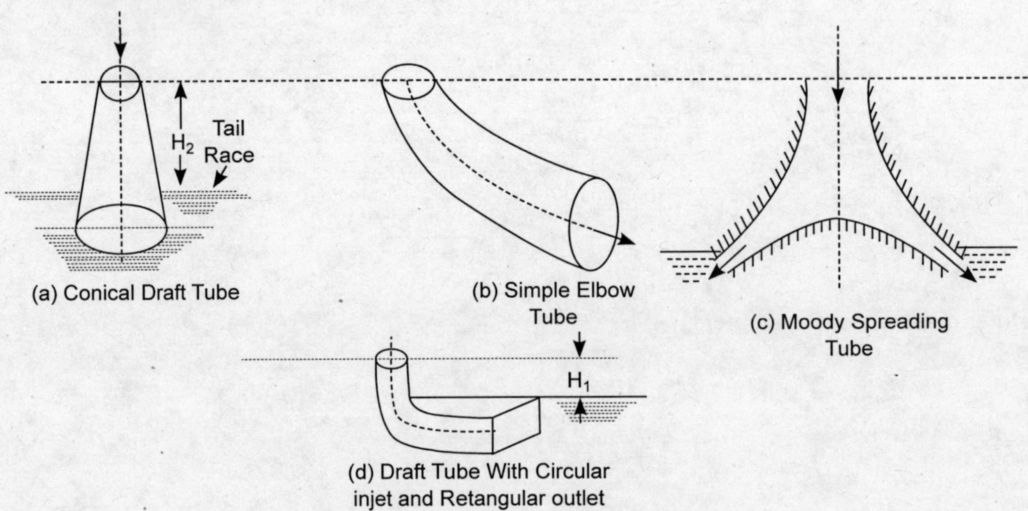

Figure 7.16: Schematic representation of some typical draft tubes. The red arrow depicts the direction of the discharge.

The outlet of the runner is connected to the tail race through the draft tube, which is basically a pipe which connects the two parts while ensuring a gradually increasing area towards the tail race. The draft tube helps in creating a suction at the outlet of the runner. It also helps in converting the outlet kinetic energy into a pressure head owing to which the turbine works across a larger pressure difference and hence it is able to generate a larger power given then inlet head is constant.

7.11 SPECIFIC SPEED

The specific speed of a turbine, *Ns*, represents the speed of a geometrically identical turbine which produces unit power (1 kW) under a unit head (1 m of water).

We can derive it as follows. We first consider the overall efficiency of the turbine as

$$\eta_o = \frac{\text{Power output}}{\text{Power input}} = \frac{P}{\rho g Q H} \tag{7.52}$$

when the numerator and denominator is expressed in kW units, we can express the above as

$$\eta_0 = \frac{P}{\frac{\rho g Q H}{1000}} \tag{7.53}$$

In the above expressions, H represents the head of water, Q represents the flow rate through the turbine, and P represents the power output. We can therefore express power as

$$P = \eta_o \frac{\rho g Q H}{1000} \tag{7.54}$$

The jet velocity, V, and the turbine tangential velocity, u, are related to the head of water at the input through

$$u \propto V \Rightarrow u \propto \sqrt{H}$$

The tangential velocity is related to the diameter and rotational speed as

$$u = \frac{RN \times 2\pi}{60} = \frac{\pi DN}{60}$$

where N is represented in revolution per minute units and therefore we have the factor of $\frac{2\pi}{60}$ in the above expression. From the above discussion we therefore have

$$\sqrt{H} \propto DN \Rightarrow D \propto \frac{\sqrt{H}}{N}$$

The discharge through the turbine can be similarly written as

$$Q = A \times V \Rightarrow Q \propto D^2 \sqrt{H} \tag{7.58}$$

Using the proportionality of D with H and N, we can therefore write down the expression for flow rate as

$$Q \propto D^2 \sqrt{H} \tag{7.59}$$

$$\propto \left(\frac{\sqrt{H}}{N}\right)^2 \sqrt{H} \tag{7.60}$$

$$\propto \frac{H^{\frac{3}{2}}}{N^2} \tag{7.61}$$

Using this expression for the discharge we can express power using equation (7.54) as

$$P \propto \frac{H^{\frac{3}{2}}}{N^2} \times H \propto \frac{H^{\frac{5}{2}}}{N^2} \tag{7.62}$$

$$\Rightarrow \qquad P = K\frac{H^{\frac{5}{2}}}{N^2}, \quad K = \text{constant} \tag{7.63}$$

In the above expressions if $P = 1$, i.e. 1 kW, and if $H = 1$, i.e. unit head, we obtain

$$1 = K\frac{1}{N_s^2} \Rightarrow N_s^2 = K \tag{7.64}$$

which implies that the specific speed is equal to the square root of the constant. We can now substitute the constant in the expression for power and obtain

$$P = N_s^2 \frac{H^{5/2}}{N^2} \Rightarrow N_s = \sqrt{\frac{N^2 P}{H^{5/2}}} \Rightarrow N_s = N\frac{\sqrt{P}}{H^{5/4}} \tag{7.65}$$

The specific speed of a Pelton turbine is typically between 6-30 RPM, while that of a Francis turbine is between 50-200 RPM. The specific speed of the Kaplan turbine is larger and lies between 200-1000 RPM.

In the following sections we will look into the performance of a turbine under varying head, speed, discharge etc. when the head of water at the turbine inlet is kept unity. These are probed under the condition where the efficiency of the turbine remains unchanged.

7.12 UNIT SPEED

The first parameter that we will look into is the unit speed, N_u. The derivation can be done much like the previous case. We have already seen that

$$u \propto V, \quad V \propto \sqrt{H}, \quad \Rightarrow u \propto \sqrt{H} \quad \text{where } u = \frac{\pi DN}{60} \tag{7.66}$$

For a given turbine, we may therefore conclude that

$$N = K\sqrt{H} \quad (K = \text{constant}) \tag{7.67}$$

In this case now a unit head yields $N = K = N_u$, where the RPM of the turbine is now simply the unit speed of the turbine, by very definition. Therefore, we can substitute the value of K in the expression to yield

$$N = N_u\sqrt{H} \Rightarrow N_u = \frac{N}{\sqrt{H}} \tag{7.68}$$

We may utilize the idea of the unit speed of a turbine to find out the relationship between the RPM of the turbine under varying conditions of inlet head. We can write

$$N_u = \frac{N_1}{\sqrt{H_1}} = \frac{N_2}{\sqrt{H_2}} \tag{7.69}$$

7.13 UNIT DISCHARGE

Unit discharge is defined as the discharge from the turbine under by the condition of a unit head. It is denoted by the symbol Q_u. We may derive it much in the methodology adopted in the earlier sections.

The discharge through a turbine is given by

$$Q = AV \tag{7.70}$$

which may be readily simplified for a choice of a given turbine as

$$Q \propto V \Rightarrow Q \propto \sqrt{H} \Rightarrow Q = K\sqrt{H} \tag{7.71}$$

where, for the case of a unit head we can write

$$Q = K = Q_u \tag{7.72}$$

thereby allowing us to further simplify the expression as

$$Q = Q_u\sqrt{H} \Rightarrow Q_u = \frac{Q}{\sqrt{H}} \tag{7.73}$$

The idea of unit discharge can be used to relate the discharge for varying heads by noting that

$$Q_u = \frac{Q_1}{\sqrt{H_1}} = \frac{Q_2}{\sqrt{H_2}} \tag{7.74}$$

7.14 UNIT POWER

Unit power, P_u, is defined as the power developed by a turbine under the application of a unit head at the inlet.

We have already seen that the expression for the power can be expressed as

$$P = \eta_o\frac{\rho g Q H}{1000} \Rightarrow P \propto QH \Rightarrow P \propto \sqrt{H}H \Rightarrow P = KH^{\frac{3}{2}} \tag{7.75}$$

where the simplification in the expression is possible for the case of a certain choice of a turbine, i.e. the diameter etc. do not change.

When there is only a unit head acting, we can see that $P = K = P_u$. Substituting the value of K in the expression for power, we obtain

$$P = P_u H^{\frac{3}{2}}, \Rightarrow P_u = \frac{P}{H^{3/2}} \tag{7.76}$$

In fact, if the idea of the unit power can be used to relate the power output for different input heads by noting that

$$P_u = \frac{P_1}{H_1^{\frac{3}{2}}} = \frac{P_2}{H_2^{3/2}} \tag{7.77}$$

Example 7.13

A turbine develops 10 MW when running at 100 r.p.m and head of 30 m. What is the speed and power developed by this turbine if the head is changed to 18 m?

Solution:

Given data:

At condition 1,

$P_1 = 10$ MW

$N_1 = 100$ r.p.m

$H_1 = 30$ m

At condition 2,

$H_2 = 18$ m

$P_2 = ?$

$N_2 = ?$

Using the relation between speed and head of the turbine we have,

$$\frac{N_1}{\sqrt{H_1}} = \frac{N_2}{\sqrt{H_2}}$$

$$\frac{100}{\sqrt{30}} = \frac{N_2}{\sqrt{18}}$$

$$N_2 = 77.46 \text{ r.p.m}$$

Now using the relation between power and head of the turbine we have,

$$\frac{P_1}{H_1^{3/2}} = \frac{P_2}{H_2^{3/2}}$$

$$\frac{10}{30^{3/2}} = \frac{P_2}{18^{3/2}}$$

$$P_2 = 4.65 \text{ MW}$$

Example 7.14

A turbine operating with 85% overall efficiency under a head of 25 m and 200 r.p.m. The discharge is 9 m^3/s. What will be the power output, r.p.m. and flow rate for 20 m head?

Solution:

Given data:

At condition-1,

Head $H_1 = 25$ m

Speed of the turbine $N_1 = 200$ r.p.m

Discharge $Q_1 = 9$ m^3/s

At condition-2,

Head $H_2 = 20$ m

$P_2 = ?$

$N_2 = ?$

$Q_2 = ?$

The overall efficiency of the turbine $\eta_o = 0.85$

Using the relation between speed and head of the turbine we have,

$$\frac{N_1}{\sqrt{H_1}} = \frac{N_2}{\sqrt{H_2}}$$

$$\frac{200}{\sqrt{25}} = \frac{N_2}{\sqrt{20}}$$

$$N_2 = 178.9 \text{ r.p.m}$$

From the overall efficiency of the turbine we have,

$$\eta_o = \frac{P_1}{\rho g Q_1 H_1}$$

$$0.85 = \frac{P_1}{1000 \times 9.8 \times 9 \times 25}$$

$$P_1 = 1876.2 \text{ kW}$$

Now using the relation between power and the head of the turbine we have,

$$\frac{P_1}{H_1^{3/2}} = \frac{P_2}{H_2^{3/2}}$$

$$\frac{1876.2}{25^{3/2}} = \frac{P_2}{20^{3/2}}$$

$$P_2 = 1342.5 \text{ kW}$$

Now using the relation between discharge and the head of the turbine we have,

$$\frac{Q_1}{\sqrt{H_1}} = \frac{Q_2}{\sqrt{H_2}}$$

$$\frac{9}{\sqrt{25}} = \frac{Q_2}{\sqrt{20}}$$

$$Q_2 = 8.05 \text{ m}^3/\text{s}$$

7.15 CHARACTERISTIC CURVES OF A TURBINE

Under varying operating conditions, characteristic curves provide us a means of determining the behaviour of turbines. In the characteristic curves, the rotational speed, N, the discharge, Q, and the head at the inlet, H are chosen as independent parameters. Apart from these three, the other parameters are the power output, P, the efficiency η_0, and the gate opening. The discussion below will shed light on the nature of the characteristic curves.

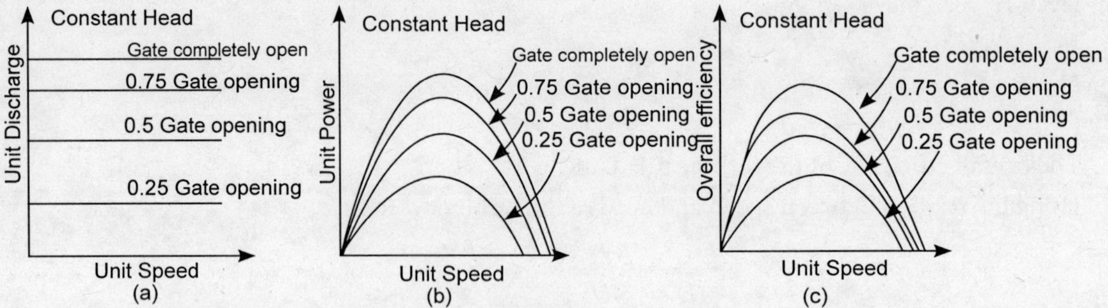

Figure 7.17: Characteristic curves.

7.15.1 Constant head curves

Constant head curves are also referred to as the main characteristic curves. These curves are experimentally obtained by maintaining a constant head at the inlet and a constant gate opening at the turbine inlet. The turbine is then loaded thereby changing the speed of the turbine. For each measured value of speed, the output power and discharge are noted. The corresponding efficiency is also then calculated. These readings then allow us to obtain the values of the unit speed, N_u, unit power, P_u, and unit discharge, Q_u. This has been discussed in the previous sections. We can then plot the locus of the unit discharge as a function of the unit speed for a fixed head. The locus can be plotted for different gate openings. A schematic of all the different curves is shown in the figure below. Similarly, the plot of unit power and efficiency as a function of the unit speed for different gate openings can also be obtained. Essentially the plots help an engineer to understand the degree of variation of the discharge, or power output or the overall efficiency as the turbine experiences different load. This can help the engineer take a decision to increase or decrease the gate opening to maintain a desired set point of a turbine which forms the part of a large hydroelectric power station.

7.15.2 Constant Speed Curves

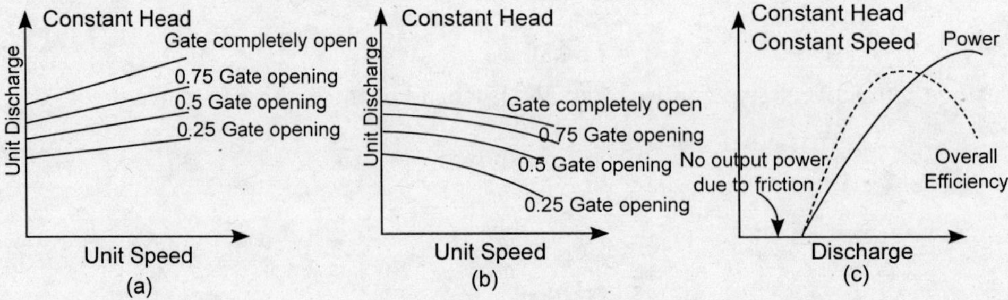

Figure 7.18: Constant speed curve of a (a) Francis turbine, (b) Kaplan Turbine, (c) Operating curve of a turbine.

Similar to the previous subsection, we can keep the turbine speed constant and find out the different characteristic curves of a turbine. Towards obtaining this, the operating speed and head are kept constant. With this the discharge is changed gradually and the parameters such as the overall efficiency and output power are plotted. An important observation is that a certain amount of discharge is required to start the turbine rotation by overcoming the friction of the various components.

7.15.3 ISO-efficiency Curves

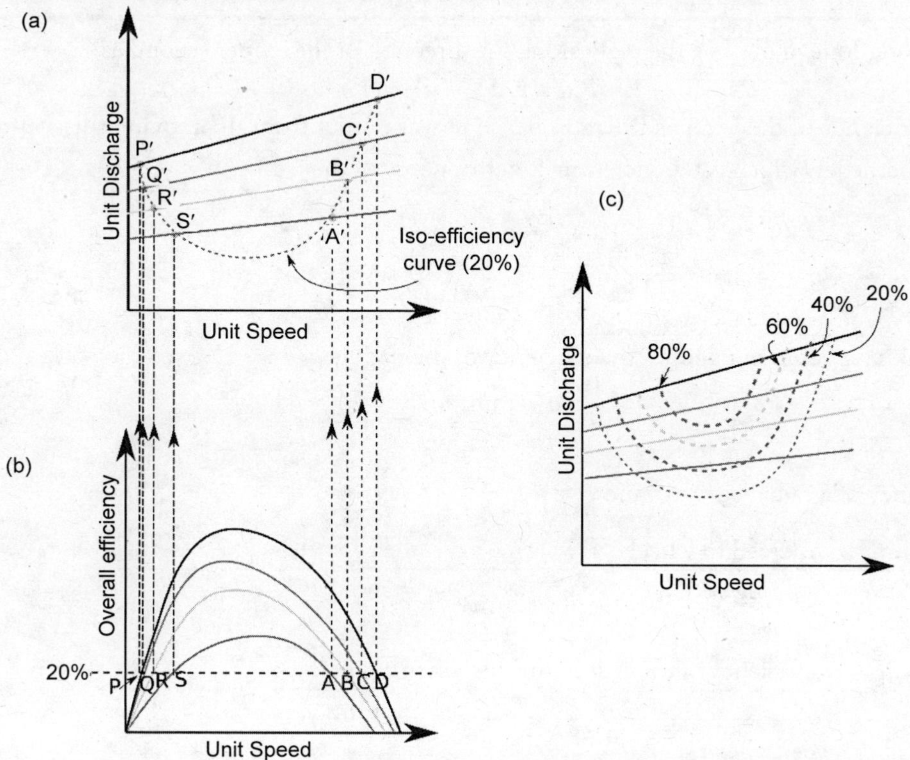

Figure 7.19: Construction of an iso-efficiency curve and iso-efficiency curves of a Kalpan turbine

Iso-efficiency curves are curves along which the overall efficiency of the turbine remains constant. These curves are obtained by keeping the inlet head constant. The speed of the turbine is once again changed by varying the load at the shaft of the turbine. Keeping a certain gate opening, we can then measure the output discharge and calculate the overall efficiency.

In Fig. 7.19 we have depicted how the iso-efficiency curve for a Kaplan turbine can be greater. Towards this the first step is to first draw the unit discharge vs unit speed curve (subplot (a)) and overall efficiency vs unit speed (subplot (b)). Let us say that we would like to plot the iso-efficiency curve at . In that case, in subplot (b) this 20 percent efficiency is a horizontal line which intersects the various curves which correspond to different gate openings. The straight line intersects the curves at two set of points. One set of points on the left, which we have denoted as P, Q, R, and S, and another set of points on the right, which we have denoted as A, B, C, and D. Now, we draw projections of these points onto the same gate opening points in the unit discharge vs unit speed curves. The points of intersection of the same corresponding gate openings, which in both the figures are denoted through the same curve color, are now marked as P', Q', R', and S' on the left, and A', B', C' and D' on the right. The curve which connects these points is shown as a dotted curve and is the locus of all points on the unit discharge vs unit speed curve which correspond to the efficiency of 20%, which is essentially the iso-efficiency curve. The same procedure can be repeated for different efficiency values. This is left as an exercise for the reader. The reader should obtain a family of curves which resemble subplot (c).

UNIT SUMMARY

- The work done in moving the plate along the direction of the jet per second as
$$P = F_x u = \rho A(V - u)^2 u(1 + \cos\theta)$$
- The efficiency of the jet, η as the ratio of the power output from the moving surface to the power input of the jet (which is the incoming kinetic energy of the jet, i.e. $\dot{m} \times V_1^2 / 2$):
$$\eta = \frac{\rho A V_{r_1}\left(V_{w_1} + V_{w_2}\right) \times u}{\frac{1}{2}\left(\rho A V_1\right) V_1^2}$$
- Jet striking a series of paddles, the efficiency of the paddle
$$\eta = \frac{\text{Power output}}{\text{Power input}} = \frac{2u(V-u)}{V^2}$$
- Efficiency of a radial curved vane
$$\eta = \frac{\text{Power transferred to wheel}}{\text{Power incoming with jet}} = \frac{\rho A V_1\left(V_{w_1} u_1 \pm V_{w_2} u_2\right)}{\frac{1}{2}\dot{m} V_1^2} = \frac{2\left(V_{w_1} u_1 \pm V_{w_2} u_2\right)}{V_1^2}$$
- The hydraulic efficiency of the Pelton turbine when $V_{r1} = V_{r2}$
$$\eta = \frac{2(V_1 - u)(1 + \cos\phi)u}{V_1^2}$$
- The relationship between the RPM of the turbine under varying conditions of inlet head
$$\frac{N_1}{\sqrt{H_1}} = \frac{N_2}{\sqrt{H_2}}$$
- The idea of unit discharge can be used to relate the discharge for varying heads by ensuring that
$$\frac{Q_1}{\sqrt{H_1}} = \frac{Q_2}{\sqrt{H_2}}$$
- The idea of the unit power can be used to relate the power output for different input heads by noting that
$$P_u = \frac{P_1}{H_1^{3/2}} = \frac{P_2}{H_2^{3/2}}$$

EXERCISES

Multiple Choice Questions

1. Water from a reservoir at a gross head h flow into a nozzle before striking a series of moving vanes. The water jet velocity at the exit of nozzle is:
   (a) Always equal to $\sqrt{2gh}$ (b) Greater than or equal to $\sqrt{2gh}$
   (c) Less than or equal to $\sqrt{2gh}$ (d) Cannot be determined from given data
2. Water jet strikes a "stationary" curved vane at one end of it in the horizontal direction. The angle of deflection between entry and exit of jet is. The magnitude of force exerted by this jet on the vane in the horizontal direction, F_x:
   (a) Increases with increase in ϕ (b) Reduces with increase in ϕ
   (c) Is non-zero with no dependency on ϕ (d) Is always zero.
3. For the previous problem (no. 2), the magnitude of force exerted by the jet on the vane in vertical direction, F_y:
   (a) Increases with increase in ϕ (b) Reduces with increase in ϕ
   (c) Is non-zero with no dependency on ϕ (d) Is always zero.
4. Water jet strikes a "stationary" symmetric curved vane in a tangential manner such that there is no shock. The angle of deflection between entry and exit of jet is. The magnitude of force exerted by this jet on the vane in the horizontal direction, F_x:
   (a) Increases with increase in ϕ (b) Reduces with increase in ϕ
   (c) Is non-zero with no dependency on ϕ (d) Is always zero.
5. For the previous problem (no. 4), the magnitude of force exerted by the jet on the vane in vertical direction:
   (a) Increases with increase in ϕ (b) Reduces with increase in ϕ
   (c) Is non-zero with no dependency on ϕ (d) Is always zero.
6. Water jet strikes one end of a "moving" curved vane in the horizontal direction. The direction of motion of vane is also horizontal. The speed of jet is V while the speed of the vane is U. The resultant force exerted by the jet on the moving vane, F_R:
   (a) Increases with increase in U
   (b) Reduces with increase in U
   (c) Is non-zero with no dependency on U
   (d) Is zero when $V = U$
7. Which of the following is/are assumed for analyzing flow through the runner of a Pelton turbine:
   (a) The relative velocity of the jet is constant throughout the runner
   (b) The vane speed is identical at the inlet and outlet
   (c) The velocity of whirl is constant throughout the runner
   (d) The absolute velocity of the jet remains constant throughout the runner.

8. Which of the following is/are assumed for analyzing flow through the runner of a Francis turbine:
   (a) The relative velocity of the flow is constant throughout the runner
   (b) The vane speed is identical at the inlet and outlet
   (c) The velocity of whirl is constant throughout the runner
   (d) The absolute velocity of the flow remains constant throughout the runner.
9. Which of the following is/are assumed for analyzing flow through the runner of a Kaplan turbine:
   (a) The relative velocity of the flow is constant throughout the runner
   (b) The blade tip speed is identical at the inlet and outlet
   (c) The velocity of whirl is constant throughout the runner
   (d) The absolute velocity of the flow remains constant throughout the runner.
10. A turbine is designed to operate with a gross head of h at an RPM of N_1 can generate a power of P_1. If another turbine of the same *specific speed* is used for generating power P_2 from a gross head h, then the relationsheip between the speed N_2 and power P_2 for the second turbine is
   (a) $N_1P_1 = N_2P_2$
   (b) $\sqrt{N_1P_1} = \sqrt{N_2P_2}$
   (c) $N_1\sqrt{P_1} = N_2\sqrt{P_2}$
   (d) $N_1 = N_2$ and $P_1 = P_2$

ANSWERS															
1.	(c)	2.	(a)	3.	(b)	4.	(a)	5.	(d)	6.	(b)	7.	(a)	8.	(d)
9.	(b)	10.	(c)												

Unsolved Questions

Level - I

1. Consider a water jet of velocity 25 m/s striking horizontally on a curved stationary plate (See figure of example 1). The jet deflects by 90° between the inlet to outlet of the curved plate. What is the resultant force for the 1 kg/s mass flow rate of jet and the direction of this resultant force with respect to the motion of vane?
2. Consider a water jet of velocity 25 m/s striking a curved moving vane in the same direction as the motion of vane. The vane moves with a velocity of 10 m/s (see figure of example 4). The jet deflects by 120° with respect to the direction of vane motion. What is the resultant force for 1 kg/s mass flow rate of jet and the direction of this resultant force with respect to the motion of vane?
3. Consider a water jet of velocity 25 m/s striking a curved moving vane at 15° with the horizontal direction. The vane also moves in the horizontal direction with a velocity of 10 m/s (see figure of example 6). The jet deflects by 120° between inlet to outlet of in the frame of reference of moving vane. What is the resultant force for 1 kg/s mass flow rate of jet and the direction of this resultant force with respect to the motion of vane?
4. A turbine develops 750 kW under a head of 100 m at 250 r.p.m. What would be expected power developed and turbine speed if the head is 50 m?
5. A turbine has to be designed for head of 10 m and discharge rate of 5 m/s. If the runner wheel has to rotate at 200 r.p.m., which type of turbine should be expected for this application?

6. A Kaplan turbine is operating with a head of 5 m with boss diameter 0.25 times the runner tip diameter. The flow ratio is 0.7 and speed ratio is 2.0. Determine the flow rate of water (m^3/s) through the runner if the runner tip diameter is 6 m.
7. Following data is for a Pelton turbine: Gross head = 100 m, diameter of jet = 100 mm, discharge at the nozzle = 0.33 m^3/s, power developed by the turbine = 250 kW. Determine the: a) maximum power (kW) that can be obtained from the water head, b) power (kW) available at the nozzle, c) power (kW) lost in runner.
8. Consider a Francis turbine with inlet absolute velocity 20 m/s, inlet guide blade angle of 30° and inlet peripheral velocity 10 m/s. Assuming radial exit from the turbine determine, (a) inlet runner vane angle and (b) exit absolute velocity.
9. A Kaplan turbine has guide blade angle of 60 and runner vane angle of 120 at the inlet. The absolute velocity at inlet is 10 m/s. The vane tip diameter is 5 m and hub diameter of 1.5 m. Determine: a) the rate of rotation (in r.p.m.) of the runner, b) water discharge rate (in m/s)
10. Consider a Francis turbine operating under a head H. The peripheral velocity at inlet is u_1, velocity of flow at the inlet is V_{f1} and the component of absolute velocity along the tangential velocity is V_{w1}. Assume velocity of flow within the vane and radial discharge at the outlet of the turbine. Neglecting any losses, derive a relationship for H in terms of velocity components given above.

Level - II

1. In a hydroelectric plant, the penstock supplies water to a Pelton wheel turbine with a gross head of 300 m. Out of this head 100 m is lost due to friction in penstock. The flow rate of water is 3 m/s. The angle of deflection of the jet in the turbine bucket is 150. Determine the: a) power generated by the turbine and b) hydraulic efficiency if the speed ratio is 0.45 and coefficient of velocity is 1.0. Assume that the relative velocity of jet remains constant throughout the bucket.
2. A Francis turbine with overall efficiency of 70% and hydraulic efficiency of 78% has to produce 150° kW of power from a head of H m. The peripheral velocity at inlet of the turbine is given by whereas the velocity of flow at inlet is. The speed of the turbine wheel is 200 r.p.m. Assume that the flow is radial at exit of the turbine and the velocity of flow is constant between inlet and exit. Determine the: (a) guide blade angle at the inlet, (b) runner vane angle at the inlet, (c) wheel diameter at the inlet and d) width of the wheel at the inlet.
3. A Francis turbine with hydraulic efficiency of 80% is operated under a head of H 10 m. The peripheral velocity at inlet of the turbine is 7.1 m/s and the velocity of flow at inlet is 13.5 m/s. The speed of the turbine wheel is 250 r.p.m with the wheel diameter at the outlet 0.5 times the wheel diameter at inlet. Assume that the flow is radial at exit of the turbine and the velocity of flow is constant between inlet and exit. Determine the: (a) guide blade angle at the inlet, (b) runner vane angle at the inlet and (c) runner vane angle at the outlet.
4. Determine the hydraulic efficiency of a Francis turbine with the runner vane angle at inlet and outlet are both 90°. The inlet guide blade angle is. Assume that the velocity of flow through the runner is constant throughout.
5. A Kaplan turbine of runner outer tip diameter of 4.5 m is rotating at 50 r.p.m. The guide blade angle at the inlet is 35° and the runner blade angle is 90° to the vane motion. At the outlet, the runner blade angle is 25 to the direction of vane motion. The total area through which water flows within the runner is 20 m^2. Assume that the velocity of flow within the runner is constant. Determine the (a) head (m) for this turbine neglecting any frictional losses, (b) hydraulic efficiency and (c) power (kW) developed by the turbine.

Answer Key

Level - I

1. 35.36 N making 45° below horizontal
2. 25.98 N making 30° below horizontal
3. 2.94 N making 60° below the horizontal
4. 265.2 kW at 177 r.p.m.
5. N_S = 131.3, therefore Francis turbine is expected
6. 183.7 m³/s
7. (a) 323.7 kW, (b) 291.37 kW, (c) 41.4 kW
8. (a) 53.8°, (b) 10 m/s
9. (a) 38.2 r.p.m., (b) 8.66 m³/s
10. $H = \frac{u_1 V_{w1}}{g} + \frac{V_{f1}^2}{2g}$

Level - II

To know more about fluid machinery

1. (a) 5437 kW, (b) 92.4%
2. (a) 31.6 b) 36.2, (c) 0.3 m, (d) 0.242 m
3. (a) 50.7 (b) 73.7, (c) 75.3
4. $\frac{2}{2 + \tan^2 \alpha}$
5. (a) 26.5 m, (b) 80.2% and (c) 3436 kW

KNOW MORE

The Pelton wheel, known for its simplicity and ease of operation, gained worldwide adoption thanks to its unique splitter principle, patented by Lester A. Pelton in 1880. Pelton built the first turbine in 1878 in Camptonville, later refining it in Nevada City. He established the Pelton Water Wheel Company in San Francisco in 1888 with A. P. Brayton and passed away in Oakland, California, in 1908.

This innovative waterwheel harnesses the momentum of water jets striking buckets on its rim, evolving from the earlier "hurdy gurdy" wheel used during the California gold rush. Significant improvements were made by William A. Doble, who patented enhancements to bucket and nozzle designs starting in 1899. A collection highlighting the Pelton waterwheel's history is housed in a mining museum at the former North Star Power Plant site, which utilized Pelton wheels for compressed air in mining operations until they were replaced by diesel power in the 1950s. The museum is operated by the Nevada County Historical Society.

[Source: https://www.asme.org/about-asme/engineering-history/landmarks/157-pelton-waterwheel-collection#:~:text=Pelton%20built%20the%20first%20Pelton,Water%20Wheel%20Company%20in%201888]

REFERENCES AND SUGGESTED READINGS

1. **Introduction to Fluid Mechanics and Fluid Machines** (3rd Edition), S K Som, Gautam Biswas, Suman Chakraborty, McGraw Hill
2. **Fluid Mechanics** (9th Edition), Frank M. White, Henry Xue, McGraw Hill

□□

8 Fluid Machinery: Pumps

UNIT SPECIFICS

In this unit we will discuss the following topics:

1. Components of a reciprocating pump
2. Components of a centrifugal pump
3. Velocity triangle analysis for centrifugal pump
4. Dimensionless pump characteristics
5. Actual pump performance curves
6. System curve and pump operating conditions
7. Cavitation in centrifugal pumps

RATIONALE

In the previous unit we discussed devices which produce work from hydraulic head. In this unit we discuss about the device which consumes mechanical/electrical energy to develop pressure head. This pressure head can be used to create flow in a pipe-network and/or lift fluid to a certain height. Such devices are called "pumps". Selection of appropriate pumps is an important task in many industries ranging from power plant to cooling applications. A detailed understanding of the principle of operation of such pumps, their mathematical analysis and performance characteristics is vital. For safe operation of centrifugal pumps, a phenomenon called "cavitation" needs to be avoided. This has practical implications on the pump installation and requires careful analysis. In all these discussions, the principles of velocity-triangle, non-dimensional analysis and fundamentals of frictional losses in pipe-network studied in previous units will be revisited.

PRE-REQUISITES

Elementary vector analysis

Trigonometric relationships

UNIT OUTCOMES

U8-O1: Various components and mathematical analysis of a reciprocating pump

U8-O2: Various components and mathematical analysis of a centrifugal pump

U8-O3: Specific speed and its application

U8-O4: Non-dimensional analysis of centrifugal pump performance characteristics

U8-O5: Variation of pump performance with discharge rate and impeller speed

U8-O6: System curve and determination of operating condition of centrifugal pumps

U8-O7: Net positive suction head in centrifugal pumps

Unit-8 Outcomes	EXPECTED MAPPING WITH PROGRAMME OUTCOMES (1 - Weak Correlation; 2 - Medium Correlation; 3 - Strong Correlation)						
	CO-1	CO-2	CO-3	CO-4	CO-5	CO-6	CO-7
U8-O1	2	-	-	3	-	-	2
U8-O2	2	-	-	3	-	-	2
U8-O3	2	-	-	3	-	-	2
U8-O4	2	-	-	3	-	-	2
U8-O5	2	-	-	3	-	-	2
U8-O6	2	-	-	3	-	-	2
U8-O7	2	-	-	3	-	-	2

8.1 INTRODUCTION

Pumps are hydraulic machines which convert mechanical energy into hydraulic energy. Typically, the hydraulic energy is obtained in the form of a pressure head. There are two primary kinds of pumps - reciprocating pumps and centrifugal pumps. The names are indicative of the process behind the operation of the pump. In the case of centrifugal pump, the pressure head is obtained through the action of centrifugal forces acting on the fluid. In the case of a reciprocating pump, the rise in pressure head is due to the liquid being drawn into a cylinder and then pressurized due to a reciprocating piston.

Let us discuss the various parts and working of the two kinds of pumps.

8.2 RECIPROCATING PUMP

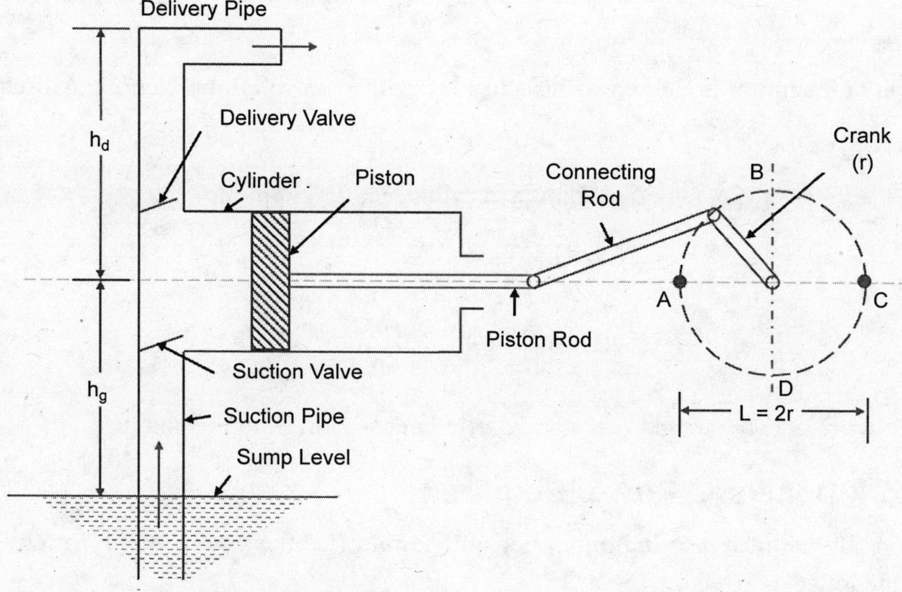

Figure 8.1: Schematic of a reciprocating pump.

The important components of a reciprocating pump are

1. Suction pipe
2. Suction valve
3. Delivery pipe
4. Delivery valve
5. Cylinder-piston arrangement

The piston in a reciprocating pump moves to and fro in a cylinder. The piston is moved to and fro by means of a slider-crank mechanism which is driven by a motor. The end of the piston are connected with the suction and delivery pipes. At the suction and delivery pipe connections to the cylinder we have non-return valves. On the suction side we have a suction valve which allows water from the sump to be drawn towards the cylinder. On the delivery side we have a discharge or delivery valve which allows the pressurized liquid to only travel up to the delivery side. The non-return valve prevents the delivery side liquid to enter into the cylinder and also prevents the pressurized liquid to exit from the suction side.

In the figure shown as the piston moves from the rightmost point towards the left, there is a suction created in the cylinder. This causes water to be drawn into the cylinder. At the same time, the discharge valve stays shut. Once the cylinder reaches the rightmost point, it now starts moving towards the left. At this point the suction valve remains shut and the delivery valve opens, thereby sending the pressurized liquid towards the delivery side. This is elucidated through the figure.

8.2.1 Discharge of a Reciprocating Pump

The discharge of a single cylinder reciprocating pump can be found out as follows: Let us consider a cylinder of diameter *D* and crank radius, *r*. Therefore, the piston area and stroke are given by

$$\text{Piston area, } A = \frac{\pi D^2}{4}, \text{ Stroke, } L = 2r \tag{8.1}$$

If the rpm of the motor is N, then we have in 1 second, $\frac{N}{60}$ revolutions. Hence the discharge of the pump in 1 second is

$$Q = \text{Volume in cylinder} \times \text{RPS of pump} \tag{8.2}$$

$$= A \times L \times \frac{N}{60} = \frac{NAL}{60} \tag{8.3}$$

$$\dot{m} = \frac{N\rho AL}{60}, \quad \dot{w} = \frac{N\rho ALN}{60} \tag{8.4}$$

where $\dot{w}$ represents the weight of water delivered per second from the pump.

8.2.2 Work Done and Power Required

The work done by the reciprocating pump per unit second (i.e. the power output in terms of power delivered to the water) is given by

$$P = \text{weight of water lifted per second} \times \text{height through which water is lifted} \tag{8.5}$$

$$= \dot{w} \times (h_s + h_d) \tag{8.6}$$

where h_s and h_d represent the height of the sump and height of the delivery reservoir respectively. Note that this is also the power input required for the pump, ignoring any losses. We can therefore write down the entire expression as

$$P = \frac{\rho g NAL \times (h_s + h_d)}{60 \times 1000} kW \tag{8.7}$$

8.2.3 Slip of a Pump

The slip of a pump is defined as the difference between the ideal or theoretical discharge and the actual discharge obtained from the pump. We therefore have

$$\text{Slip} = Q_{th} - Q_{act} \tag{8.8}$$

which, in terms of percentage slip can be expressed as

$$\%\ \text{Slip} = \frac{Q_{th} - Q_{act}}{Q_{th}} \times 100 \tag{8.9}$$

$$= (1 - C_d) \times 100 \tag{8.10}$$

where $C_d = \frac{Q_{act}}{Q_{th}}$ represents the coefficient of discharge of the pump. It is important to note that the slip may be positive or negative.

8.3 CENTRIFUGAL PUMP

The principle of working of a centrifugal pump is essentially opposite to that of a Francis turbine. Unlike Francis turbine, where fluid flows radially inwards through the runner, here fluid flow through the rotating vanes is radially outwards. As the fluid flows outwards, it gains energy provided to it by the rotating vanes (also called impeller). This mechanical energy gets converted to pressure head and kinetic energy of flow at the discharge of the pump.

8.3.1 Components of a Centrifugal Pump

A centrifugal pump comprises of the following components

1. Impeller
2. Casing
3. Suction pipe
4. Foot valve
5. Delivery pipe

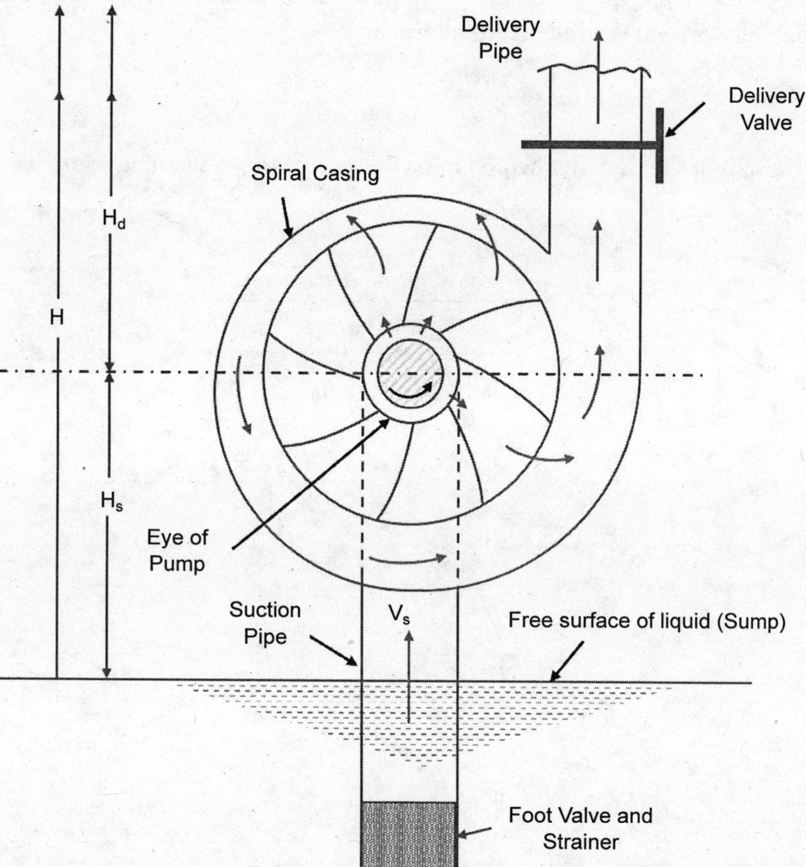

Figure 8.2: Schematic of a centrifugal pump.

The impeller is the rotating part of the pump and it consists of backward facing vanes. The impeller shaft is powered through a motor. Depending on the application and availability, either electric motor or fossil fuel powered motors are used. The casing of the centrifugal pump is the passage around the impeller which allows for the water to transfer its kinetic energy into a pressure head, or pressure energy, while diverting the fluid to the delivery pipe. Typically, the casing has a very specific shape. Some commonly found casings are the volute casing, the vortex casing, and the guide vane casing. An important observation is that the centrifugal pump may be thought of as the opposite of a Francis turbine.

8.3.2 Work Done by a Centrifugal Pump

The basic idea is to use the concepts behind the reaction turbine, in particular the Francis turbine, to derive the work done by a centrifugal pump. Water enters radially at the inlet which allows for maximum efficiency. In what follows, we will analyze only pumps with outlet vane angle <90°, generally referred to as backward facing pumps. In such a case, the radially incoming water is perpendicular to the direction of the motion. As a consequence $\alpha = 90°$; and correspondingly, $V_{w1} = 0$. The reader is advised to familiarize oneself with the convention of symbols explained earlier in the discussion of turbines. Let D_1 be the diameter of the impeller at the inlet and let N be the rotational speed of the impeller in RPM. The corresponding impeller speed at the inlet is therefore

$$u_1 = \frac{\pi D_1 N}{60} \tag{8.11}$$

Similarly, at the exit of the impeller, which has a diameter , the tangential velocity is

$$u_2 = \frac{\pi D_2 N}{60} \tag{8.12}$$

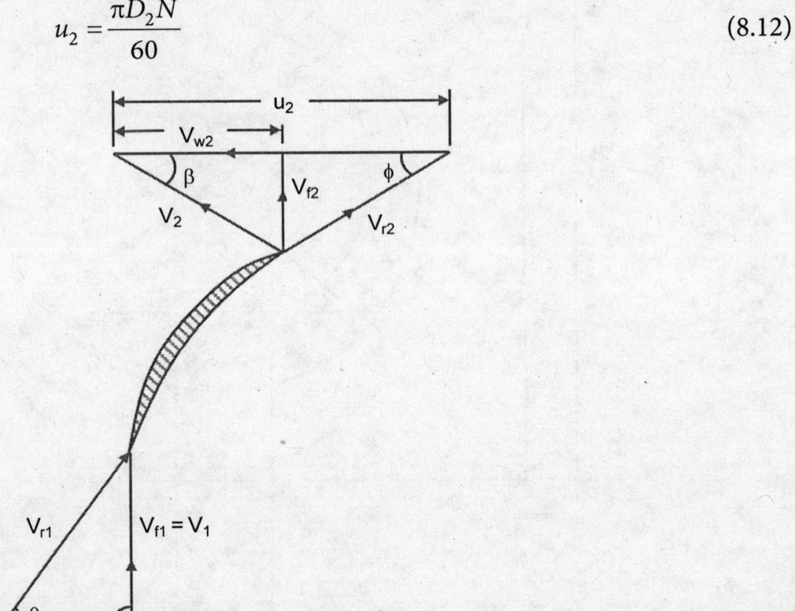

Figure 8.3: Velocity diagram at the inlet and exit of an impeller of a centrifugal pump.

If we just reverse the analysis of a Francis turbine, we can show that the work done by the pump on water per unit time per unit weight is given by

$$\frac{P}{\dot{m}g} = -\frac{P_{turb}}{\dot{m}g} = -\left(\frac{1}{g}\left(V_{w_1}u_1 - V_{w_2}u_2\right)\right) = \frac{1}{g}\left(V_{w_2}u_2 - V_{w_1}u_1\right) \tag{8.13}$$

where P_{turb} represents the power calculated if the velocity triangles corresponded to that of a reaction turbine. We note that since the velocity component $V_{w1} = 0$ at the inlet, the above expression may be simplified to

$$\frac{P}{\dot{m}g} = \frac{1}{g}V_{w_2}u_2 \Rightarrow P = \dot{m}V_{w_2}u_2 = \rho Q V_{w_2}u_2 \tag{8.14}$$

where the flow rate is determined from

$$Q = \pi D_1\, B_1\, V_{f1} \tag{8.15}$$

where V_{f1} represents the flow velocity at the inlet, and B_1 represents the width of the impeller at the inlet.

Example 8.1

The external impeller diameter of a centrifugal pump is 400 mm and it rotates at 1500 r.p.m. The pump delivers water flow rate of 0.5 m³/s at a height of 30 m. The vane width at the outlet is 50 mm. Manometric efficiency of the pump is 80%. Assume radial entry to the pump. Determine the vane angle at the outlet.

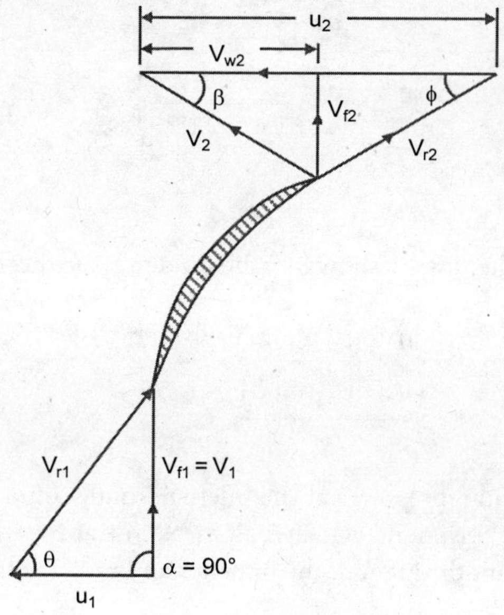

Solution:

Given data

$d_2 = 400$ mm

$b_2 = 50$ mm

$Q = 0.5$ m³/s

H = 30 m

N = 1500 rpm

$V_{w1} = 0$

$\eta_{man} = 0.8$

The tangential velocity at the pump inlet is given by:

$$u_2 = \frac{\pi D_2 N}{60} = \frac{\pi \times 0.4 \times 1500}{60} = 31.41 m/s$$

Using the definition of manometric efficiency,

$$\eta_{man} = \frac{gH}{u_2 V_{w2}} \Rightarrow 0.8 = \frac{9.81 \times 30}{31.41 \times V_{w2}}$$

$\Rightarrow$ $V_{w2} = 11.71$ m/s

The volume flow rate,

$$Q = \pi d_2\, b_2\, V_{f2}$$

$$0.5 = \pi \times 0.4 \times 0.05 \times Vf_2$$

$$V_{f2} = 7.96 \text{ m/s}$$

From outlet velocity triangle,

$$\tan\phi = \frac{V_{f2}}{u_2 - V_{w2}}$$

$$\phi = tan^{-1}\left(\frac{7.96}{31.41 - 11.71}\right) = 22^\circ$$

Example 8.2

Neglecting all frictional and other losses, show that the fluid pressure head difference $\frac{p_2 - p_1}{\rho g}$ across the inlet and outlet of an impeller of a centrifugal pump is given by:

$$\frac{\left(u_2^2 - V_{f2}^2 cot^2\phi\right)}{2g}$$

Where p_1 and p_2 are the fluid pressures at the inlet and outlet of impeller. ϕ is the vane angle, Vf_2 the velocity of flow and u_1 the peripheral velocity; all measured at the impeller outlet. Assume that the velocity of flow remains constant throughout the impeller and radial entry into the impeller.

Solution

Given data:

$V_{w1} = 0$

$$V_{f1} = V_{f2}$$

Losses are neglected; hence, applying the modified Bernoulli's equation across the impeller of a centrifugal pump we get:

$$p_1 + \frac{\rho V_1^2}{2g} + \rho g h_1 + \Delta P_{pump} = p_2 + \frac{\rho V_2^2}{2g} + \rho g h_2$$

Neglecting any height change between inlet and outlet of pump (i.e.) we can simplify the above relationship to:

$$\left(\frac{V_1^2}{2g} - \frac{V_2^2}{2g}\right) + \frac{\Delta p_{pump}}{\rho g} = \frac{p_2 - p_1}{\rho g}$$

$$\left(\frac{V_1^2}{2g} - \frac{V_2^2}{2g}\right) + H_{man} = \frac{p_2 - p_1}{\rho g}$$

$$\frac{p_2 - p_1}{\rho g} = \left(\frac{V_1^2}{2g} - \frac{V_2^2}{2g}\right) + \eta_{man}\left(\frac{u_2 V_{w2}}{g}\right)$$

Neglecting all losses, the pressure difference in fluid across the impeller is:

$$\frac{p_2 - p_1}{\rho g} = \left(\frac{V_1^2}{2g} - \frac{V_2^2}{2g}\right) + \left(\frac{u_2 V_{w2}}{g}\right)$$

From inlet velocity triangle we get:

$$V_1 = Vf_1$$

From outlet velocity triangle:

$$\tan\phi = \frac{V_{f2}}{u_2 - V_{w2}}$$

$\Rightarrow$ $$V_{w2} = u_2 - V_{w2} \cot\phi$$

Also, we get

$$V_2^2 = V_{f2}^2 + V_{w2}^2$$

$$V_2^2 = V_{f1}^2 + V_{w2}^2$$

$$V_2^2 = V_1^2 + V_{w2}^2$$

Substituting the expression for we obtain the expression for manometric head developed across the pump:

$$\frac{p_2 - p_1}{\rho g} = \left(\frac{V_1^2}{2g} - \left(\frac{V_1^2 + V_{w2}^2}{2g}\right)\right) + \left(\frac{u_2 V_{w2}}{g}\right)$$

$$\frac{p_2 - p_1}{\rho g} = \frac{u_2 V_{w2}}{g} - \frac{V_{w2}^2}{2g}$$

Substituting the expression for V_{w2} we obtain

$$\frac{p_2 - p_1}{\rho g} = \left[\frac{u_2\left(u_2 - V_{f2}\cot\phi\right)}{g}\right] - \left[\frac{u_2^2 + V_{f2}^2 cot^2\phi - 2u_2 V_{f2}\cot\phi}{2g}\right]$$

$$\frac{p_2 - p_1}{\rho g} = \frac{2u_2^2 - 2u_2 V_{f2}\cot\phi - u_2^2 - V_{f2}^2 cot^2\phi + 2u_2 V_{f2}\cot\phi}{2g}$$

$$\frac{p_2 - p_1}{\rho g} = \left[\frac{u_2^2 - V_{f2}^2 cot^2\phi}{2g}\right]$$

8.3.3 Specific Speed of a Centrifugal Pump

The specific speed, *Ns*, is defined as the speed of a geometrically similar pump which delivers a flow rate of 1 m^3/s flow rate against a head of 1 m.

The flow rate of a centrifugal pump, as has been shown before, is given by

$$Q = \text{flow velocity} \times \text{area} = \pi D B V_f \tag{8.16}$$

where the flow velocity, V_f, is at a section having an impeller diameter of D and width B. It is worth noting that the flow velocity is related to the manometric head at the inlet through $V_f \propto \sqrt{H_m}$

Therefore we have

$$Q \propto DBV_f \tag{8.17}$$

However, we have the relation that $D \propto B$ thereby allowing us to write

$$Q \propto D_2\ V_f \tag{8.18}$$

Moreover, we have the tangential velocity given by

$$u = \frac{\pi DN}{60} \Rightarrow u \propto DN \tag{8.19}$$

which means that the tangential velocity can be written down in terms of the head as

$$u \propto V_f \Rightarrow u \propto \sqrt{H_m} \Rightarrow DN \propto \sqrt{H_m} \Rightarrow D \propto \frac{\sqrt{H_m}}{N} \tag{8.20}$$

Upon substituting the proportionality of D in the expression for the flow rate, Q, we obtain

$$Q \propto \frac{H_m}{N^2}\sqrt{H_m} \Rightarrow Q \propto \frac{H_m^{\frac{3}{2}}}{N^2} \tag{8.21}$$

For the case of the specific speed we can substitute Q = 1m^3/s and $H_m = 1$ while $N = N_s$ and obtain

$$1 \propto \frac{1}{N_s^2} \tag{8.22}$$

Dividing equations (8.21) and (8.22) we obtain

$$Q = \frac{N_s^2}{N^2} H_m^{\frac{3}{2}} \Rightarrow N_s = \frac{N\sqrt{Q}}{H_m^{3/4}} \tag{8.23}$$

8.4 DIMENSIONLESS PUMP PERFORMANCE CURVES

We now proceed to the dimensionless pump performance parameters. Let us say we have a given pump design. In that case, the parameters such as the head, H, and horsepower, P, depend on the discharge, Q, impeller diameter, D, and the shaft speed, n. Apart from these parameters, the performance should also depend on some more parameters. Two fluid parameters, density, ρ, and viscosity, μ, play an important role in assessing pump performance. Apart from this, the surface roughness of the pump, $\in$, also plays an important role in deciding the aforementioned. In this case, we can represent the head or horsepower as a functional form

$$gH = f_1(Q, D, n, \rho, \eta, \in) \quad P = f_2(Q, D, n, \rho, \mu, \in) \tag{8.24}$$

Now, it is not convenient to express the head and power in terms of all these parameters, but rather it is expedient to represent the variables on the right hand side in terms of combination of terms through

$$\frac{gH}{n^2 D^2} = g_1\left(\frac{Q}{nD^3}, \frac{\rho n D^2}{\mu}, \frac{\in}{D}\right), \frac{P}{\rho n^3 D^5} = g_2\left(\frac{Q}{nD^3}, \frac{\rho n D^2}{\mu}, \frac{\in}{D}\right) \tag{8.25}$$

The reader may confirm that all the terms appearing in the expressions above are dimensionless. This process of obtaining a non-dimensional group of terms is extremely useful for experiments in fluid mechanics as it reduces the number of experiments required to obtain a functional relationship. The observant reader will note that the parameters Reynolds number and roughness ratio are seen, i.e.

$$\text{Re} = \frac{\rho n D^2}{\mu}, \frac{\in}{D} \tag{8.26}$$

Therefore we can represent the parameters appearing in the parenthesis in the expressions above as pump parameters

$$\text{Capacity coefficient, } C_Q = \frac{Q}{nD^3} \tag{8.27}$$

$$\text{Head coefficient, } C_H = \frac{gH}{n^2 D^2} \tag{8.28}$$

$$\text{Power coefficient, } C_P = \frac{P}{\rho n^3 D^5} \tag{8.29}$$

If there are two pumps, which we denote by pump 1 and pump 2, then we can relate their flow rate ratios, head ratios, and power ratios, as

$$\frac{Q_2}{Q_1} = \frac{n_2}{n_1}\left(\frac{D_2}{D_1}\right)^3 \tag{8.30}$$

$$\frac{H_2}{H_1} = \left(\frac{n_2}{n_1}\right)^2 \left(\frac{D_2}{D_1}\right)^2 \tag{8.31}$$

$$\frac{P_2}{P_1} = \frac{\rho_2}{\rho_1}\left(\frac{n_2}{n_1}\right)^3 \left(\frac{D_2}{D_1}\right)^5 \tag{8.32}$$

These are known as similarity rules and are used to determine the influence of changing fluid, rotational speed, and size of a turbomachinery for a geometrically similar pair of machines. Typically we expect that for two turbomachines which are similar, the efficiencies would be equal. However, it is noted that large diameter pumps with a large Reynolds number and low roughness ratios and clearance ratios yield higher efficiency.

Example 8.3

The internal and external diameter of an impeller of a centrifugal pump running at 1500 r.p.m are 100 mm and 200 mm respectively. The volume flow rate of water through the pump is 0.02 m³/s. The velocity of flow is constant throughout the impeller and is equal to 1 m/s. Vane angle at the outlet is 45° and the inlet flow is radial. The pipeline at the suction of pump has 150 mm diameter and that at the discharge has 100 mm diameter. The pressure head at the suction is 3 m absolute and at the discharge is 15 m absolute. Neglect any static head difference between the suction and discharge pipelines. This pump consumes 5 kW electrical work. Determine: (i) the inlet vane angle, (ii) Manometric efficiency of the pump, (iii) Overall efficiency of the pump.

Solution

Given data

$d_1 = 100$ mm

$d_2 = 200$ mm

N = 1500 rpm

$\phi = 45°$

Q = 0.02 m³/s

$V_{f1} = Vf_2 = 2$ m/s

— = 15 – 3 = 12 m

$d_{in} = 150$ mm

$d_{out} = 100$ mm

$W_{elec} = 5$ kW

The tangential velocities at the impeller inlet and outlet are given by:

$$u_1 = \frac{\pi d_1 N}{60} = \frac{\pi \times 0.1 \times 1500}{60} = 7.85 m/s$$

$$u_2 = \frac{\pi d_2 N}{60} = \frac{\pi \times 0.2 \times 1500}{60} = 15.71 m/s$$

The volume flow rate Q is given by,

$$Q = \frac{\pi}{4} \times d_{in}^2 \times V_{in} = \frac{\pi}{4} \times d_{out}^2 \times V_{out}$$

$$V_{in} = \frac{4Q}{\pi d_{in}^2} = 1.132\, m/s$$

$$V_{out} = \frac{4Q}{\pi d_{out}^2} = 2.547\, m/s$$

Here V_{in} and V_{out} the velocity of water at the suction and discharge pipelines.

(i) From inlet velocity triangle:

$$V_{f1} = V_1 = 2 \text{ m/s, and } V_{w1} = 0$$

$$\theta = \tan^{-1}(0.2547) = 14.29°$$

(ii) Applying Bernoulli's principle between the suction and discharge of the pump:

$$\frac{p_1}{\rho g} + \frac{V_{in}^2}{2g} + h_1 + H_{man} = \frac{p_2}{\rho g} + \frac{V_{out}^2}{2g} + h_2$$

$$H_{man} = \left(\frac{p_2}{\rho g} + \frac{V_{out}^2}{2g} + h_2\right) - \left(\frac{p_1}{\rho g} + \frac{V_{in}^2}{2g} + h_1\right)$$

Neglecting height difference between suction and discharge of pump, we obtain:

$$H_{man} = \left(\frac{p_2}{\rho g} + \frac{V_{out}^2}{2g}\right) - \left(\frac{p_1}{\rho g} + \frac{V_{in}^2}{2g}\right) = \frac{\Delta p}{\rho g} + \frac{2.547^2 - 1.132^2}{2 \times 9.81} = 12.27\, m$$

From outlet velocity triangle:

$$\tan\phi = \frac{V_{f2}}{u_2 - V_{w2}}$$

$$V_{w2} = u_2 - V_{w2} \cot\phi = 15.71 - 2 \times \cot 45° = 13.71 \text{ m/s}$$

$$\eta_{man} = \frac{gH_{man}}{u_2 V_{w2}} = \frac{9.8 \times 12.27}{15.71 \times 13.71} = 55.89\%$$

(ii) Overall efficiency is given by,

$$\eta_{overall} = \frac{\rho g Q H_{man}}{W_{elec}} = \frac{1000 \times 0.02 \times 9.81 \times 12.27}{5000} = 48.13\%$$

Example 8.4

A one-eighth scale model of a single-stage centrifugal pump is tested in a laboratory. The model pump has a speed of 1000 r.p.m and the head developed is 8 m. The power consumed by the model pump is 1.5 kW. The field pump has to develop 25 m. Determine the: (i) working speed of field pump, (ii) power consumed by field pump and (iii) ratio of flow rates for the two pumps.

Solution:

Given data

$D_1/D_2 = 1/8$

$H_1 = 8$

$N_1 = 1000$ rpm

$P_1 = 1.5$ kW

$H_2 = 25$ m

Subscript 1 is used for model pump and subscript 2 is used for the field pump,

(i)

$$\frac{\sqrt{H_1}}{D_1N_1} = \frac{\sqrt{H_2}}{D_2N_2}$$

(ii)

$$\frac{\sqrt{8}}{D_1 \times 1000} = \frac{\sqrt{25}}{8 \times D_1 \times N_2} = 221\,rpm$$

$$\frac{P_1}{D_1^5N_1^3} = \frac{P_2}{D_2^5N_2^3}$$

$$P_2 = P_1(\frac{D_2}{D_1})^5(\frac{N_2}{N_1})^3 = 530.3\,kW$$

(iii)

$$\frac{Q_1}{D_1^3N_1} = \frac{Q_2}{D_2^3N_2}$$

$$\frac{Q_2}{Q_1} = (\frac{D_2}{D_1})^3\left(\frac{N_2}{N_1}\right) = 8^3\,\frac{221}{1000} = 113.15$$

8.5 PUMP CHARACTERISTIC CURVES

A pump's performance can be assessed with the help of various performance curves. We describe below the general idea of performance curves for centrifugal pumps. Performance curves are curves, which are typically plotted for constant impeller speed, N. The independent quantity is the discharge, Q. The dependent variable is taken to be the head at discharge, H, the output power, P, or the power required, P , or the efficiency .

A typical head vs discharge rate curve for a constant impeller speed of a centrifugal pump is shown in Fig. 8.4. This curve is obtained by conducting experiments on an actual pump and it depends on the overall pump design. At zero discharge rate the head developed by the pump is maximum, i.e. $Q = 0$ for $H = H_{max}$. This pressure is also known as the "shut-off" head for the pump. From the figure we also observe that the head developed by the pump continuously decreases as the pump discharge rate is increased. At zero head, we have the maximum discharge zero, i.e. $H = 0$ for $Q = Q_{max}$. However, this maximum discharge rate (Q_{max}) is practically not achievable due to phenomenon called "choking". Therefore, this portion of curve is denoted by dotted line in Fig. 8.4.

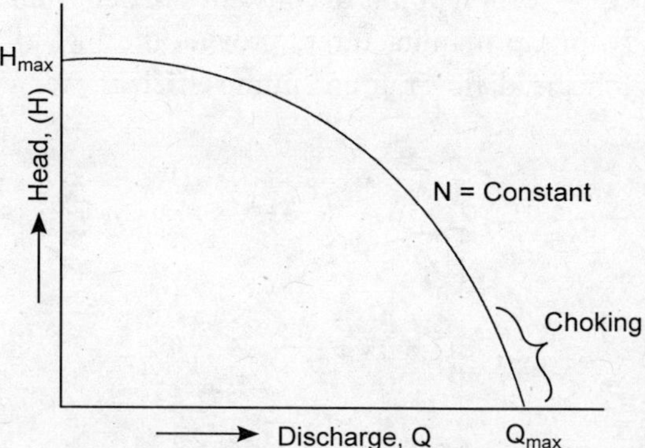

Figure 8.4: Typical head vs discharge rate curve for a centrifugal pump at constant impeller speed.

Fig. 8.5 shows the variation of power input to the pump, the power output (from the impeller to water) and the pump efficiency with discharge rate for a fixed impeller speed. The "power input" to the pump increases as the discharge rate is increased. Interestingly, the "power output" from the pump i.e. the power transferred by the impeller to water is maximum at certain discharge rate and then it starts to decrease. This is due to the hydraulic losses incurred as water flows through the various components of the pump. A detailed discussion on these losses is beyond the scope of this textbook. The point at which pump operates with maximum efficiency is known as the best efficiency point (BEP). This is typically obtained for a discharge of .

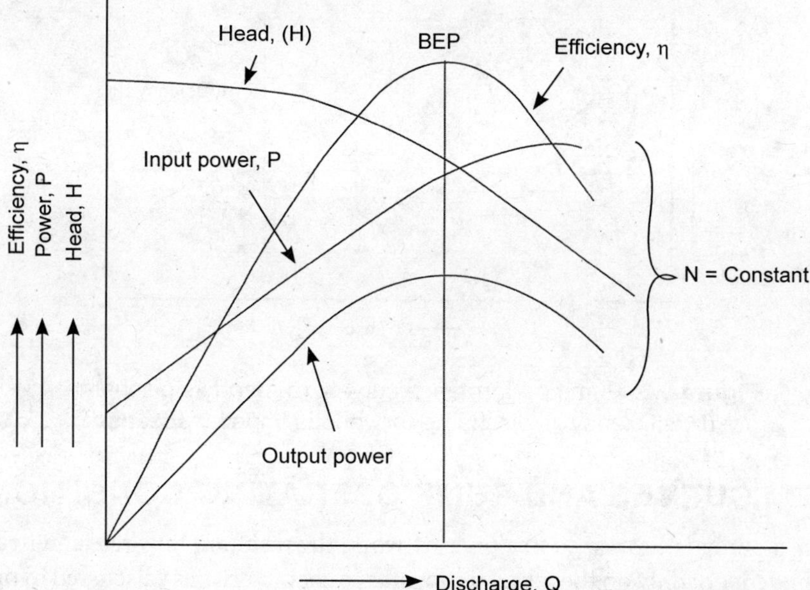

Figure 8.5: Pump performance parameters vs discharge rate at constant impeller speed.

Fig. 8.6 (a) shows the family of head vs discharge curves whereas Fig. 8.6 (b) shows the family of efficiency vs discharge curves for various impeller speeds. Here, the impeller speeds $N_4 > N_3 > N_2 > N_1$. The locus of constant-efficiency points in the head vs discharge plot (Fig. 8.6a) is obtained by orthographic projection of the intersection points of constant-efficiency line with the family of efficiency curves in Fig. 8.6 b). Typically, pump manufacturers provide the Fig. a) as a part of their data sheet through which one can predict the head, flow rate and pump efficiency for any given operating condition.

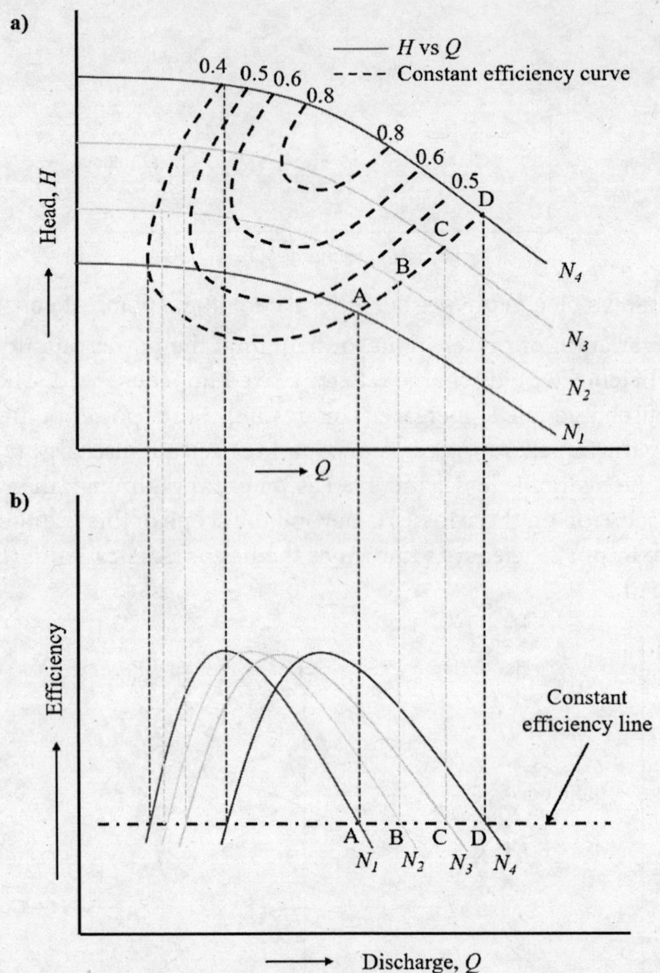

Figure 8.6: Pump performance curves: (a) head vs discharge, (b) efficiency vs discharge for various impeller speeds.

8.6 SYSTEM CURVES AND PUMP OPERATING CONDITIONS

When a centrifugal pump is connected to a pipe-network, the resultant flow rate and head developed by the pump is dependent not only on the pump performance characteristics (discussed in previous section) but also on the pipeline dimensions (length and diameter), number of bends, presence of fittings and connectors (like elbows, tees, etc.) as well as valves.

8.6.1 System curves

As we have already studied in chapter 5, the total head loss for flow through a pipe-network is divided into two broad categories: (a) Major loss (which includes the pressure drop due to friction in a straight pipe) and (b) Minor losses (which includes losses due to sudden area changes, pipe connectors like elbows, tees and valves). When a centrifugal pump is connected to this pipe-network, it has to match this total head loss (H_{sys}) for a desired flow rate (Q). A plot between total head loss (H_{sys}) vs flow rate (Q) for the pie-network is often called as the "system curve". A simplified form of mathematical expression H_{sys} for in terms of an effective loss coefficient K_{eff} is written as: $H_{sys} = K_{eff}\left(\frac{V^2}{2g}\right)$ Here K_{eff} includes all the loss coefficients described in chapter 5. If the pipe-network is complex and involves several components and/or several combinations of serial and parallel flow paths, then K_{eff} needs to be derived appropriately. These concepts have already been discussed in details in chapter 5. As the flowrate through the pipeline $Q \propto V$, the above expression can be rewritten as: $H_{sys} = C_{sys}\, Q^2$ Here C_{sys} is a constant which is directly proportional to the effective loss coefficient of the pipe-network. In addition to frictional losses, if the pipe-network also involves hydrostatic head difference (h_{hyd}) between the flow inlet and outlet, then the new system curve equation will be: $H_{sys} = C_{sys}\, Q^2 + h_{hyd}$ Thus, a plot between head loss in the pipe-network vs flowrate will be parabolic in nature (Fig. 8.7). Here, the slope of the parabola depends on the numerical value of C_{sys} and the apex of the parabola is dependent on h_{hyd}.

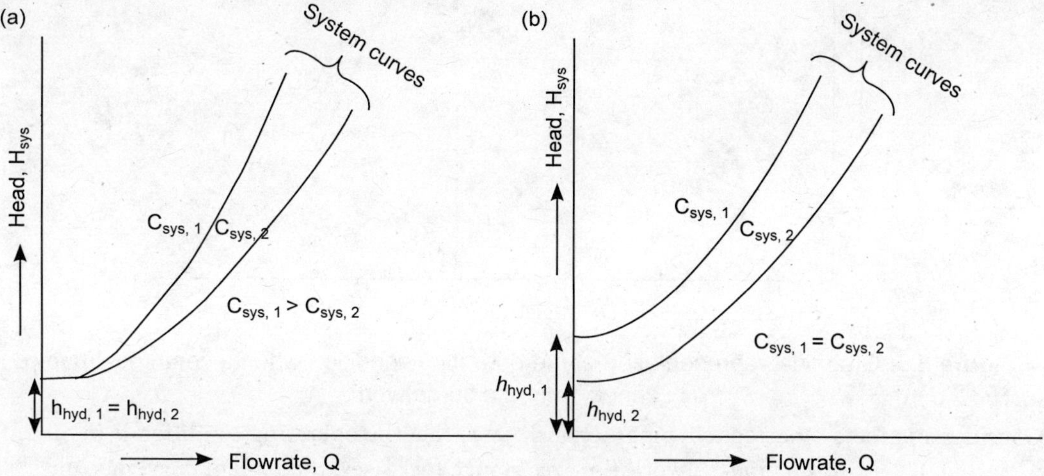

Figure 8.7: Variation of system curves with (a) change in flow resistance, (b) change in hydrostatic head difference of the pipe-network.

If the effective loss coefficient for the pipe-network increases, the parabola becomes steeper in nature as depicted in Fig. 8.7 a). On the contrary, if the hydrostatic head difference between the inlet and outlet (h_{hyd}) of pipe network is increased/decreased then the system curve simply shifts up/down without any change in slope of the system curve. This effect is shown in Fig. 8.7 b).

Practically, after the installation of a pipe-network, head difference does not vary. Similarly, pipe lengths, elbows, bends, tees and other connectors do not vary in a pipe-network. However, presence of valve in the pipe-network allows the control on the effective loss coefficient and thereby the slope of the system curve (C_{sys}). Therefore, valves are common in any pipe-network as it allows the user to vary the slope of the system curve. The effect of this variation on the overall flowrate through the pipe-network will be discussed in next sub-section.

8.6.2 Pump Operating Conditions

A centrifugal pump connected to any pipe-network will lead to an operating flowrate (Q_{op}) and a certain discharge pressure head (H_{op}) will be developed. Mathematically, this operating condition (head and flowrate) is such that it satisfies both the system curve as well as the pump characteristic equations simultaneously. Therefore, on the plot of *Hvs Q*, the operating point is an intersection of these two curves. This is shown in Fig. 8.8. Any variation to the pipe-network effective flow coefficient or change in pump will lead to a new operating condition.

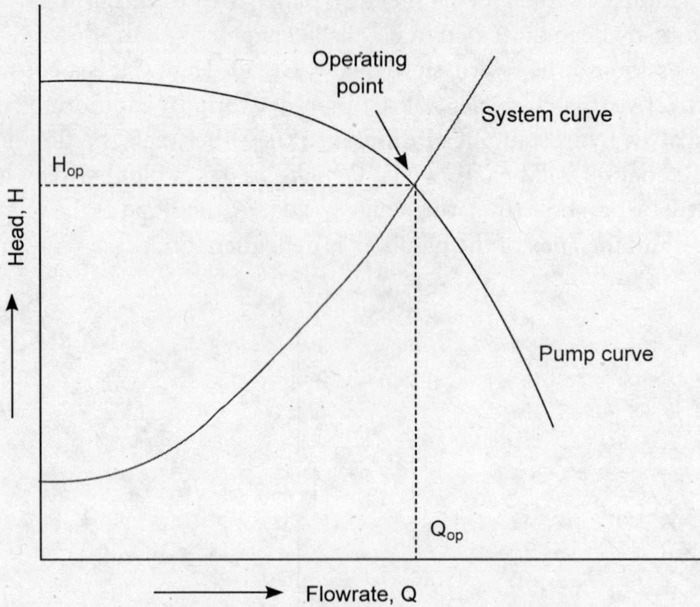

Figure 8.8: Graphical representation of the operating condition when a centrifugal pump is connected to a pipe-network.

Consider a pump connected to a pipe-network, such that the operating condition is represented by point-1 in Fig. 8.9 a). If the pipe-network flow resistance increases, the system curve becomes steeper and it shifts from curve-I to curve-II. The same pump curve remains identical to previous case. However, due to the shift in system curve, the operating condition shifts from point-1 to point-2 (see Fig. 8.9 (a). When compared to point-1, the flowrate is lower but the pressure head is higher at this new operating condition, point-2. Conversely, when the pipe-network flow resistance decreases, the system curve shifts from curve-I to curve-III. For the same pump curve, the operating point shifts to point-3. At this new operating condition, the flowrate is higher than point-1 while the pressure head has decreased. These changes in operating conditions are depicted in Fig. 8.9 (a). Thus, even without change of pump characteristics, if one changes the system curve (for example by opening or closing a valve), the discharge rate and pressure head changes. The change in flowrate and pressure head are inversely related.

Fig. 8.9 (b) represents the scenario where the pipe-network resistance remains unchanged and only the impeller speed is varied. Decreasing the impeller speed will shift the pump curve downwards (curve-II). This shifts the operating condition from point-1 to point-2 decreasing the operating flow as well as the pressure head. Conversely, if the impeller speed is increased the pump curve shifts upwards thereby shifting the operating condition to point-3. At this condition the flowrate as well as pressure head is higher than point-1. It is evident from the above discussions that the change in impeller speed also changes the operating conditions and the change in flowrate and pressure head are directly related.

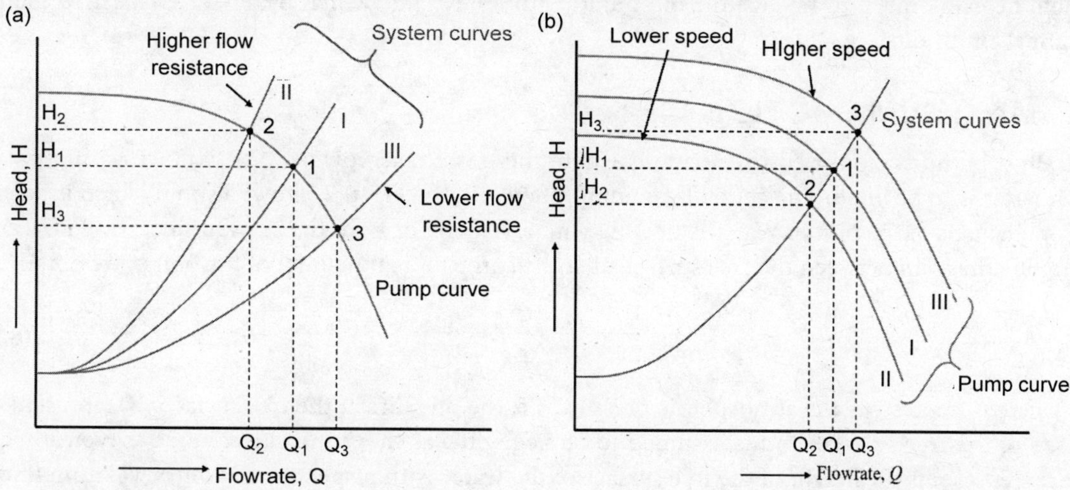

Figure 8.9: Change in operating conditions with variation of: (a) pipe-network resistance, (b) pump impeller speed.

8.6.3 Pump Selection

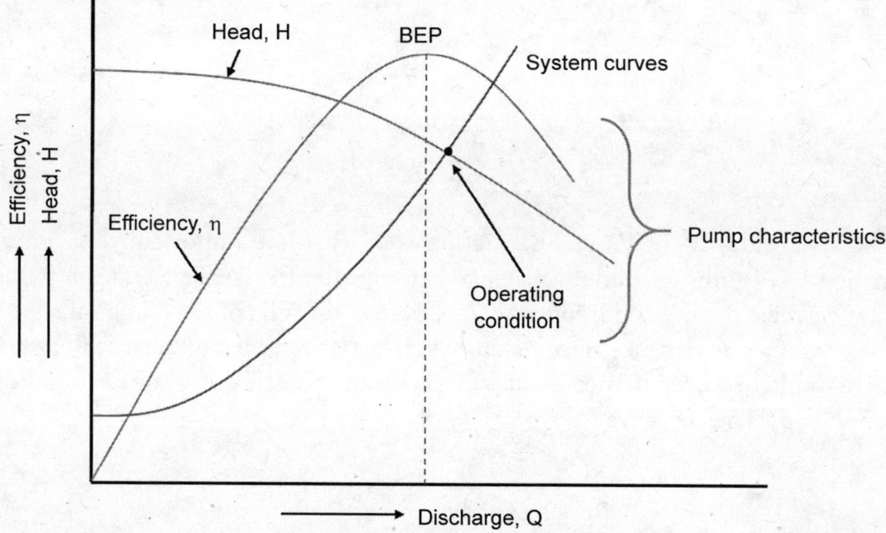

Figure 8.10: Selection of pump such that the operating condition is close to the BEP.

For appropriate pump selection for any pipe-network, the system curve needs to be determined. Several pump models available commercially from various pump-manufacturers, each having unique characteristic curves. Typically, an appropriate pump is chosen such that the desired operating range is close to its BEP (see Fig.8.10). Minor changes in the operating condition can be made by either changing the system curve (using a valve in the pipe-network) or changing the impeller speed (using a variable speed motor). However, major deviation from the initial operating condition can lead to significant reduction in pump efficiency thereby increasing power consumption and operational cost. Sometimes, it might be economical to replace the old pump with a new one, which has its BEP closer to the new operating condition.

8.7 MAXIMUM SUCTION LIFT

Typically a pump is employed in situations where there is a reservoir of water, known as sump, from which water is to be lifted. The centrifugal pump, shown in the Fig. 8.2, shows a sump which is a height H_s from the axis of the pump. Water is flowing with a velocity of V_s in the suction side. If we now apply Bernoulli's theorem between the free surface of at the sump and at the inlet of the pump, we obtain

$$\frac{p_a}{\rho g}+\frac{V_a^2}{2g}+z_a=\frac{p_1}{\rho g}+\frac{V_1^2}{2g}+z_1+H_L \tag{8.33}$$

where p_a represents the atmospheric pressure, i.e. the pressure at the free surface. V_a represents the velocity of the free surface, and is assumed to be negligible as the sump surface area is typically quite large. z_a represents the height of the free surface of the water with respect to an arbitrary datum line; we can choose $z_a = 0$ as the datum line for convenience. p_1 and V_1 represent the pressure and velocity at the inlet of the pump. z_1 represents the distance of the axis of the centrifugal pump from the arbitrary datum. Given that the datum of the system is chosen at the free surface, we have $z_1 = H_s$. H_L represents the head loss due to the foot valve and any other causes of head losses (such as bends etc.).

Substituting these values, we obtain

$$\frac{p_a}{\rho g}+0+0=\frac{p_1}{\rho g}+\frac{V_1^2}{2g}+H_s+H_F \tag{8.34}$$

$$\frac{p_1}{\rho g}=\frac{p_a}{\rho g}-\left(\frac{V_1^2}{2g}+H_s+H_F\right) \tag{8.35}$$

We can now substitute $V_1 = V_s$, i.e. the suction velocity. More importantly, in order to find the maximum suction lift, i.e. the maximum distance between the free surface and the pump which the pump can draw, we must put the condition at the pressure at the inlet of the pump must be just equal to the vapour pressure of water. Because, if the distance between the pump and surface were to increase any more, then there would be a drop in pressure below the vapour pressure, leading to cavitation. Therefore we must ensure $p_1 = p_v$. Thus we have

$$\frac{p_v}{\rho g}=\frac{p_a}{\rho g}-\left(\frac{V_s^2}{2g}+H_s+H_F\right) \tag{8.36}$$

$$\Rightarrow \qquad H_v = H_a - \left(\frac{V_s^2}{2g} + H_s + H_F\right) \qquad (8.37)$$

where $H_v = \frac{p_V}{\rho}$ represents the vapour pressure head. The above equation can be recast as

$$H_s = H_a - H_v - \frac{V_S^2}{2g} - H_F \qquad (8.38)$$

This equation represents the maximum suction lift for a centrifugal pump. The suction height of a centrifugal pump must be necessarily kept less than this value in order to avoid cavitation at the inlet of the pump.

8.8 NET POSITIVE SUCTION HEAD (NPSH)

The idea of a maximum suction lift and net positive suction head are closely related; the latter being more utilized in industry. The *net positive suction head (NPSH)* represents the difference between the sum of absolute pressure head at the inlet and velocity head and the of vapour pressure head. Basically we have

$$NPSH_{\text{available}} = \frac{p_1}{\rho g} - \frac{p_v}{\rho g} + \frac{V_s^2}{2g} \qquad (8.39)$$

We use the result of the previous section to obtain

$$\frac{p_1}{\rho g} = \frac{p_a}{\rho g} - \left(\frac{V_s^2}{2g} + H_s + H_F\right) \qquad (8.40)$$

Using this equation in the definition of the NPSH, we obtain

$$NPSH_{\text{available}} = \left[\frac{p_a}{\rho g} - \left(\frac{V_s^2}{2g} + H_s + H_F\right)\right] - \frac{p_v}{\rho g} + \frac{V_s^2}{2g} \qquad (8.41)$$

$$\Rightarrow \qquad \text{NPSH}_{\text{available}} = H_a - H_v - H_s - H_F \qquad (8.42)$$

The above expression provides us an estimate of the NPSH available for the pump for a certain suction height, flowrate and pipe dimensions. In order to prevent cavitation the NPSH available should always be greater than the minimum NPSH required by the pump. The numerical value of minimum NPSH required by a pump is obtained through experiments. One such expression for minimum NPSH required for safe operation of pump is

$$\frac{NPSH_{\text{min,required}}}{H_m} = 1.03 \times 10^{-3} \times \left[\frac{N\sqrt{Q}}{H_m}\right]^{4/3} \qquad (8.43)$$

Here N represents the impeller speed, H_m is the manometric head developed by the pump and Q is the flow rate delivered by the pump.

Example 8.5

Consider a pump which is placed 3 m above the sump datum level. The atmospheric pressure is 101 kPa and vapor pressure of water is 3 kPa. The pressure drop in the suction line due to frictional losses is 10 kPa. What is the NPSH (m) available at the pump suction? (Assume density of water as 1000 kg/m)

Solution:

Given data

$$H_a = \frac{P_a}{\rho g} = \frac{101\times10^3}{1000\times9.81} = 10.3 \text{ m}$$

$$H_v = \frac{P_v}{\rho g} = \frac{3\times10^3}{1000\times9.81} = 0.306 \text{ m}$$

$h_s = 3$ m

$$h_f = \frac{10\times1000}{1000\times9.81} = 1.02 \text{ m}$$

$$\text{NPSH}_{\text{available}} = (H_a - h_s - h_f) - H_v$$

$$\text{NPSH}_{\text{available}} = (10.3 - 3 - 1.02) - 0.306 = 5.974 \text{ m}$$

Example 8.6

Consider a long pipeline with several bends, pipefittings and valves. The total minor loss coefficient is 20 and major loss coefficient (due to pipe friction) is 10. Consider a pump is whose characteristics equation is of the form: $=A_1 Q^2 + A_2$. The pump is capable to provide a maximum pressure head of 30 m at no flow condition (shut-off condition) and a maximum flow rate of 0.0025 m³/s at zero head condition. When this pump is connected to the above-mentioned pipeline a flow rate of 0.00125 m³/s is obtained. Now a valve in this pipeline is opened such that the new minor loss coefficient reduces to 5. Determine the new flow rate through the pump. Assume that the major loss remains unchanged even after opening the valve.

Solution:

Pump:

The pump characteristics equation:

$$\text{H} = A_1 Q^2 + A_2$$

Shut-off case (Q = 0) the head developed is 30 m

$$\Rightarrow \quad 30 = 0 + A_2 \Rightarrow A_2 = 30$$

For no head condition (H = 0) we have maximum flow rate of 0.0025 m³/s

$$0 = 0.0025\times A_1^2 + A_2$$

$$A_1 = -\frac{30}{0.0025^2} = -4.8\times10^6$$

Hence, the pump curve is

$$H_{\text{Pump}} = -(4.8\times10^6)\, Q^2 + 30$$

Pipeline:

For the pipeline total head loss,

$$H_{pipe} = \left(K_{major} + \Sigma K_{minor}\right)\frac{V^2}{2g}$$

Volume flow rate Q is directly proportional to the velocity of fluid through the pipeline, V. Hence, the head loss through the pipeline can be rewritten as:

$$H_{pipe} = (\Sigma K_{major} + K_{minor})CQ^2$$

where C is a constant.

Case 1:

$K_{major,\,1} = 10$

$\Sigma K_{minor,\,1} = 20$

$Q_1 = 0.00125\ m^3/s$

$$H_{pipe,1} = \left(K_{major,1} + \Sigma K_{minor,1}\right)CQ_1^2$$

$$H_{pipe,1} = (10+20)CQ_1^2 = 30CQ_1^2$$

At operating condition, the head developed and flow rate delivered by the pump is identical to the frictional head loss and flow rate through the pipeline. Therefore,

$$H_{Pump} = H_{pipe,\,1}$$

$$\Rightarrow$$

$$-\left(4.8\times10^6\right)Q_1^2 + 30 = 30CQ_1^2$$

$$\Rightarrow$$

$$-(4.8\times10^6)\ 0.00125^2 + 30 = 30 \times C \times 0.00125^2$$

$$\Rightarrow$$

$$C = 0.48\times10^6$$

Case 2:

$$K_{major,\,2} = K_{major,\,1} = 10$$

$$\Sigma K_{minor,\,2} = 5$$

$$H_{pipe,2} = \left(K_{major,2} + \Sigma K_{minor,2}\right)CQ_2^2$$

$$H_{pipe,2} = (10+5)CQ_2^2 = 15CQ_2^2$$

At operating condition of this new pipeline curve with the same pump as above, we get:

$$-\left(4.8\times10^6\right)Q_2^2 + 30 = \left(15\times0.48\times10^6\right)Q_2^2$$

$$\Rightarrow \quad Q^2 = 0.00158\ m^3/s$$

UNIT SUMMARY

- The power input required for the reciprocating pump

$$P = \frac{\rho g N A L \times (h_s + h_d)}{60 \times 1000} kW$$

- Specific speed of centrifugal pump

$$N_s = \frac{N\sqrt{Q}}{H_m^{3/4}}$$

- Capacity coefficient, $C_Q = \frac{Q}{nD^3}$

- Head coefficient, $C_H = \frac{gH}{n^2 D^2}$

- Power coefficient, $C_P = \frac{P}{\rho n^3 D^5}$

- If there are two pumps, which we denote by pump 1 and pump 2, then we can relate their flow rate ratios, head ratios, and power ratios, as

$$\frac{Q_2}{Q_1} = \frac{n_2}{n_1}\left(\frac{D_2}{D_1}\right)^3$$

$$\frac{H_2}{H_1} = \left(\frac{n_2}{n_1}\right)^2\left(\frac{D_2}{D_1}\right)^2$$

$$\frac{P_2}{P_1} = \frac{\rho_2}{\rho_1}\left(\frac{n_2}{n_1}\right)^3\left(\frac{D_2}{D_1}\right)^5$$

- Expression for minimum NPSH required for safe operation of pump is:

$$\frac{NPSH_{\text{min,required}}}{H_m} = 1.03 \times 10^{-3} \times \left[\frac{N\sqrt{Q}}{H_m}\right]^{4/3}$$

EXERCISES

Multiple Choice Questions

1. In a reciprocating pump, which of the following statements is/are TRUE:
(a) Discharge flow is continuous
(b) Discharge flow is fluctuating
(c) Increase in stroke length increases the discharge flow rate
(d) Increase in stroke length reduces the discharge flow rate

2. For a centrifugal pump, which of the following statements is/are TRUE:
(a) Discharge flow is continuous
(b) Discharge flow is fluctuating
(c) Increase in impeller width increases the flow rate
(d) Increase in impeller width decreases the flow rate

3. For a centrifugal pump with radial entry of liquid into the impeller
(a) The absolute velocity vector at the inlet makes 90 degrees with the peripheral velocity
(b) The relative velocity vector at the inlet makes 90 degrees with the peripheral velocity
(c) The velocity of swirl at inlet is non-zero
(d) The velocity of swirl at inlet is zero.

4. For a centrifugal pump with velocity of flow constant throughout the impeller:
(a) The impeller width at the inlet and outlet is identical
(b) The impeller width at inlet is smaller than at the outlet
(c) The impeller width at outlet is smaller than at the inlet
(d) Can't comment on the impeller width based on the provided data

5. Consider a centrifugal pump with impeller of diameter D rotating at speed of N. The head developed across the impeller is:
(a) Proportional to D^2 (b) Proportional to D^3
(c) Proportional to N^2 (d) Proportional to N^3

6. A centrifugal pump operating at speed N_1 consumes power P_1. If the speed of the pump is made $2N_1$, then the power consumed will be
(a) P_1 (b) $2P_1$
(c) $4P_1$ (d) $8P_1$

7. A fixed speed centrifugal water pump is sized for a certain pipeline layout to deliver a flow rate of Q_1 at an estimated gross head of H_1. However, after actual installation of the pump, the flow rate is measured as Q_2 which is different from Q_1. Which of the following statements is true?
(a) If $Q_1 < Q_2$ then $H_2 > H_1$ (b) If $Q_2 > Q_1$, then $H_2 < H_1$
(c) Cannot comment.

8. Which of the following case(s) is unlikely to cause cavitation in a pump?
(a) Using small suction pipe lengths
(b) Using small suction pipe diameter
(c) Using a fluid with high vapor pressure
(d) Keeping the pump below the free surface of fluid at the suction

ANSWERS															
1.	(b)	2.	(a)	3.	(a)	4.	(c)	5.	(a)	6.	(d)	7.	(d)	8.	(d)

Level - I

1. The internal and external impeller diameters of a centrifugal pump are 200 mm and 400 mm respectively. The impeller inlet and outlet vane angles are 15 degree and 45 degree respectively. The impeller rotates at 1000 r.p.m. Assume velocity of flow is constant throughout the vane and radial entry to the pump blade. What is the work transferred by the impeller per unit mass of water?
2. The internal impeller diameter of a centrifugal pump is 150 mm. The impeller inlet and outlet vane angles are 20 degree and 40 degree respectively. The impeller rotates at 1500 r.p.m. Work transferred by the impeller per unit mass of water is 434.6 kJ/kg. Assume velocity of flow is constant throughout the vane and radial entry to the pump blade. What is the impeller external diameter?
3. The external impeller diameter of a centrifugal pump is 100 mm and it rotates at 3000 r.p.m. The pump delivers water flow rate of 120 liters per minute at a height of 20 m. The vane width at the outlet is 5 mm. Vane angle at the outlet is 60 degrees. Assume radial entry to the pump. Determine the manometric efficiency of the pump.
4. For the above problem (no. 3), if the inlet vane width is 20 mm and velocity of flow remains constant throughout the vane, then determine: (a) internal diameter of vane and (b) inlet vane angle.
5. What is the specific speed of a pump delivering 90 liters per minute at 20 m head? The pump is operated by a motor with 1500 r.p.m.
6. A one-fifth scale model of a single-stage centrifugal pump has a speed of 3000 r.p.m and the head developed is 5 m. The power consumed by the model pump is 2.0 kW. The prototype pump operates with 1000 r.p.m. Determine the: (a) head developed by the prototype pump, (b) power consumed by prototype pump and (c) ratio of flow rates for the two pumps.
7. Consider a pump which is placed 4 m above the free surface datum level. The atmospheric pressure is 101 kPa and vapor pressure of water is 2.5 kPa. The pressure drop in the suction line due to frictional losses is 15 kPa. Will cavitation occur if the minimum NPSH required is 3 m?
8. The head developed by a pump is given by $H = C_0 N^2 + C_1 NQ + C_2 Q^2$, where N is the RPM of the pump and Q is the flow rate delivered by the pump. C_0, C_1, and C_2 are constants. Obtain an expression for the
   (a) Maximum head developed
   (b) Maximum ideal flow rate developed

Level - II

1. A centrifugal pump is running at 800 r.p.m. The outlet vane angle of the impeller is 60 degrees and velocity of flow is 2 m/s. The discharge flow rate is 150 litres/s when the pump is operating against a gross head of 25 m. The monomeric efficiency of the pump is 70%. Determine: (a) outlet impeller diameter, (b) impeller width at the outlet.
2. Neglecting all frictional and other losses in the impeller, show that the pressure rise in the impeller of a centrifugal pump is given by

$$\frac{V_{f1}^2 + u_2^2 - V_{f2}^2 cosec^2\phi}{2g}$$

where V_{f1} and V_{f2} are the flow velocities at the inlet and outlet respectively. u_2 repersents the tangential velocity of the impeller at the outlet and ϕ represents the vane angle at the outlet. Assume radial entry to the impeller.

3. A centrifugal pump with impeller outer diameter of 600 mm and inner diameter of 300 mm rotates at 600 r.p.m. The impeller width at the outlet as 53 mm. The pump delivers water a flow rate of 500 liters per second at a gross head of 10 m. The absolute velocity at the inlet makes an angle of 75 degrees with the tangential velocity. The outlet vane angle is 45 degrees. Assuming constant velocity of flow through the impeller, determine: (a) manometric efficiency and (b) work input to the pump in kW.
4. A centrifugal pump rotating at 1500 RPM delivers 200 litres/s of water against a gross head of 25 m. The pump is installed where the atmospheric pressure is 10^5 Pa (abs.) and the vapour pressure of water is 3 kPa. Determine the maximum allowable height of this pump from the free surface of the water in the sump. Note that the expression of the minimum NPSH required for safe operation of the pump is given by

$$\frac{NPSH_{min}}{H_{gross}} = 1.03 \times 10^{-3} \times \left[\frac{N\sqrt{Q}}{H_{gross}}\right]$$

where N is the RPM of the impeller.

Answer Key

Level - I

1. 397.7 kJ/kg
2. 300 mm
3. 83.5%
4. (a) 25 mm, (b) 18
5 6.14 RPM
6. (a) 13.89 m, (b) 231.5 kW, (c) 41.67
7. No. Since available NPSH = 4.51 m which is greater than the minimum NPSH required = 3 m.
8. (a) $C_0 N_2$, (b) $\dfrac{-C_1 N + \sqrt{C_1^2 N^2 - 4C_2 C_0 N^2}}{2C_2}$

Level - II

1. (a) 461 mm, (b) 51.8 mm
2. NA
3. (a) 53.1%, (b) 92.34 kW
4. 7.82 m

To know more about pumps

KNOW MORE

The earliest classified pump was a mud lifting machine detailed in a late 15th-century treatise by Italian engineer Francesco di Giorgio Martini. About 200 years later, French inventor Denis Papin developed the first true centrifugal pump, which utilized straight vanes for local drainage.

In contrast, the reciprocating pump has deeper historical roots, originating around 200 BC with Greek inventor Ctesibius, who created a water organ. This device featured an air pump with valves, a water tank, and a row of pipes, laying the groundwork for the modern reciprocating pump.

[Source: https://news.ipecsales.com/the-difference-between-centrifugal-and-reciprocating-pumps]

REFERENCES AND SUGGESTED READINGS

1. **Introduction to Fluid Mechanics and Fluid Machines** (3rd Edition), S K Som, Gautam Biswas, Suman Chakraborty, McGraw Hill
2. **Fluid Mechanics** (9th Edition), Frank M. White, Henry Xue, McGraw Hill

CO AND PO ATTAINMENT TABLE

Course outcomes (COs) for this course can be mapped with the programme outcomes (POs) after the completion of the course and a correlation can be made for the attainment of POs to analyze the gap. After proper analysis of the gap in the attainment of POs necessary measures can be taken to overcome the gaps.

Table for CO and PO attainment

Course Outcomes	Expected Mapping with Programme Outcomes (1- Weak Correlation; 2- Medium Correlation; 3- Strong Correlation)						
	PO-1	PO-2	PO-3	PO-4	PO-5	PO-6	PO-7
CO-1	-	-	-	-	-	-	-
CO-2	-	-	-	-	-	-	-
CO-3	-	-	-	-	-	-	-
CO-4	-	-	-	-	-	-	-
CO-5	-	-	-	-	-	-	-
CO-6	-	-	-	-	-	-	-
CO-7	-	-	-	-	-	-	-

The data filled in the above table can be used for gap analysis.

INDEX

❑❑